Green Chemistry and Sustainable Technology

Aims and Scope

The series *Green Chemistry and Sustainable Technology* aims to present cutting-edge research and important advances in green chemistry, green chemical engineering and sustainable industrial technology. The scope of coverage includes (but is not limited to):

– Environmentally benign chemical synthesis and processes (green catalysis, green solvents and reagents, atom-economy synthetic methods etc.)
– Green chemicals and energy produced from renewable resources (biomass, carbon dioxide etc.)
– Novel materials and technologies for energy production and storage (bio-fuels and bioenergies, hydrogen, fuel cells, solar cells, lithium-ion batteries etc.)
– Green chemical engineering processes (process integration, materials diversity, energy saving, waste minimization, efficient separation processes etc.)
– Green technologies for environmental sustainability (carbon dioxide capture, waste and harmful chemicals treatment, pollution prevention, environmental redemption etc.)

The series *Green Chemistry and Sustainable Technology* is intended to provide an accessible reference resource for postgraduate students, academic researchers and industrial professionals who are interested in green chemistry and technologies for sustainable development.

More information about this series at http://www.springer.com/series/11661

Sourav Mondal • Mihir Kumar Purkait
Sirshendu De

Advances in Dye Removal Technologies

Springer

Sourav Mondal
Mathematical Institute
University of Oxford
Oxford, UK

Mihir Kumar Purkait
Department of Chemical Engineering
Indian Institute of Technology Guwahati
Guwahati, Assam, India

Sirshendu De
Department of Chemical Engineering
Indian Institute of Technology Kharagpur
Kharagpur, West Bengal, India

ISSN 2196-6982 ISSN 2196-6990 (electronic)
Green Chemistry and Sustainable Technology
ISBN 978-981-10-6291-9 ISBN 978-981-10-6293-3 (eBook)
DOI 10.1007/978-981-10-6293-3

Library of Congress Control Number: 2017951626

Printed on acid-free paper

This Springer imprint is published by Springer Nature
The registered company is Springer Nature Singapore Pte Ltd.
The registered company address is: 152 Beach Road, #21-01/04 Gateway East, Singapore 189721, Singapore

Preface

With increasing textile industries, the toxic and hazardous dye effluent is increasing in the ecological system due to unavailability of energy-efficient and inexpensive treatment methods. The book presents a detailed understanding of various advanced dye removal technologies. The application of the developed technologies in the context of textile effluent treatment is presented. With stringent environmental norms, the need for efficient and economic treatment option is warranted. The proposed monograph describes the rationale behind the application of the treatment technologies compared to conventional processes like microbial degradation, coagulation, ion exchange, etc. Depending on the type and concentration of dyes in the effluent, different technologies may be appropriate. The effect of various treatment conditions and operating conditions are key factors in determining the efficiency of the removal process. The mass transfer mechanism behind the separation process is also analyzed and presented in this book.

Experimental investigations have been carried out to adsorb different dyes from aqueous medium using commercial activated charcoal as an adsorbent and are described in Chap. 2. Synthetic solution of three dyes, chrysoidine, eosin, and Congo red, in aqueous medium is used for the adsorption study. Effects of adsorbent dose, initial dye concentration, contact time, pH, and temperature have been studied for the adsorption of each dye under stirred batch condition. Breakthrough curve for the column adsorption has been generated for the adsorption of eosin dye. Standard adsorption isotherms have been considered to explain the experimental equilibrium data. The rate of adsorption has been described by both first- and pseudo-second-order kinetic models. A two-resistance mass transfer model for batch adsorption process is presented that includes a film mass transfer coefficient and an internal effective diffusivity controlling the internal mass transport process based on the pore diffusion mechanism. Validation of model is checked with the literature adsorption data.

Chapter 3 describes a detailed study of the adsorption of dyes. Three systems are selected for the adsorption study. Aqueous synthetic solutions of crystal violet and methylene blue in a single-component system are selected for the primary study.

Actual textile effluent containing a mixture of two reactive dyes is used for further study. Aqueous synthetic solutions of the reactive dyes in a single-component system as well as two-component mixture of the dyes are studied. Experimental studies consist of preparation of the adsorbent, equilibrium, and kinetic studies in each system. Adsorption equilibrium data are fitted with various isotherms. A mass transfer model having a film mass transfer coefficient and an internal effective diffusivity is used for the prediction of the concentration profiles of the dyes in batch adsorption for both single-component and two-component systems.

Chapter 4 describes a systematic study of the regeneration of spent carbon (obtained after adsorption of different toxic dyes) using a novel, in situ, low-energy surfactant-enhanced carbon regeneration technique. Experimental investigations have been carried out for the regeneration of spent carbon applying surfactant-enhanced carbon regeneration technique using different types of cationic and anionic surfactants. An empirical kinetic model for the regeneration of adsorbent is presented.

The nanofiltration of the dyes is discussed in Chap. 5. An aqueous synthetic solution of crystal violet and methylene blue in a single-component system is used for the initial study. Using a suitable molecular weight cutoff (MWCO) membrane, a fractionation study is carried out with the aqueous synthetic mixture of crystal violet and methylene blue. Actual textile effluent (containing a mixture of the reactive dyes) is used for the final study. Nanofiltration is carried out in two different process configurations: an unstirred batch system and a steady-state cross flow system. Experiments are conducted in each system to observe the effects of various process parameters on the permeate concentration and flux profile. Theoretical model for unstirred batch system is developed for the single-component as well as the two-component system and tested with the data. Evaluation of the model parameters, viz., diffusivity and real retention, is carried out using profile optimization of the experimental data. Film theory is used to explain the experimental data in the cross flow system.

The application of the proposed combination processes involving (i) adsorption followed by nanofiltration and (ii) advance oxidation followed by nanofiltration is represented in Chap. 6. A detailed parametric study is carried out for both the processes (combination method). The performance of different combinations of separation technologies is discussed in detail.

In Chap. 7, micellar-enhanced ultrafiltration of dye is discussed. Aqueous synthetic solution of dye is used as feed of the micellar-enhanced ultrafiltration system. The micellar-enhanced ultrafiltration experiments are carried out in two process configurations: an unstirred batch system and a steady-state cross flow system. All the experiments are conducted to observe the effects of various process parameters on the retention of solute and permeate flux. A resistance in series model is used to quantify the flux decline in the batch system. The different resistances and growth kinetics of the gel layer have been investigated as function of the operating conditions. A two-step chemical treatment process for the recycling of surfactant from the permeate and retentate stream has been proposed, and its performance has been discussed in this chapter.

Cloud point extraction of toxic dyes is represented in Chap. 8. An aqueous synthetic solution of chrysoidine, Congo red, and eosin in a single-component system is used for the initial study. Effects of different operating parameters, e.g., concentration of the feed mixture (both dye and surfactant), pH, temperature, and the presence of mono- and divalent salt on the extraction of both the dye and surfactant, have been studied in detail. The performance of two nonionic surfactants is investigated to extract dye from the synthetic dye solution. From the experimental data, a solubilization isotherm is developed to quantify the amount of Congo red and eosin solubilization. A method has been proposed to design a cloud point extractor for the separation of Congo red and eosin dye. To test the efficiency of surfactant recovery from the dilute phase, a solvent extraction technique has been adopted and explained in detail in the same chapter.

Electrocoagulation of dyes is a powerful method of disintegration of dyes in wastewater. The electrochemical pathway leading to degradation and the effects of various operating conditions (strength of current, electrode material, electrolytes, etc.) are discussed elaborately in Chap. 9. The economics of the process is also discussed. Emulsion liquid membrane is a relatively new technology used for the extraction and separation of metal ions and bioactive materials. Chapter 10 deals with extraction of textile dyes from water using a double emulsion. The relative effect of surfactant, alkali, feed concentration, etc., on the separation efficiency is described in detail.

We hope that the book will be useful to engineers and researchers working in the field of chemical engineering, environmental studies, water technology, civil engineering, and industrial engineering. We have tried our best to make this book comprehensive and hope that it will ignite further research and development in the relevant domains of engineering. We sincerely hope that the readers will be benefitted from the book and inspired to implement them in practice. Although we have put in our best efforts to organize and present all information regarding dye removal technologies, any suggestion for improvements will be gratefully acknowledged from the readers.

The authors would like to acknowledge all dear and near ones, who have directly or indirectly inspired and influenced in preparing the contents of this book. Dr. Mondal would also like to acknowledge that this work was completed while he was at the Department of Chemical Engineering, IIT Kharagpur, during his PhD, and thank his supervisor and colleagues for their support during the same.

Oxford, UK — Sourav Mondal
Guwahati, Assam, India — Mihir Kumar Purkait
Kharagpur, West Bengal, India — Sirshendu De

Contents

List of Figures

List of Tables

About the Authors

Dr. Sourav Mondal has done his Ph.D. in Chemical Engineering from the Indian Institute of Technology Kharagpur. Presently he is working as a post-doctoral researcher in the University of Oxford. His research work is dedicated to various modeling and mathematical analysis of the transport phenomena in different separation processes including membrane-based filtration of wastewater treatment, fruit juice clarification, dye separation, bio-product separation, mixed matrix membrane filtration, etc. He is also working on modeling of adsorption, microfluidic mass transport, computational fluid dynamics of multiphase separation processes and other fluid dynamical processes as well as environmental challenges. He has been the recipient of Shastri Indo-Canadian research mobility award in 2012-2013 as a part of the scientific and innovative exchange program between the two nations. He has already authored two books, one book chapter, one patent, more than 30 publications in international journals of repute and presented at several international conferences.

Prof. Mihir Kumar Purkait is a Professor in the Department of Chemical Engineering at the Indian Institute of Technology Guwahati (IITG). Prior to joining as faculty in IITG, he has received his Ph.D. and M.Tech. in Chemical Engineering from the Indian Institute of Technology Kharagpur (IITKGP) after completing his B.Tech. and B.Sc. (Hons) in Chemistry from University of Calcutta. He has received several awards like: Dr. A.V. Rama Rao Foundation's Best Ph.D. Thesis and Research Award in Chemical Engineering from IIChE (2007), BOYSCAST Fellow award (2009–2010) from the DST, Young Engineers Award in the field of chemical engineering from the Institute of Engineers (India, 2009), and Young Scientist Medal award from the Indian National Science Academy (INSA, 2009). His current research activities are focused in four distinct areas viz.: i) membrane synthesis and separation technology (both ceramic and polymeric), ii) water pollution control, iii) smart material and iv) adsorption. In each of the areas, his goal is the synthesis of responsive materials and to develop a more fundamental understanding of the factors governing the performance of these technologies.

He has more than 12 years of experience in academics and research and published 135 papers in different reputed journals of importance. He has submitted 7 patents and was involved with 14 sponsored projects from various funding agencies. Dr. Purkait has guided 13 Ph.D. students and 7 are yet to get their Ph.D. degree.

Prof. Sirshendu De is a Professor in the Department of Chemical Engineering, Indian Institute of Technology Kharagpur, India. He has obtained his B.Tech. (1990), M.Tech. (1992) and Ph.D. (1997) degrees from the Indian Institute of Technology Kanpur. His major field of interest is membrane-based separation processes: design, modeling and fabrication of flat sheet and hollow fiber membranes. He has already authored 8 books, 10 book chapters, 14 patents and more than 250 publications in national and international journals of repute. He has also transferred three technologies for commercialization. Other fields of research of Dr. De are adsorption, transport phenomena, modeling of flow through microchannel, etc. He has been the recipient of several prestigious awards, including Shanti Swarup Bhatnagar Award in 2011 for fundamental contribution and innovation in basic science and technology (Engineering Science category). He has also received Innovation Award in 2016 from Indian Desalination Association; 5th National Award for Technology Innovation in various fields of petrochemicals and downstream plastic processing industry from Central Institute of Plastic Engineering and Technology in 2015; CSMCRI Distinguished Speaker Award in CHEMCON 2015 and 2013 from Indian Institution of Chemical Engineers; NASI-Reliance Industries Platinum Jubilee Award, 2013, from National Academy of Science, Allahabad; Silver Jubilee Young Engineer Award, 2012, from Indian National Academy of Engineering, New Delhi; DAE-SRC Outstanding Investigator Award, 2012, from Department of Atomic Energy, Govt. of India; Herdillia Award in 2010 for excellence in basic research in chemical engineering; VNMM Award in 2009 for excellence in innovative applied research; Young Engineer Award from Indian National Academy of Engineering in 2001 for excellence in engineering research; and Amar Dye Chem Award in 2000 for excellence in chemical engineering research. He is the fellow and Chair Professor of Indian National Academy of Engineering, New Delhi, and fellow of National Academy of Sciences, Allahabad, for his contribution in engineering and research.

Chapter 1
Introduction

Abstract The presence of colored wastewater in the aquatic streams is hazardous to the environmental ecology affecting the aquatic life severely. With the growth of textile industry, dye-containing effluent is a potential water treatment issue. Moreover, dye-containing wastewater (even in trace amounts) is toxic to human consumption. There have been several conventional water treatment technologies based on adsorption which purifies dye-containing wastewater. The present chapter explains an overview to the different processes, types of dye compounds, and the safe discharge limits of different types of contamination-containing dyes.

Keywords Color • Wastewater • Dye structures • Effluent • Textile industry

Color has always fascinated mankind, for both esthetic and social reasons. Throughout history, dyes and pigments have been the major articles of commerce. The use of dyes for coloring dates back to the prehistoric era of more than 5000 years. However, most of these dyes are from natural resources particularly with plant extracts. It is only in 1856 William Henry Perkin serendipitously discovered the first synthetic dye, mauveine. Since then, more than 10,000 different variants of synthetic dyes have been produced and are available commercially (Hunger 2003; Husain 2006). Synthetic dyes quickly replaced the traditional natural dyes as they are inexpensive and easy to prepare, have a wide range of new colors, and impart better properties to the dyed materials. The classification of dyes is based on the chemical properties and applications. The different types of dyes currently available are described in Table 1.1.

The production of dyes has increased many folds in the past decade, and it is mostly dominated (both consumption and production) by the South-Asian markets. The average annual global production is around 10^6 metric tonnes (Ghaly et al. 2014). According to the world trade export data of 2012, China is major exporter of textiles followed by European Union and India (WTO 2012), as shown in Fig. 1.1.

The major end-use application of dyestuff is the textile industry, which in turn indirectly depends on the population growth and private spending limits. With enormous growth in the textile industry, treatment of dye-contaminated effluent is a serious environmental concern. Approximately 10^4 tonnes of dyes per year are discharged into waste streams by the textile industries (Choy et al. 1999). In a typical dye processing unit, approximately, 80 liters of water is consumed per kg of

S. Mondal et al., *Advances in Dye Removal Technologies*, Green Chemistry and Sustainable Technology, DOI 10.1007/978-981-10-6293-3_1

Table 1.1 Categories of dye based on dyeing process and applications

Type	Chemical property	Applications
Acid	Water soluble – anionic	Dyeing of silk, wool, nylon, etc.
Basic	Water soluble – cationic	Coloring of acrylic fiber and paper
Direct	Dyeing is carried with addition of sodium salts in mild alkaline bath	Used on cotton, paper, leather, wool, silk, and nylon. Also used for pH indicators and biological staining
Mordant	Requires a mordant	Used for the navy or black shades in wool
Vat	Insoluble; reduction in alkaline environment produces water-soluble metal salt which binds to the textile fiber; subsequent oxidation reforms the original insoluble dye	Highly effective for the color of indigo
Reactive	Use a chromophore to directly react with fiber substrate	Dyeing cotton and other cellulose fibers at home or in the art studio in normal temperature and pH
Disperse	Water-insoluble nonionic dyes used for hydrophobic fibers	Dyeing cellulose acetate fibers, polyester, nylon, acrylic fibers
Azo	Insoluble dye is prepared in situ in the fiber by treating with both diazoic and coupling components	Useful for dyeing cotton; however, it is obsolete due to toxic nature of diazo compounds
Sulfur	Treating the fabric in a solution of an organic compound and sulfide compounds	Inexpensive; used for dark shades in dyeing cotton

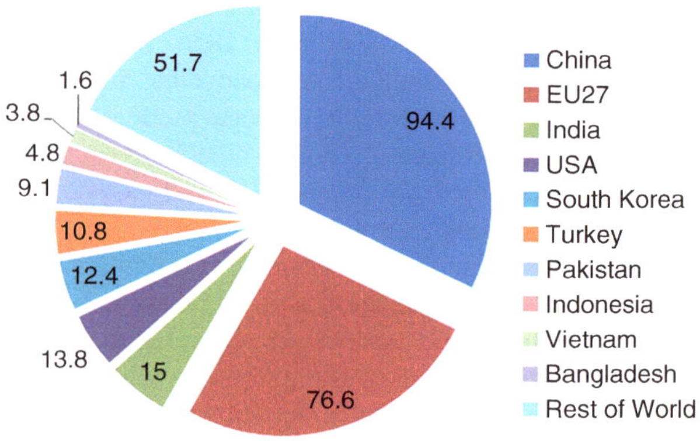

Fig. 1.1 Major textile exporters in the world in 2012

product, as illustrated in Fig. 1.2. The big consumers of dyes other than textile industries are dyeing, paper and pulp, tannery, and paint industries, and hence the effluents of all these industries as well as those from the dye-manufacturing plants contain dyes in sufficient quantities. Dyes are considered as objectionable pollutant because of their general toxicity (Combes and Havelandsmith 1982; Christie 2007; Bae and Freeman 2007) in oral ingestion and inhalation, skin and eye irritation, and skin sensitization leading to problems, like skin irritation and skin sensitization, and

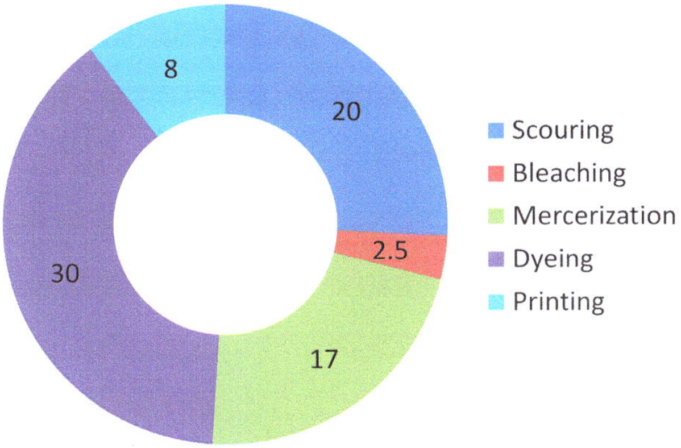

Fig. 1.2 Typical amount of water consumed (in m^3/1000 liters of product) in a conventional continuous process

also due to carcinogenicity (Hatch and Maibach 1999; Rai et al. 2005). Dye-contaminated colored effluent restricts transmission of light into the aquatic environment and thereby disrupts the aquatic ecosystem significantly (Kuo 1992). With stringent environmental norms, many textile-processing units, particularly in China, have been shut down in recent years. Thus, the treatment of colored effluent discharges in most countries is the need of the hour, giving rise to the need for the development of new and innovative separation technologies that are highly efficient as well as economic. The technological target would be to reduce the cost of the treated water for recycling than using freshwater to make the process economically sustainable.

1.1 Environmental Problems Associated with the Colored Industrial Effluent

The effluent from textile industries contains huge amount of toxic liquid wastes containing inorganic and organic compounds (Labanda et al. 2009). During the process of dyeing, all the dyes applied to the fabrics are not fixed on them, and there is a significant portion of unbound dyes that is washed out. Various process industries such as textile, pulp and paper, dyeing, paint, ink, etc. use large quantity of dyes for coloring their products. The typical range of various dye fixations onto the fibers and the unfixed amounts are presented in Fig. 1.3. The fraction of unbound dyes are high in most textile effluents (Shiou et al. 2008). The quantity of water used and subsequently released is dependent on the type of fabric dyed in relation to its affinity for the fixation of the dye molecule. With respect to the production volumes, azo dyes are the largest group of colorants, constituting 60–70% of all organic dyes produced in the world (Bafana et al. 2011). Some of the azo dyes can show toxic effects, especially carcinogenic and mutagenic manifestations (Chung and Cerniglia 1992; Pinheiro et al. 2004). Often the effluents from such industries, which are rich in color, are discharged into the water bodies without proper treatment leading to extensive environmental pollution. The textile dyeing industry consumes large quantities of water and results in huge amount of

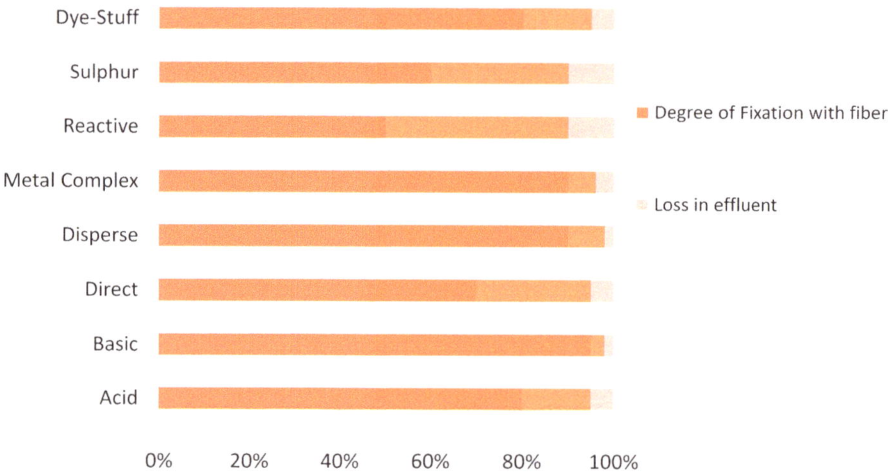

Fig. 1.3 The typical amounts of fixed and unfixed dye concentration used for dyeing fibers (EWA 2005)

wastewater at their various steps of dyeing and finishing processes. A typical textile unit generates different types of wastewater in various magnitudes and qualities. Approximately, 0.08–0.10 m^3 of water is used to produce 1 kg of fabrics (Yaneva and Georgieva 2012). Pagga and Brown (1986) and Al-Kdasi et al. (2004) estimated that around 1000–3000 m^3 of water is discharged to process about 12–20 tonnes of textiles per day. The effluents being enriched in dyes and toxic chemicals, many of which are nonbiodegradable and carcinogenic, pose a major threat to the public health and the environmental ecology if left untreated.

Dyes are generally synthetic organic aromatic compounds, embodied with various functional groups and heavy metals. The aromatic amines present in the azo dyes cause increased risk of bladder cancer, skin irritation, contact dermatitis, chemosis, lacrimation, exophthalmos, permanent blindness, rhabdomyolysis, acute tubular necrosis supervene, vomiting gastritis, hypertension, vertigo, and, upon ingestion, edema of the face, neck, pharynx, tongue, and larynx along with respiratory distress (Puvaneswari et al. 2006). Dyes increase the biochemical oxygen demand (BOD) of the receiving water and reduce the reoxygenation process, therefore affecting the development of the photoautotrophic organisms. The colored dye effluents are considered to be highly toxic to the aquatic organisms and disturb the natural balance in the aquatic life by inhibiting the photosynthetic activities (Venkatamohan et al. 2002). Decreasing light penetration through water limits the photosynthetic activity, generating oxygen deficiency and deregulating the biological cycles of aquatic life. Basic dyes are also toxic and can cause allergic dermatitis, skin irritation, mutations, and even cancer. Cationic dyes can cause increased heart rate, shock, vomiting, cyanosis, jaundice, quadriplegia, heinz body formation, and tissue necrosis in humans (Nohyneka et al. 2004). The ingestion of the dye-containing wastewater causes several symptoms like severe headaches, skin irritation, vomiting, and acute diarrhea (Gupta et al. 2000). The presence of heavy metals in the complex framework of dye induces chronic toxicity, leading to ulceration of the skin and mucous membranes, failure of kidney, etc. Besides being toxic, chemicals present in the dye effluent are also teratogenic (disruption to normal embryonic development), mutagenic, and carcinogenic to several living organisms, particularly to the aquatic life (Mathur and Bhatnagar 2007; Novotny et al. 2006). This makes the problem graver, as the effect is not expressed immediately and makes it vulnerable for changing the genetic material (Vogel 1982). It has been reported in biological studies that some specific forms of nitro and azo dyes, which are generally present in textiles, get converted into carcinogenic amines or other toxic compounds inside the bodies of living organisms (Weisburger 2002; Weber and Wolfe 1987; Umbuzeiro et al. 2005; Chen 2006). Moreover, there are several instances of dye being produced from known carcinogens – benzidine (Baughman and Perenich 1988). In some physiological research, it is investigated that specific forms of the azo dyes are related to the splenic sarcomas, cancer in human bladder, and irregularities in the chromosomal material of the mammalian cells (Percy et al. 1989; Medvedev et al. 1988).

Therefore, treatment of these effluents is essential. The allowable wastewater environmental discharge limits decided by different nations are presented in the subsequent section. Separation and reuse of the costly dyes from the effluent mixtures can be an added advantage considering the economy of the process.

1.2 Toxicity Levels and Allowable Limits of Various Dye Concentrations in Streams

Toxicity and allowable limits of different dyes in aqueous stream in various countries are presented in Table 1.2.

1.3 Existing Processes for the Separation of Dye from Wastewater

In the past, municipal treatment systems were used for the purification of textile dye effluents. This method was effective for removing the water-insoluble dyes (e.g., disperse and vat dyes), which exhibited good exhaustion properties. However, with the increasing use of water-soluble dyes (reactive dyes), conventional biological treatment processes failed to achieve adequate color removal. When these dyes were released into aquatic systems, anaerobic bacteria in the sediment could not mineralize the dyes completely, resulting in the formation of toxic amines (Banat et al. 1996; Robinson et al. 2002). Shaw et al. (2002) reported the results on the possible advantages of anaerobic/aerobic metabolism in sequenced redox reactors to treat a synthetic textile effluent. In that work, the wastewater contained polyvinyl

Table 1.2 Environmental discharge limits of the various developing nations across the globe

Parameter	CCME	China	BIS	Hong Kong	EPA	Thailand	Philippines	Indonesia	Bangladesh	Sri Lanka
pH	5.5–8.5	6–9	5.5–9	6–10	6–9	5–9	6–9	6–9	6.5–9	6–8.5
Temperature (°C)	30	–	50	45	40	–	40	–	40–45	40
Color (Pt-co)	100	80	–	1	7	–	100–200	–	–	30
TDS (mg/l)	2000	–	2100	–	2000	2000–5000	1200	–	2100	2100
TSS (mg/l)	40	150	100	800	30	30–150	90	60	100	500
Sulfide (μg/l)	200	1000	2000	1000	200	–	–	–	1000	2000
Free chlorine (μg/l)	1000	–	1000	–	1000	–	1000	–	–	–
COD (mg/l)	80	200	250	2000	80	120–400	200–300	250	200	600
BOD_5 (mg/l)	50	60	30	800	50	20–60	30–200	85	150	200
Oil and Grease (mg/l)	–	–	10	20	10	300	5–15	5	10	30
Dissolved oxygen (μg/l)	6000	–	–	4000	–	–	1000–2000	–	4500–8000	–
Nitrate (μg/l)	13,000	–	10,000	–	20,000	–	–	–	10,000	45,000
Ammonia (μg/l)	0.1	–	–	500	0.2	–	–	–	5000	60
Phosphate (μg/l)	<4000	1000	5000	5000	5000	–	–	2000	–	2000
Calcium (μg/l)	–	–	–	–	20,000	–	200,000	–	–	240,000
Magnesium (μg/l)	200,000	–	–	–	200,000	–	500–5000	–	–	150,000

Chromium (μg/l)	1000	–	2000	1000	< 1000	5000	–	5000	2000	50
Aluminum (μg/l)	5	–	–	–	< 1000	–	1000	–	–	–
Copper (μg/l)	< 1000	1000	3000	1000	5	1000	1000–5000	2000	500	3000
Manganese (μg/l)	5	2000	2000	500	20,000	5000	1000–20,000	–	5000	500
Iron (μg/l)	300	–	3000	1500	< 10,000	–	5000–10,000	5000	2000	1000
Zinc (μg/l)	30	500	5000	600	0.05	–	5	5000	5000	10,000
Mercury (μg/l)	0.025	–	0.01	1		5		–	10	1

alcohol from desizing and an azo dye (Remazol Black). Color removal was 94%, but dye metabolites caused reactor instability. Aromatic amines from the anaerobic breakdown of the azo dyes were not completely mineralized by the aerobic phase.

The existing processes for removal of color from wastewater can broadly be divided into two categories: (i) chemical methods and (ii) physical methods. Chemical methods include mainly coagulation-flocculation and chemical oxidation processes, whereas adsorption and membrane-based separation processes belong to the physical treatment processes. These methods are discussed briefly in the subsequent section.

In the coagulation process, the charge on colloids and suspended solids are destabilized using a coagulant. The coagulant material disperses through the effluent by flash mixing (Kawamura 2000). Metallic salts like alum (Chu 2001), lime (Georgion et al. 2003), magnesium salts (Panswed and Wangchaisuwan 1986; Haitan et al. 2000), etc. are commonly used. Flocculation is the gentle mixing phase that follows the rapid dispersion of coagulant by the flash mixing unit. Its purpose is to accelerate the rate of particle collisions, causing the agglomeration of electrolytically destabilized colloidal particles into settleable and filterable sizes. The terms coagulation and flocculation are sometimes used interchangeably and used frequently by the industries for the treatment of colored wastewater. The performance of these processes is dependent on the final floc formation and its settling quality. Although this method is shown to be effective in decolorizing both soluble and insoluble dyes, one major disadvantage of the process is sludge generation that needs to be disposed (Robinson et al. 2001).

In the work of Chu (2001), recycled alum sludge (RAS) generated by the coagulation process itself was used for the separation of one hydrophobic and one hydrophilic dye, thus reducing the consumption of fresh alum dosage. RAS was found to be a good way of removing hydrophobic dye in wastewater. However, the use of RAS was not recommended for the removal of hydrophilic dyes, since the high solubility characteristics of such dyes can cause deterioration in water quality during recycling.

Ferrous sulfate and lime were used by Georgiou et al. (2003), for the treatment of a textile wastewater. Treatment with lime alone was proved to be very effective. Treatment with lime in the presence of increasing doses of ferrous sulfate was tested successfully. However, the process was very costly mainly due to extensive precipitation of solids.

Another most commonly used method for dye removal is wet oxidation. In this process, dye is chemically oxidized in the presence of air or some oxidants like hydrogen per oxide. Often decolorization varies depending on the way in which the peroxide is activated (Slokar and Le Marechal 1997). Oxidation using ozone is capable of removing various types of dyes and other pollutants. The use of ozone was first pioneered in the early 1970s and proved to be a very good oxidizing agent due to its high instability. One major advantage of this process is that ozone can be applied in its gaseous state and, therefore, does not increase the volume of wastewater and sludge. Since this process can achieve very high conversion rates, the wet air oxidation process requires much less space. However, the process of ozonation is expensive. Due to the short half-life of ozone, continuous ozonation is required

(Robinson et al. 2001). The summary of various chemical, physical, and biological treatment processes is described in Table 1.3.

1.4 Molecular Structures of Commonly Found Dyes in Wastewater

Molecular structures of common dyes in wastewater is presented in Table 1.4

Table 1.3 Brief summary on different types of dye removal treatment processes

Treatment methodology	Treatment stage	Advantage	Limitations
Chemical treatment			
Precipitation, coagulation-flocculation	Pretreatment	Short detention time and low capital costs	Agglomerate separation and treatment. Selected operating condition
Electrokinetic coagulation	Pretreatment	Economically feasible	High sludge production
Fenton process	Pretreatment	Effective for both soluble and insoluble colored contaminants. No alternation in volume.	Sludge generation; problem with sludge disposal. Prohibitively expensive.
Ozonation	Main treatment	Effective for azo dye removal. Applied in gaseous state: no alteration of volume	Not suitable for dispersed dyes. Releases aromatic dyes. Short half-life of ozone (20 min)
Oxidation with NaOCl	Posttreatment	Low-temperature requirement. Initiates and accelerates azo bond cleavage	Cost-intensive process. Release of aromatic amines
Photochemical means	Posttreatment	No sludge production	Formation of byproducts
Electrochemical oxidation	Pretreatment	No additional chemicals required and the end products are non-dangerous/ hazardous	Cost-intensive process; mainly high cost of electricity
Ion exchange	Main treatment	Regeneration with low loss of adsorbents	Specific application; not effective for all dyes
Advanced oxidation	Main treatment	Complete mineralization ensured. Growing number of commercial applications. Effective pretreatment methodology in integrated systems and enhances biodegradability	Cost intensive

(continued)

Table 1.3 (continued)

Treatment methodology	Treatment stage	Advantage	Limitations
Photocatalysis	Posttreatment	Process carried out at ambient conditions. Inputs are not toxic and inexpensive. Complete mineralization with shorter detention times	Effective for small amount of colored compounds. Expensive process
Enzymatic treatment	Posttreatment	Effective for specifically selected compounds	Enzyme isolation and purification is cumbersome. Efficiency curtailed due to the presence of interferences
Physical treatment			
Membrane filtration	Main treatment	Removes all dye types; recovery and reuse of chemicals and water	Concentrated sludge production
Irradiation	Posttreatment	Effective oxidation at lab scale	Requires a lot of dissolved oxygen (O2)
Adsorption with			
Activated carbon	Posttreatment	Economically attractive. Good removal efficiency of wide variety of dyes	Very expensive; cost-intensive regeneration process
Peat	Pretreatment	Effective adsorbent due to cellular structure. No activation is required	Surface area is lower than activated carbon
Coal ashes	Pretreatment	Economically attractive. Good removal efficiency	Larger contact times and huge quantities are required. Specific surface area for adsorption are lower than activated carbon
Wood sawdust	Pretreatment	Effective adsorbent due to cellular structure. Economically attractive. Good adsorption capacity for acid dyes	Long retention times and huge quantities are required
Silica gels	Pretreatment	Effective for basic dyes	Side reactions prevent commercial application
Biological treatment			
Aerobic	Posttreatment	Partial or complete decolorization for all classes of dyes	Expensive
Anaerobic	Main treatment	Resistant to a wide variety of complex colored compounds. Biogas produced is used for stream generation	Longer acclimatization phase
Single cell (fungal, algal, bacterial)	Posttreatment	Good removal efficiency for low volumes and concentrations. Very effective for specific color removal	Culture maintenance is cost intensive. Cannot cope up with large volumes of wastewater

Table 1.4 Molecular structure of various commercially available dyes

Dyes	Molecular structure	C.I.
Acid dyes		
Acid black 1	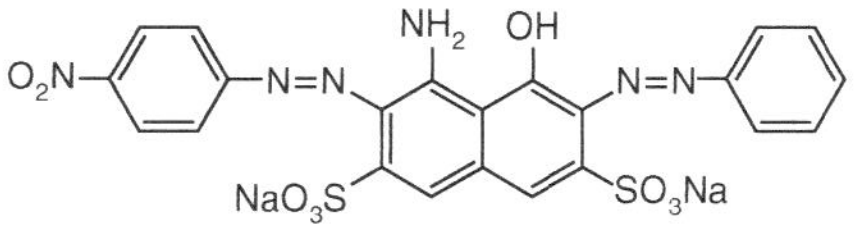	20,470
Acid black 24	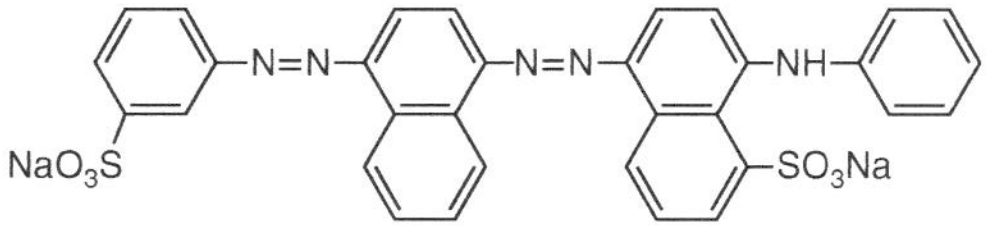	26,370
Acid black 48	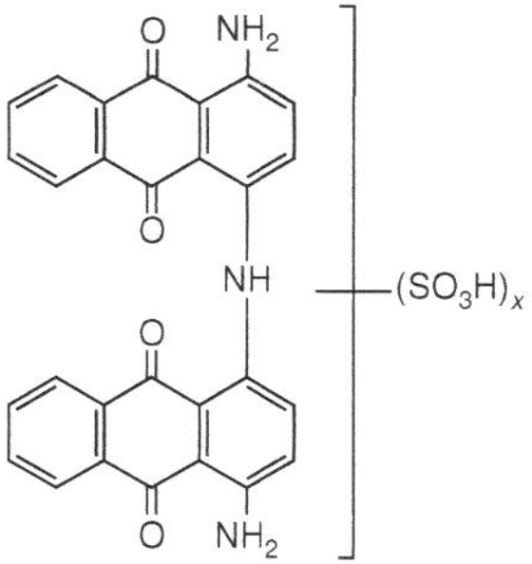	65,005

(continued)

Table 1.4 (continued)

Dyes	Molecular structure	C.I.
Acid black 180	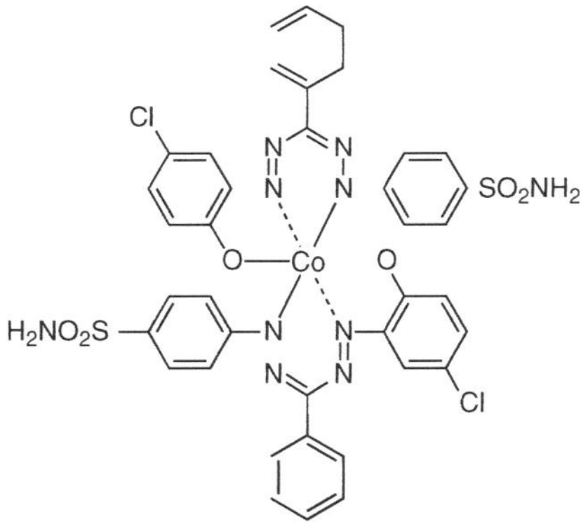	13,710
Acid blue 25	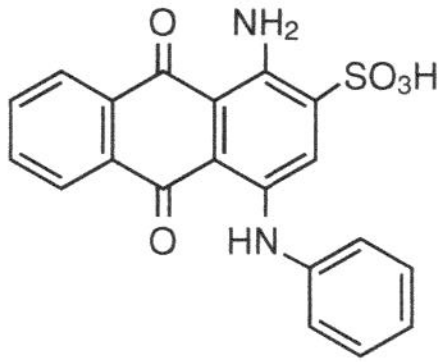	62,055
Acid blue 40	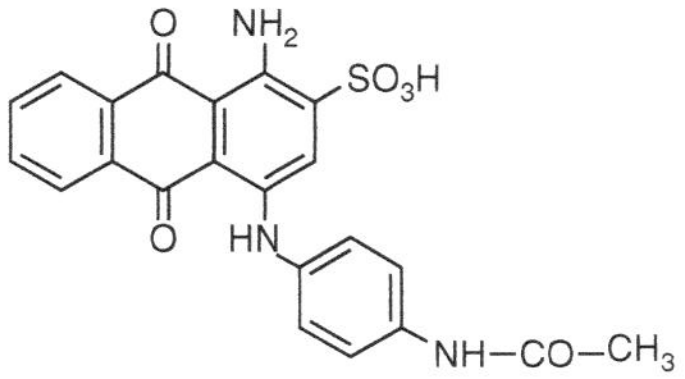	62,125

Acid blue 92	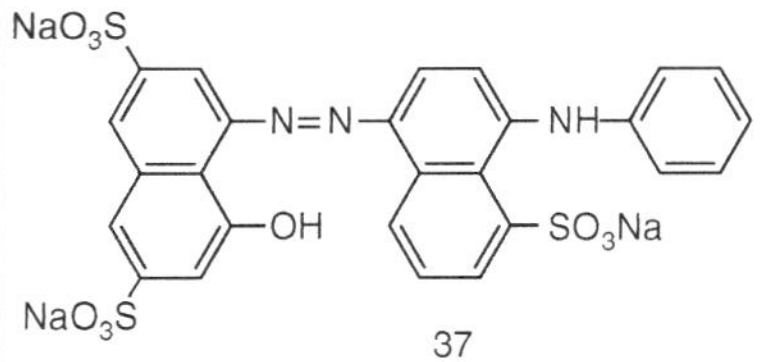 37	13,390
Acid brown 20	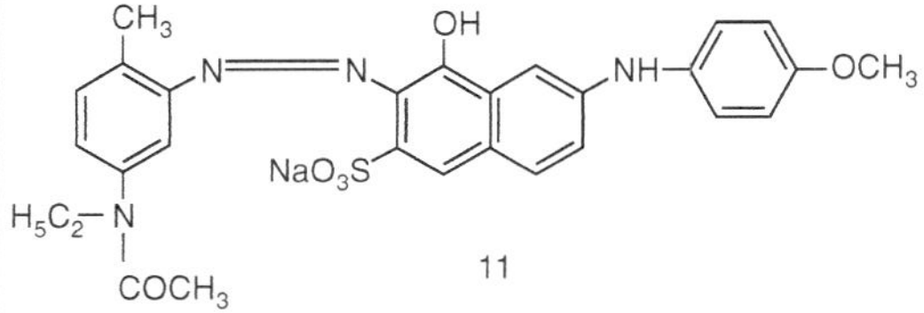 11	17,640
Acid brown 123	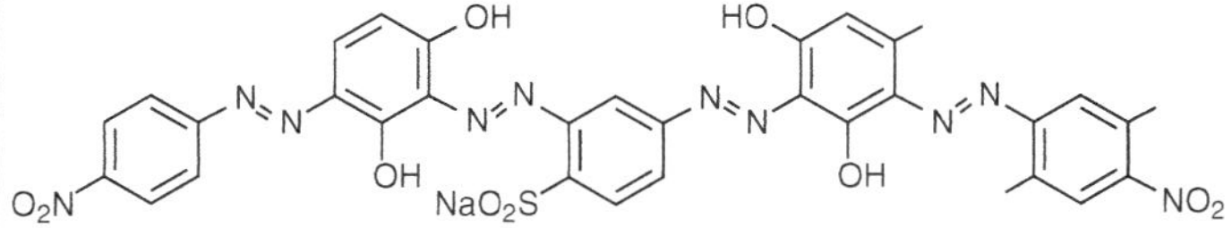	35,030
Acid green 1	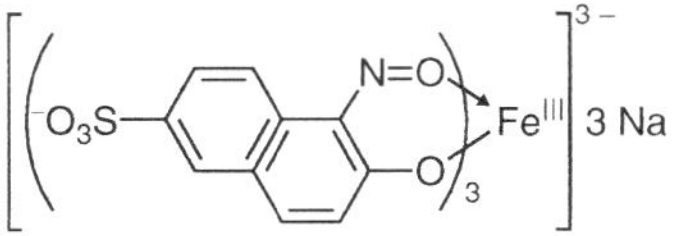	10,020
Acid green 41	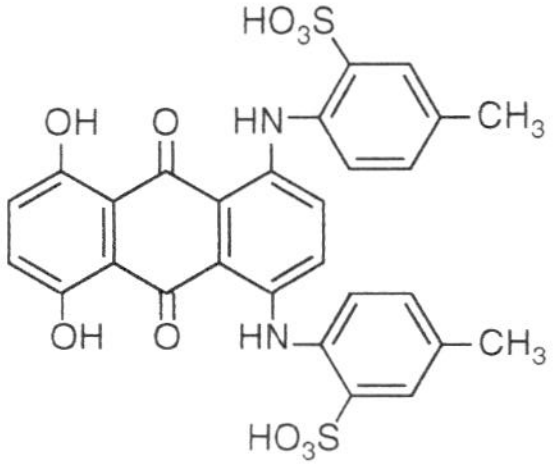	62,560

(continued)

Table 1.4 (continued)

Dyes	Molecular structure	C.I.
Acid orange 3	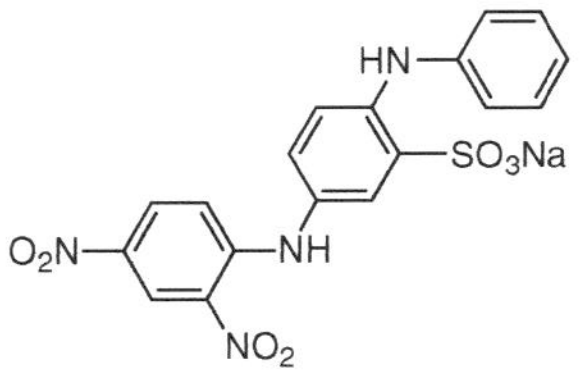	10,385
Acid orange 7	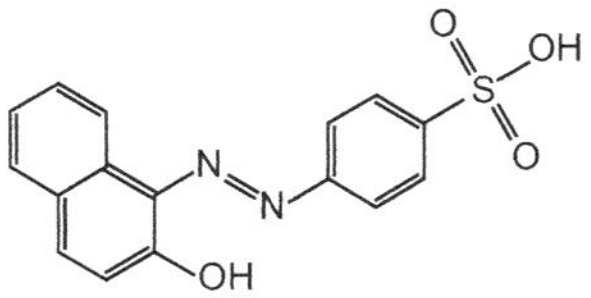	15,510
Acid orange 74	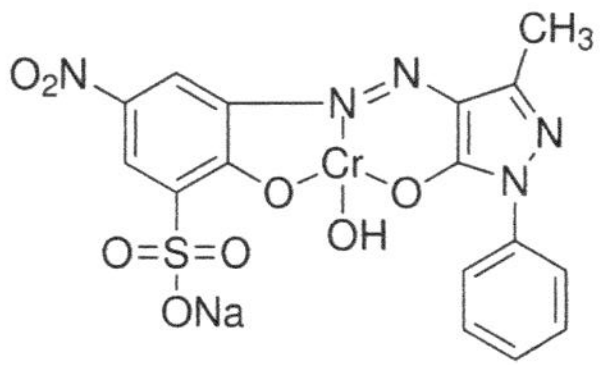	18,745
Acid red 1	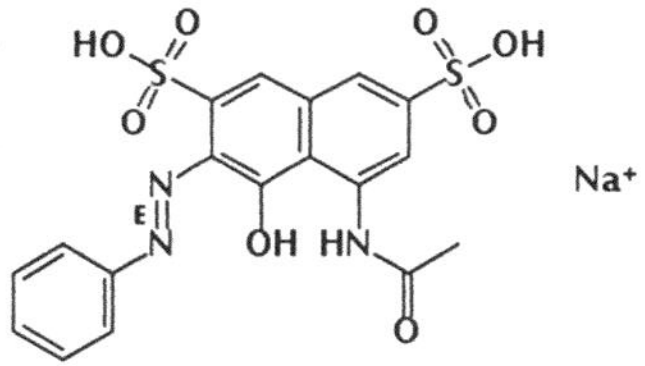	18,050

Acid red 14	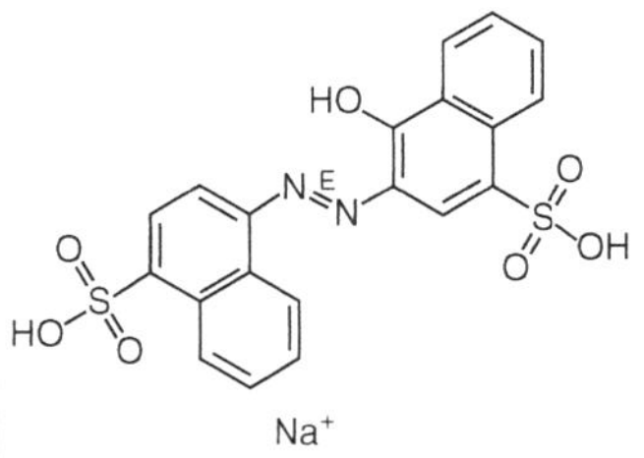	14,720
Acid red 52	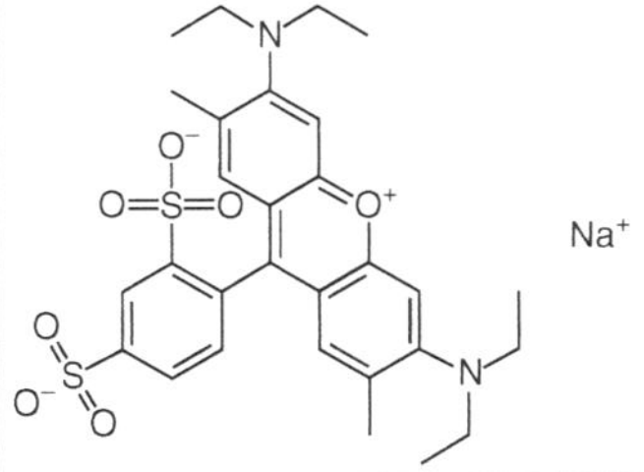	45,100
Acid red 134	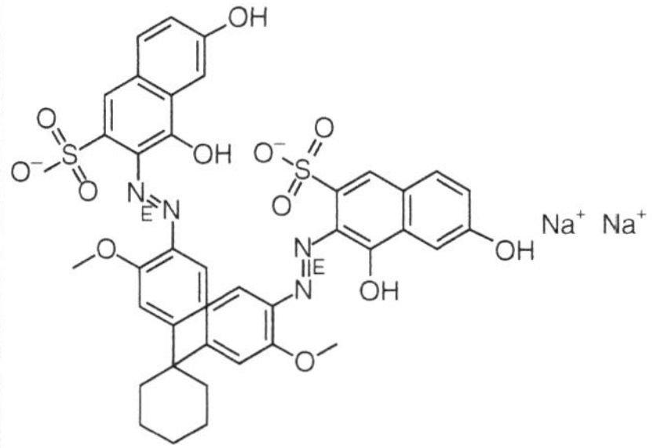	24,810

(continued)

Table 1.4 (continued)

Dyes	Molecular structure	C.I.
Acid red 249	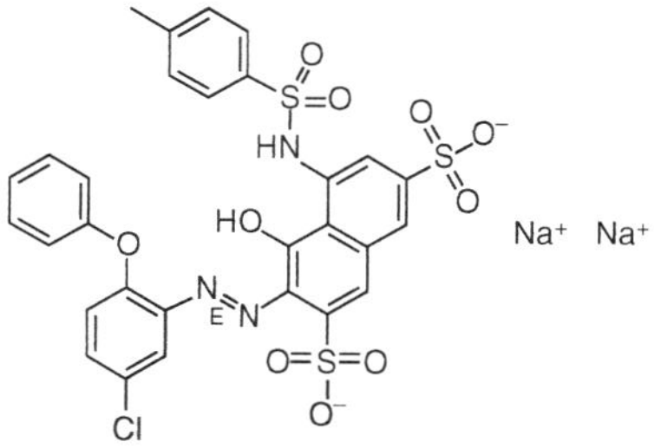	18,134
Acid violet 14	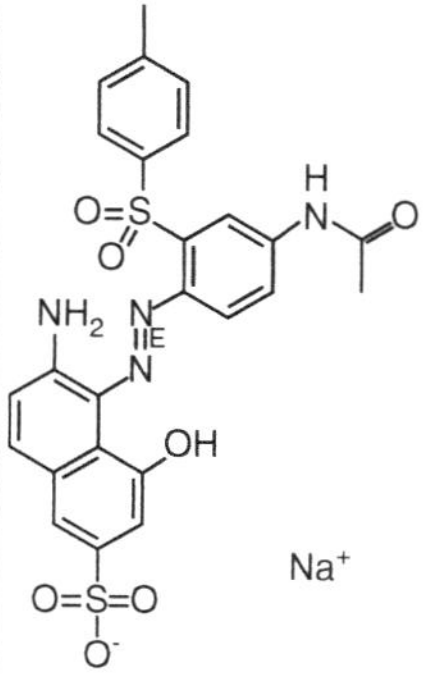	17,080

Acid violet 17	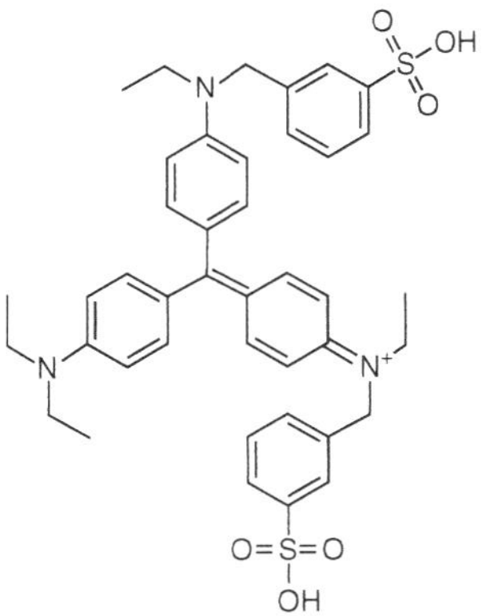	42,650
Acid violet 42	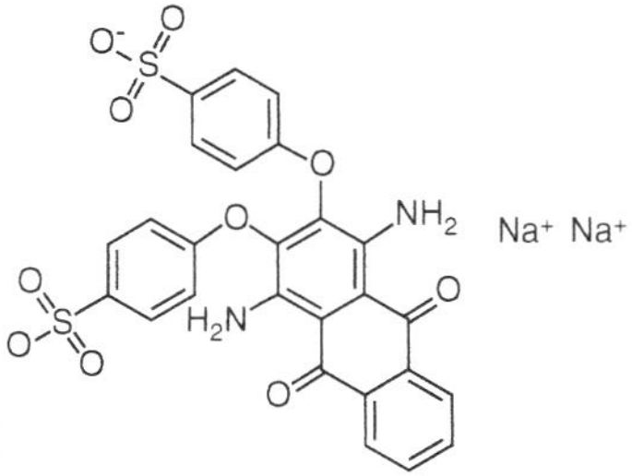	62,026
Acid yellow 1	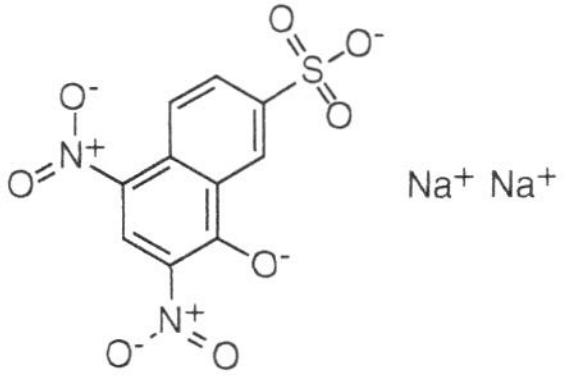	10,316

(continued)

Table 1.4 (continued)

Dyes	Molecular structure	C.I.
Acid yellow 11	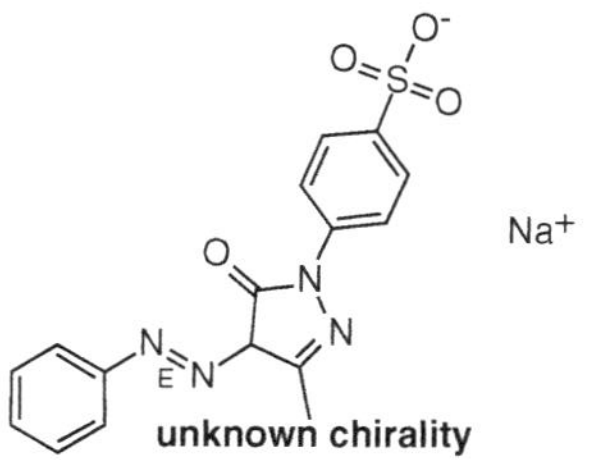	18,820
Acid yellow 38	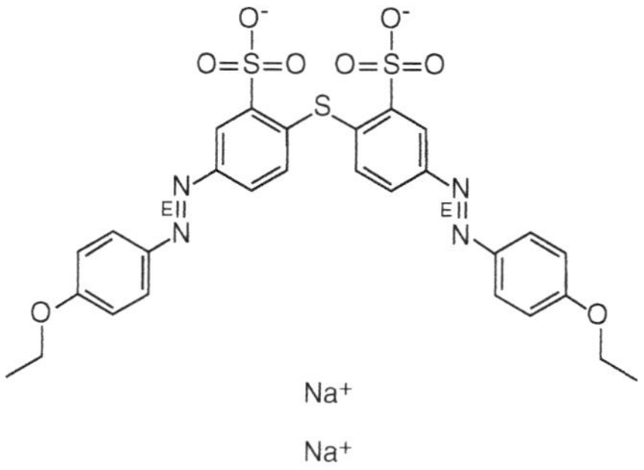	25,135
Acid yellow 76	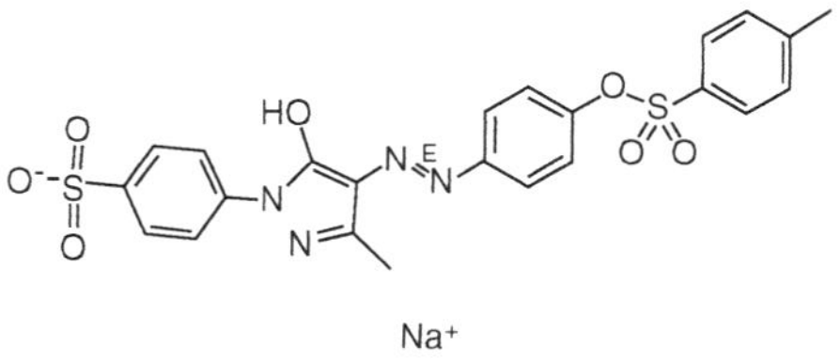	18,850

Acid yellow 127	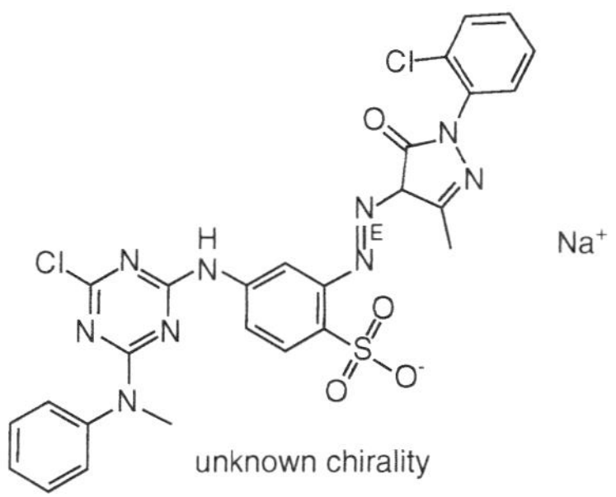	18,888
Basic dyes		
Basic blue 3	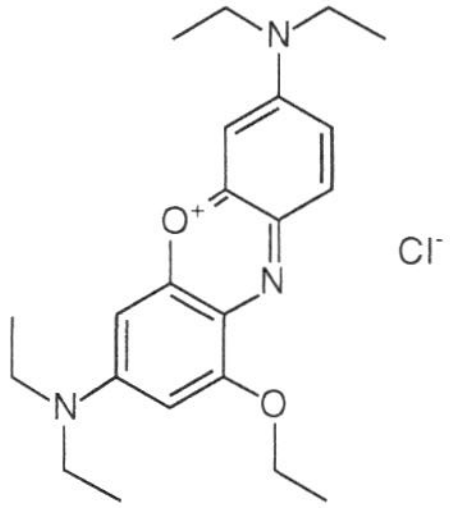	51,005
Basic blue 64	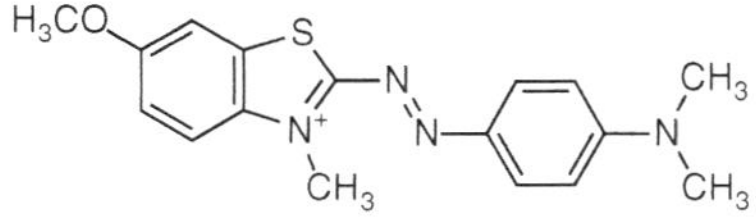	

(continued)

Table 1.4 (continued)

Dyes	Molecular structure	C.I.
Basic blue 140	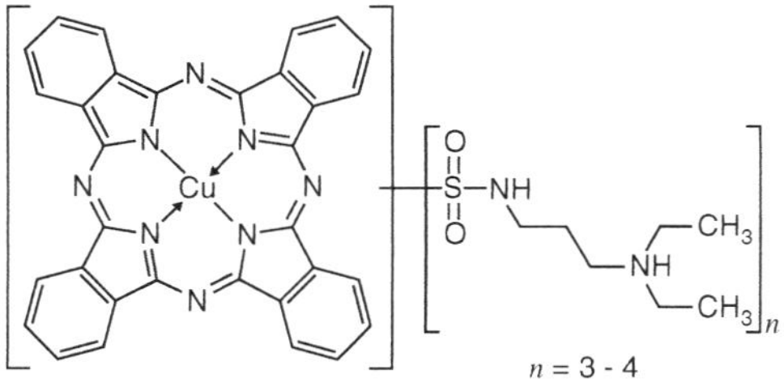	
Basic green 4	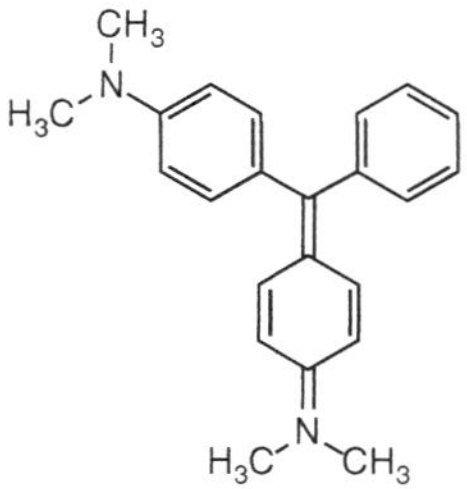	42,000
Basic orange 1	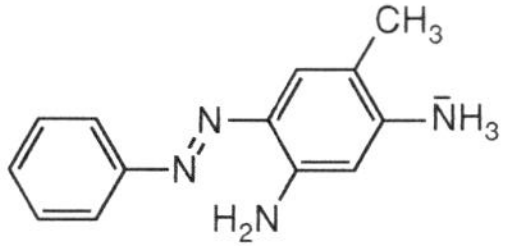	11,320
Basic red 18	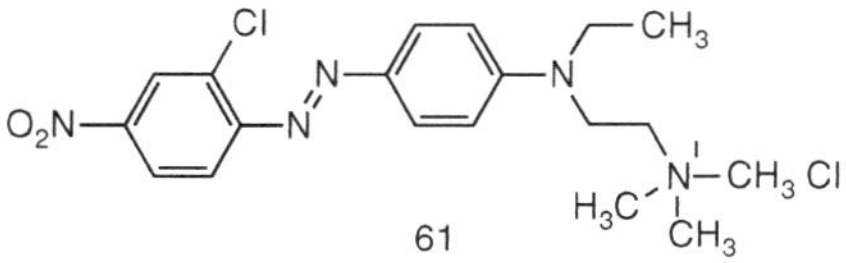 61	11,085

Basic red 46	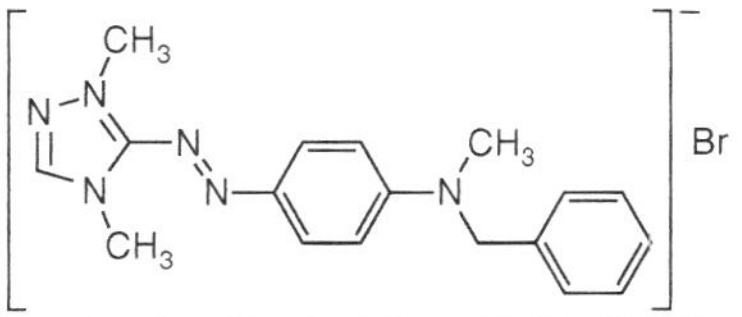	110,825
Basic violet 16	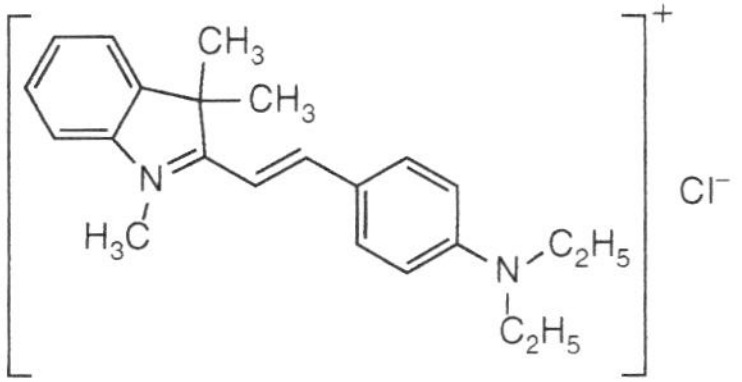	48,013
Basic yellow 15	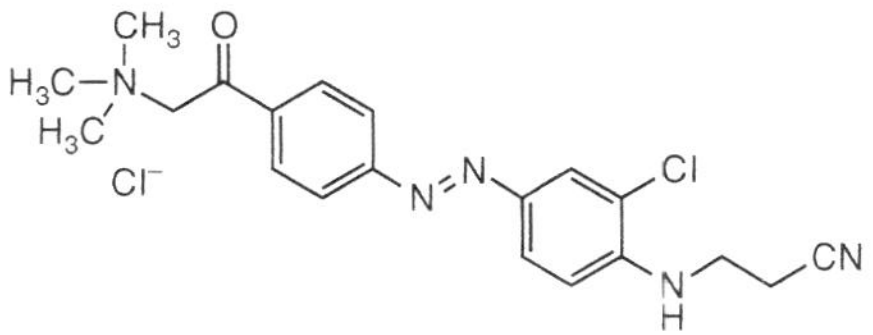	11,087
Basic yellow 51	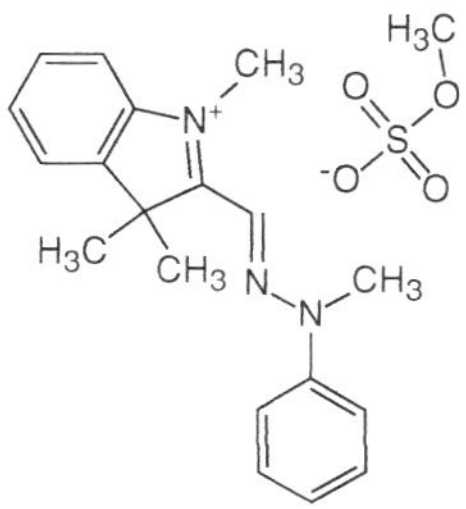	480,538

(continued)

Table 1.4 (continued)

Dyes	Molecular structure	C.I.
Direct dyes		
Direct black 22	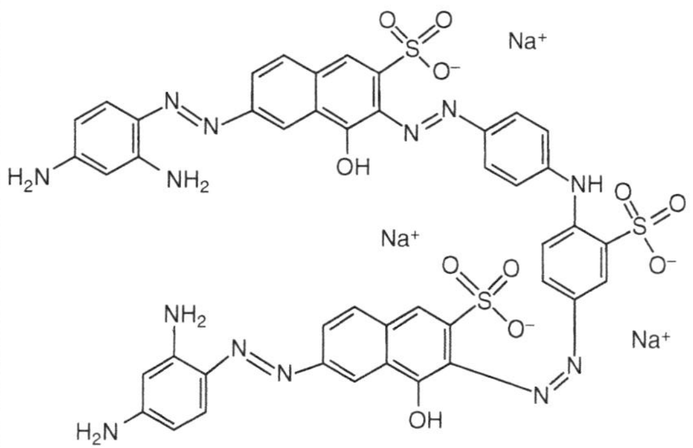	35,435
Direct black 168	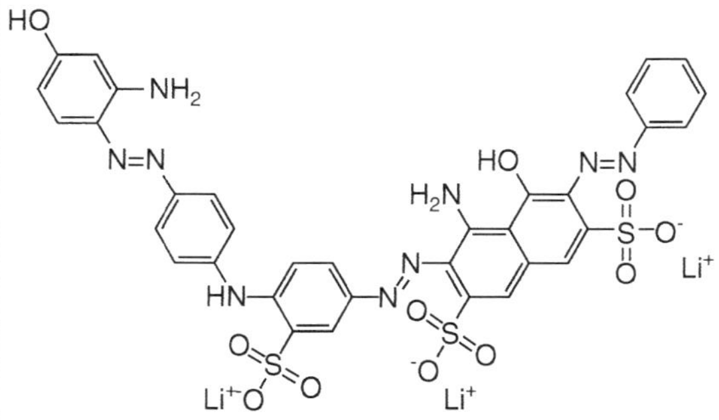	30,410

Direct blue 78	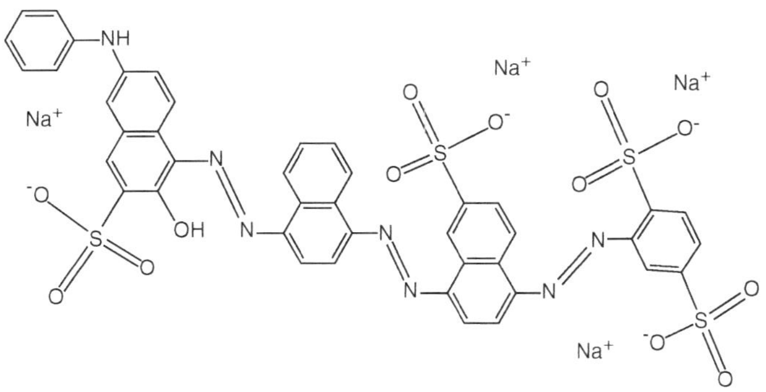	34,200
Direct blue 86	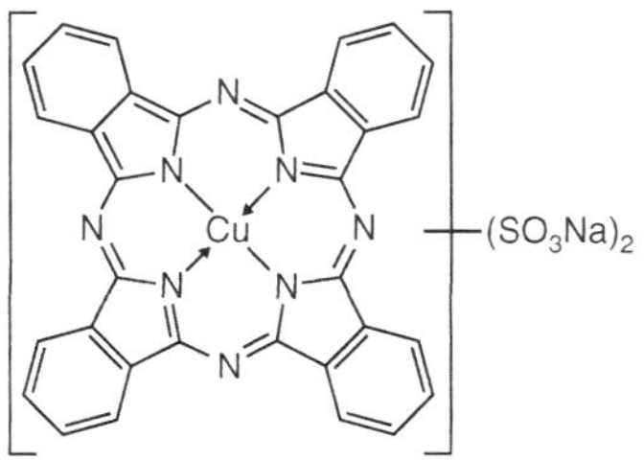	74,180
Direct blue 93	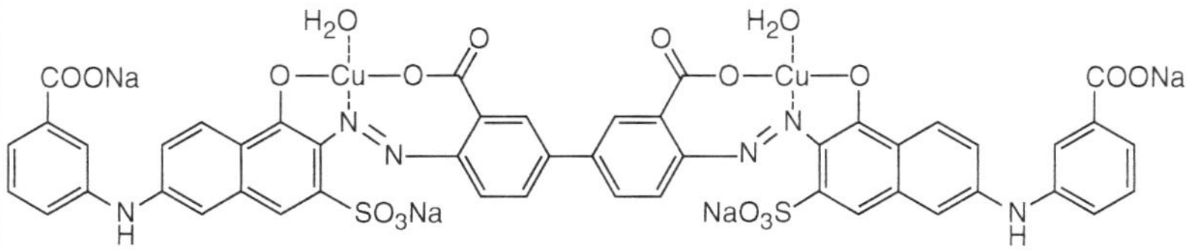	22,810

(continued)

Table 1.4 (continued)

Dyes	Molecular structure	C.I.
Direct green 26	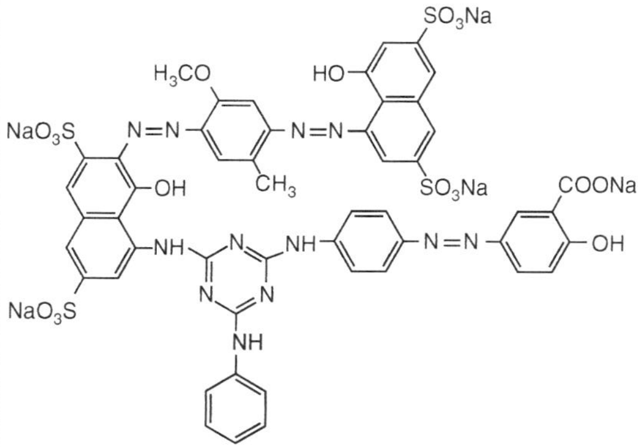	34,045
Direct orange 102	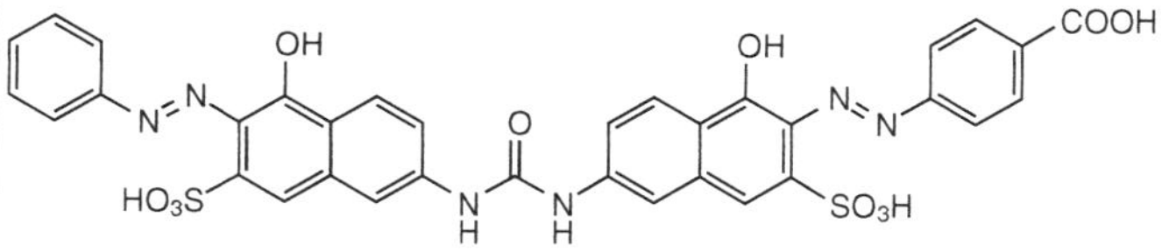	29,156
Direct red 81	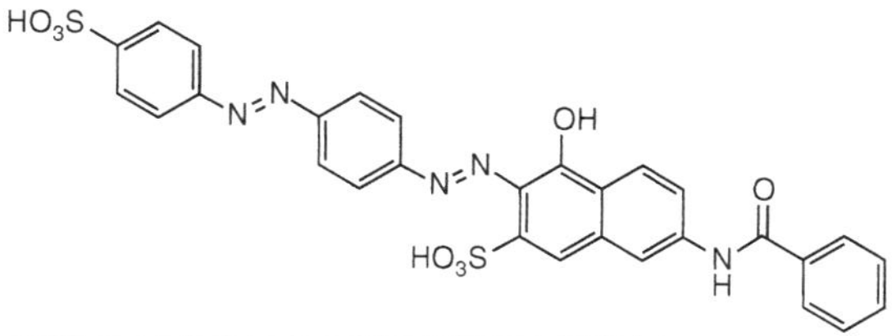	28,160

Direct violet 51	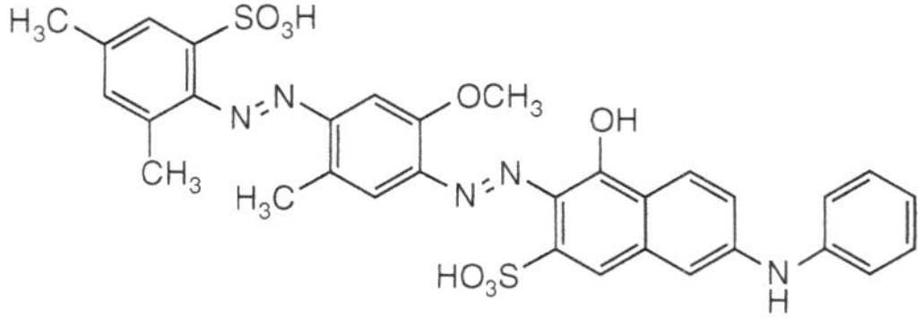	27,855
Direct yellow 12	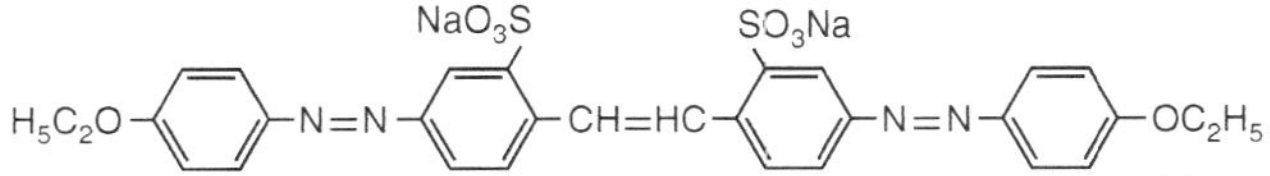	24,895
Direct yellow 28	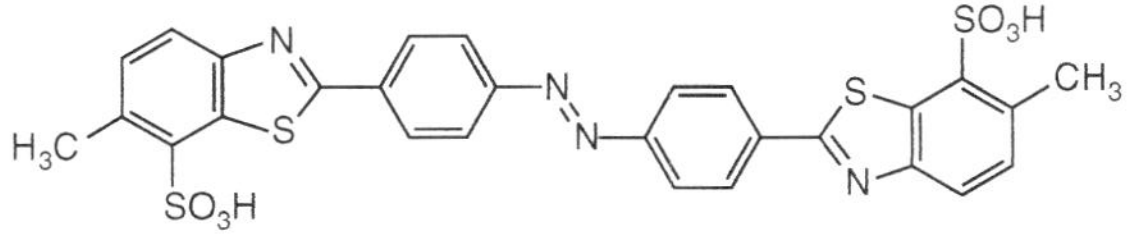	19,555
Disperse dyes		
Disperse blue 7	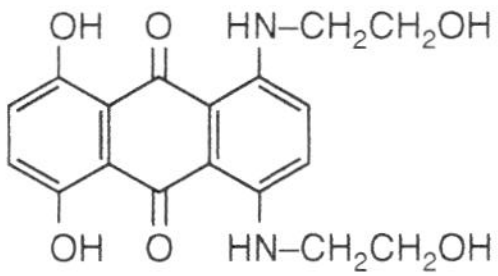	62,500
Disperse blue 14	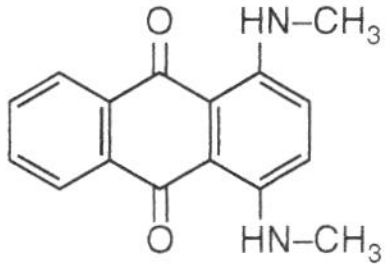	61,500

(continued)

Table 1.4 (continued)

Dyes	Molecular structure	C.I.
Disperse blue 56	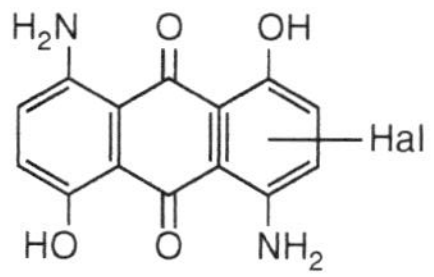	63,285
Disperse blue 73	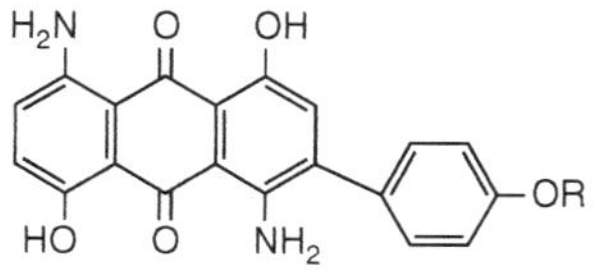	63,265
Disperse blue 148	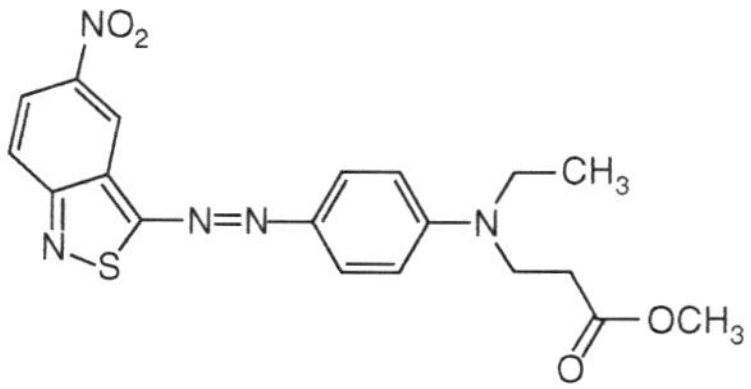	11,124
Disperse blue 354	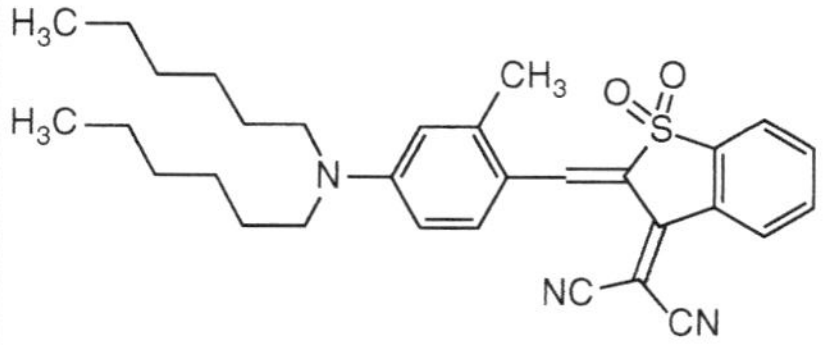	48,480

Disperse green 9	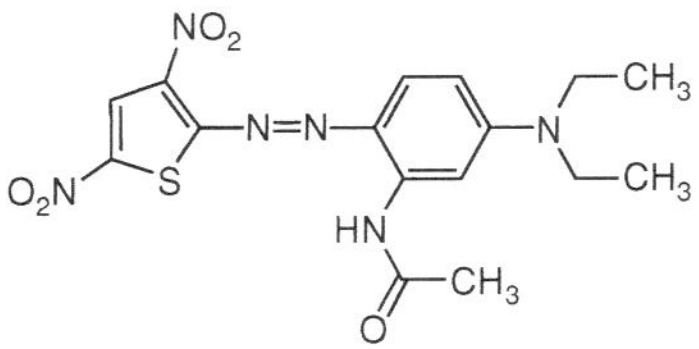	110,795
Disperse orange 29	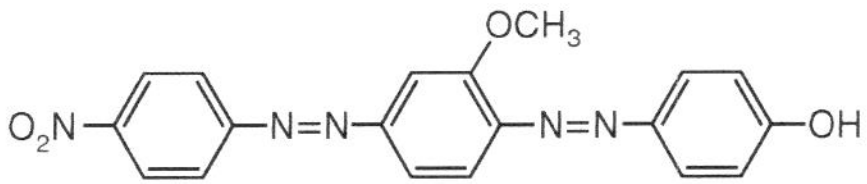	26,077
Disperse orange 56	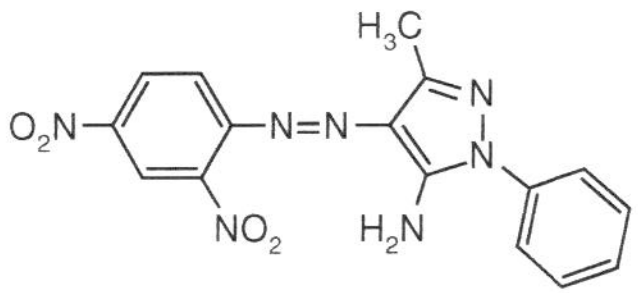	–
Disperse red 17	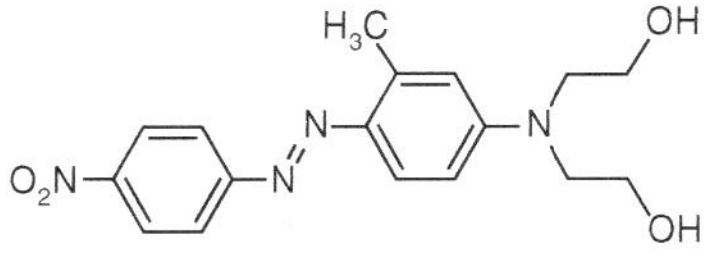	11,210
Disperse red 177	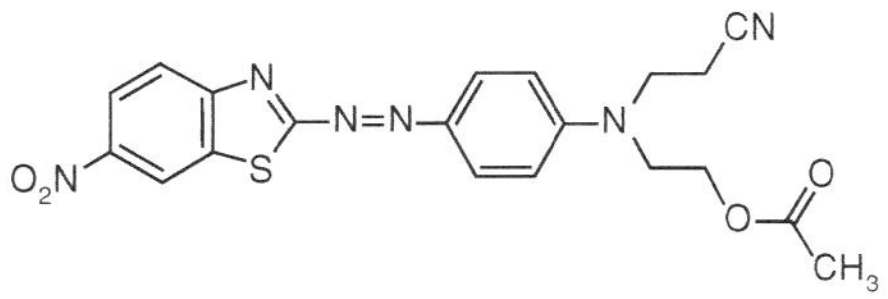	11,122

(continued)

Table 1.4 (continued)

Dyes	Molecular structure	C.I.
Disperse red 338	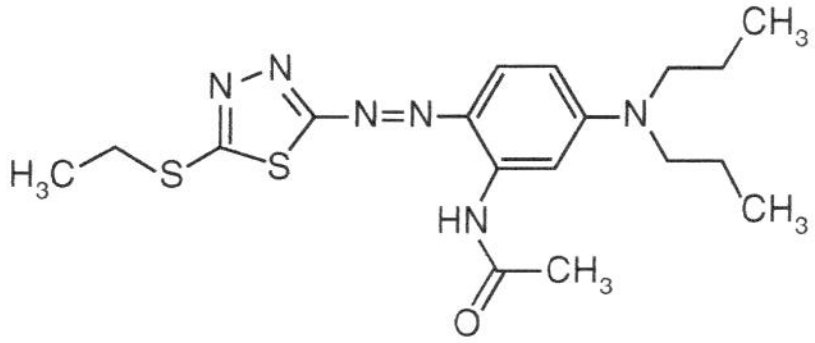	111,430
Disperse violet 28	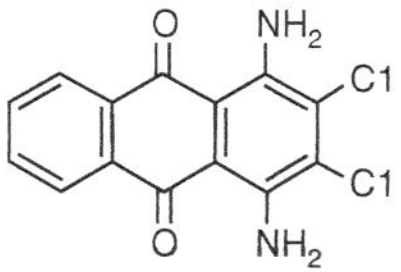	61,102
Disperse yellow 5	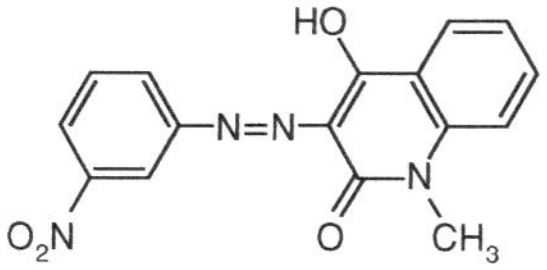	12,790
Disperse yellow 42	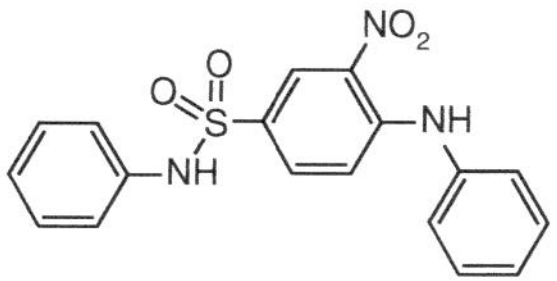	10,338
Disperse yellow 64	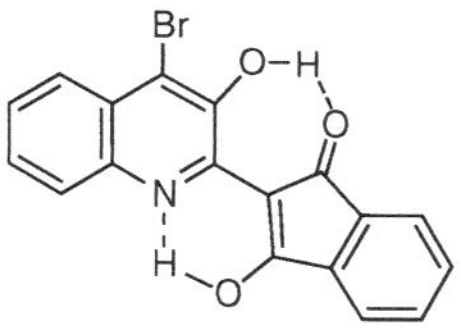	42,023

Disperse yellow 211	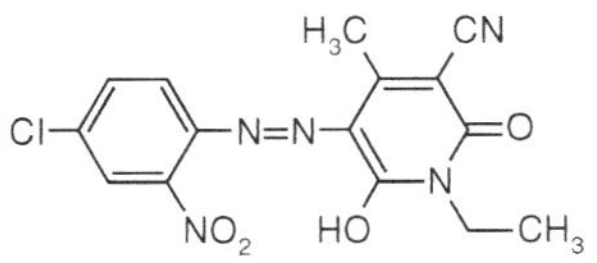	12,755
Disperse yellow 332	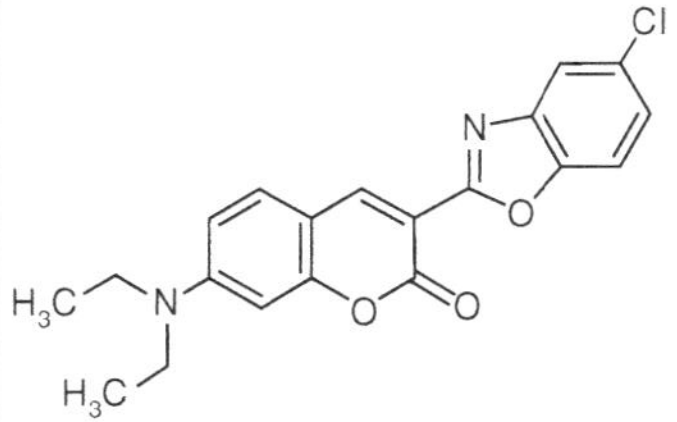	55,165
Mordant dyes		
Mordant black 3	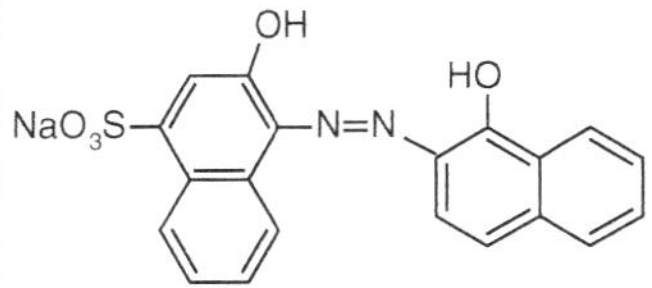	14,640
Mordant blue 7	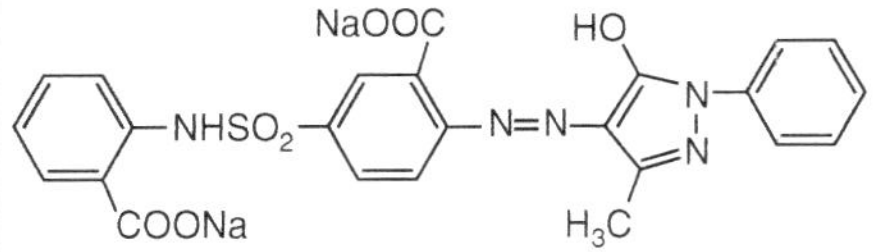	17,940
Mordant brown 33	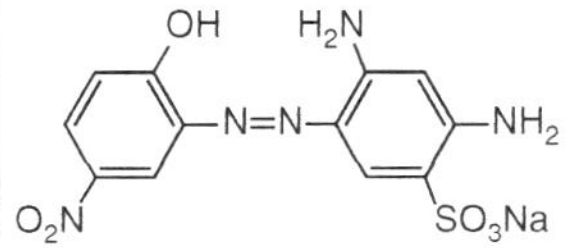	13,250

(continued)

Table 1.4 (continued)

Dyes	Molecular structure	C.I.
Mordant red 7	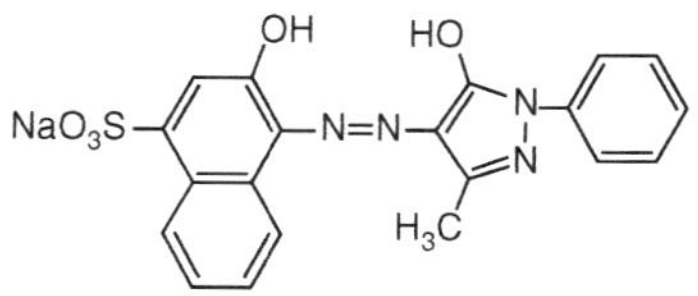	18,760
Mordant yellow 5	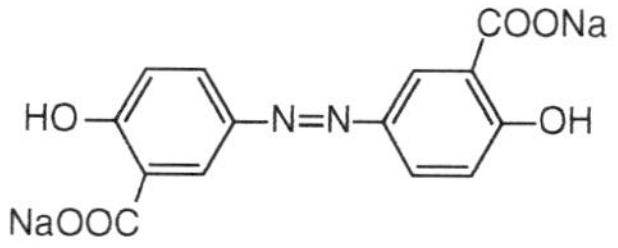	14,130
Reactive dyes		
Reactive black 5	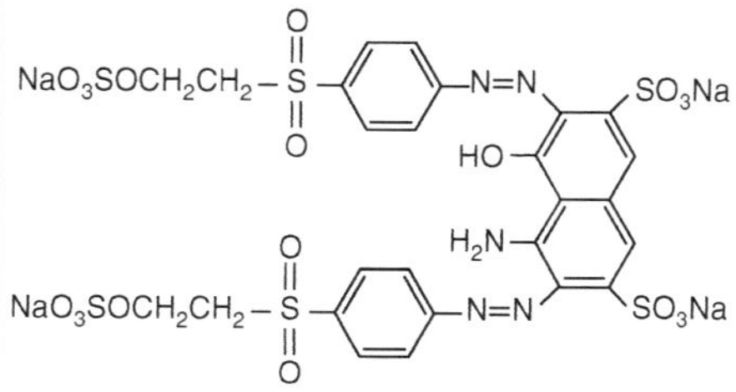	20,505

Reactive black 31	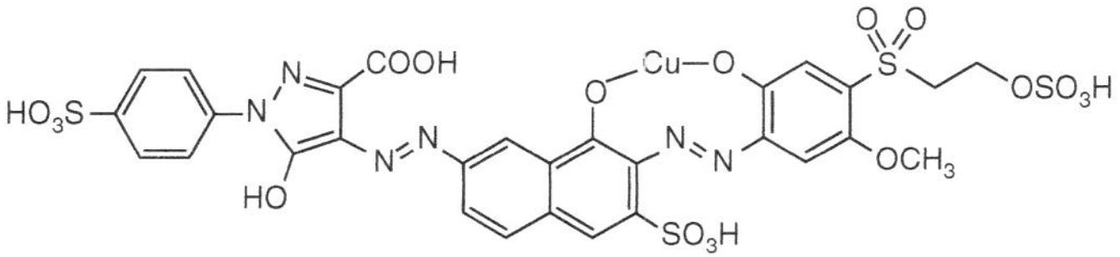	
Reactive blue 15	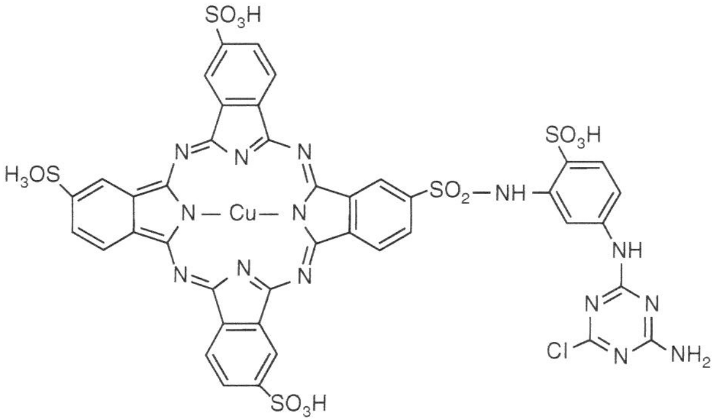	74,459
Reactive blue 19	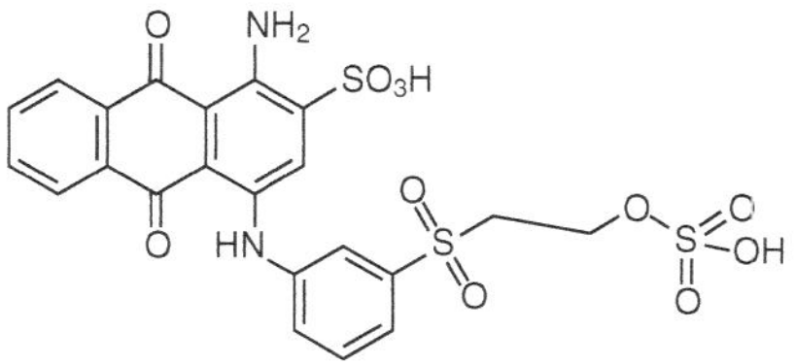	61,200

(continued)

Table 1.4 (continued)

Dyes	Molecular structure	C.I.
Reactive blue 216	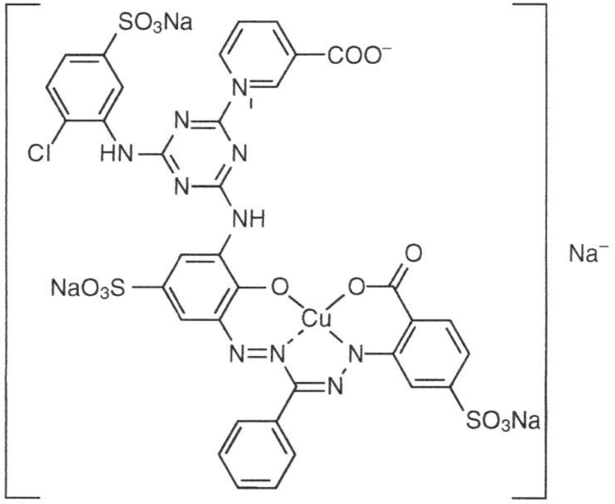	137,155
Reactive yellow15	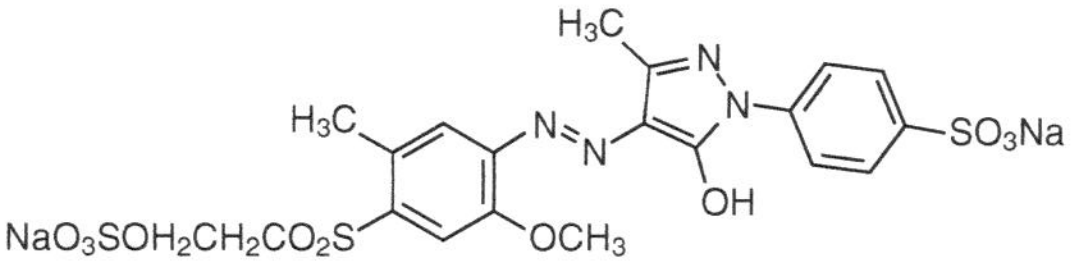	
Reactive red 23	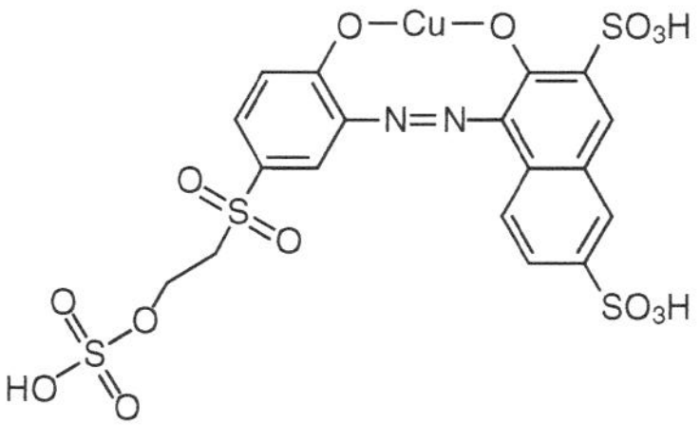	16,202

Reactive red 24	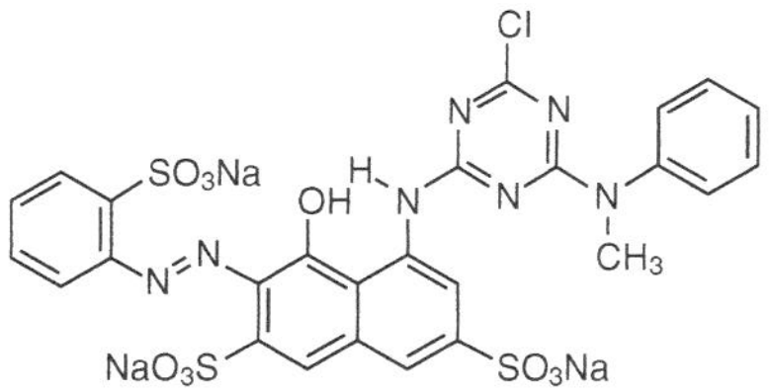	
Reactive brown 10	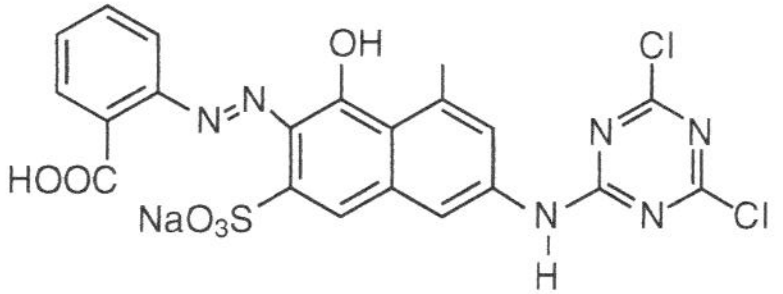	
Reactive violet 4	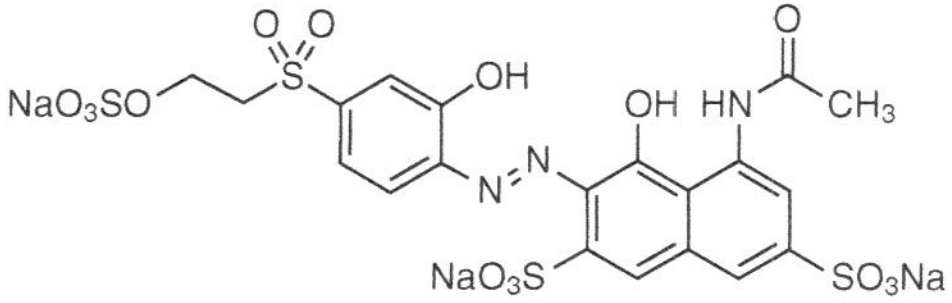	17,965
Solvent dyes		
Solvent black 5	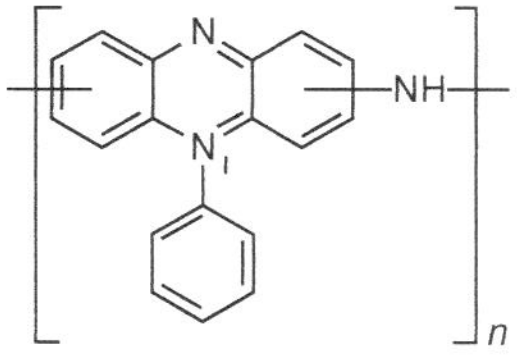	50,415

(continued)

Table 1.4 (continued)

Dyes	Molecular structure	C.I.
Solvent blue 66	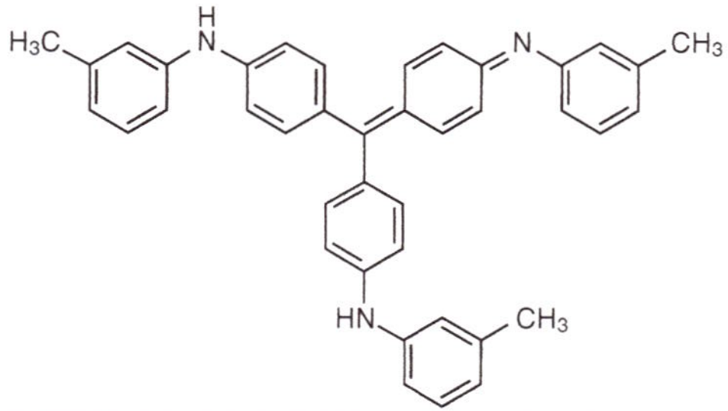	42,799
Solvent orange 5	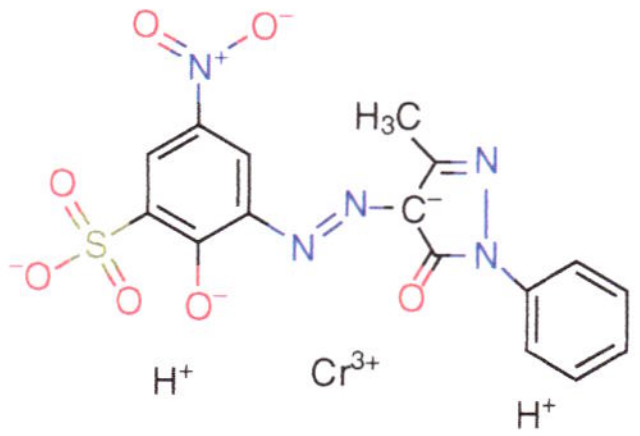	518,745:1
Solvent yellow 16	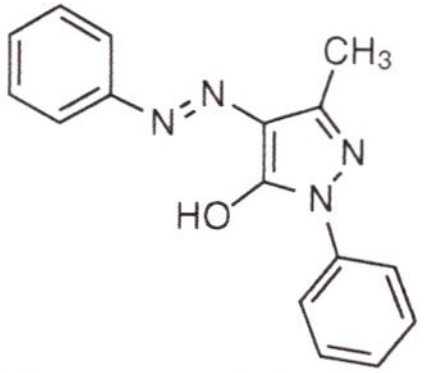	12,700

Solvent yellow 32	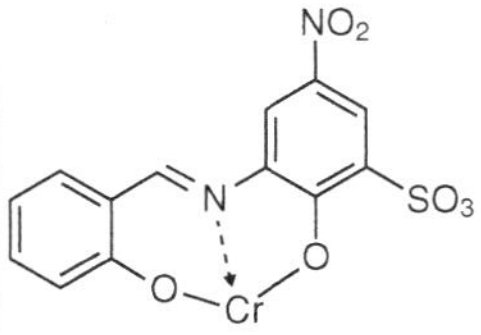	48,045
Solvent yellow 56	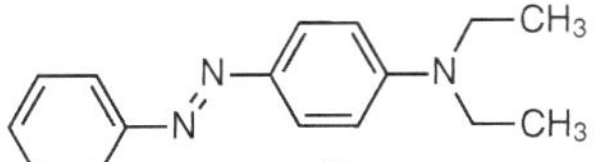 10	11,021
Solvent green 3	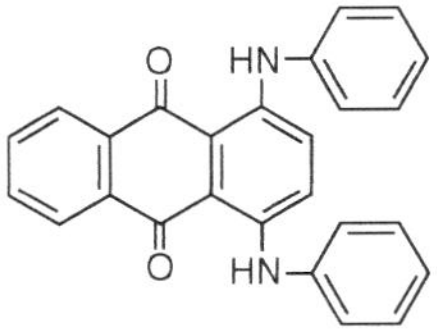	61,565
Solvent violet 13	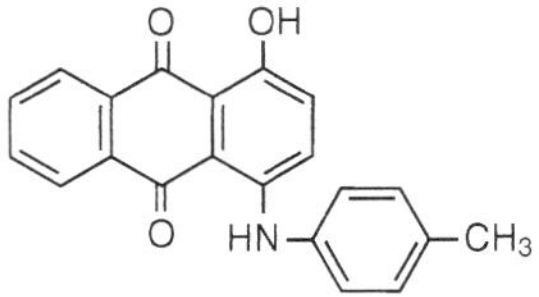	60,725
Solvent red 8	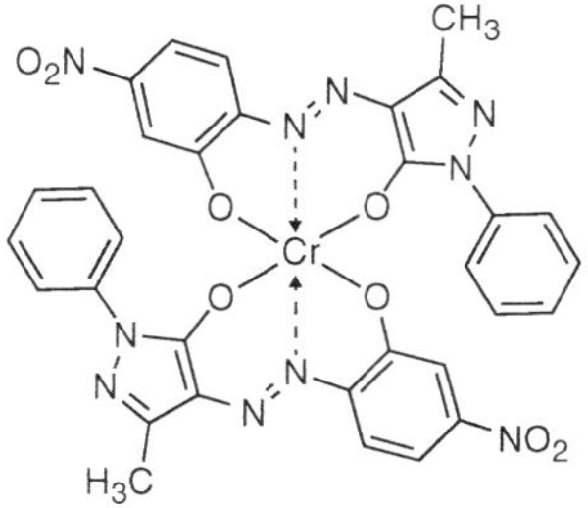	12,715

(continued)

Table 1.4 (continued)

Dyes	Molecular structure	C.I.
Solvent red 52	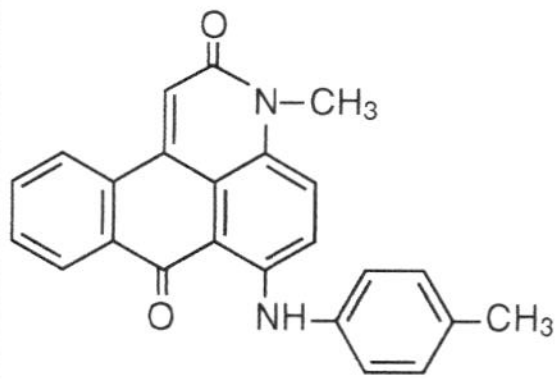	68,210
Solvent red 111	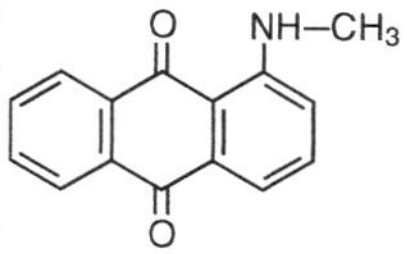	60,505
Vat dyes		
Vat yellow 10	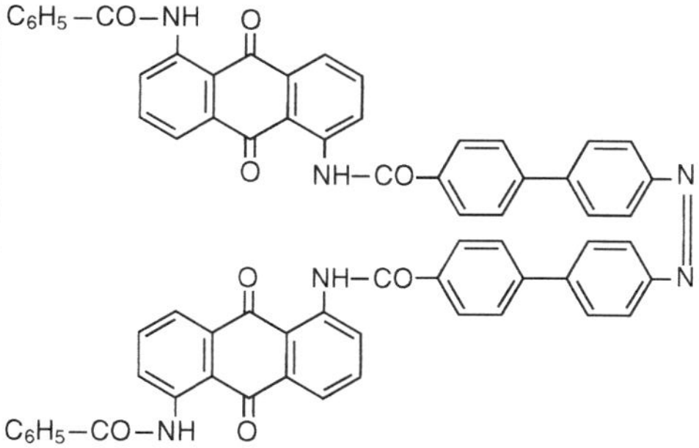	65,430

Vat yellow 12	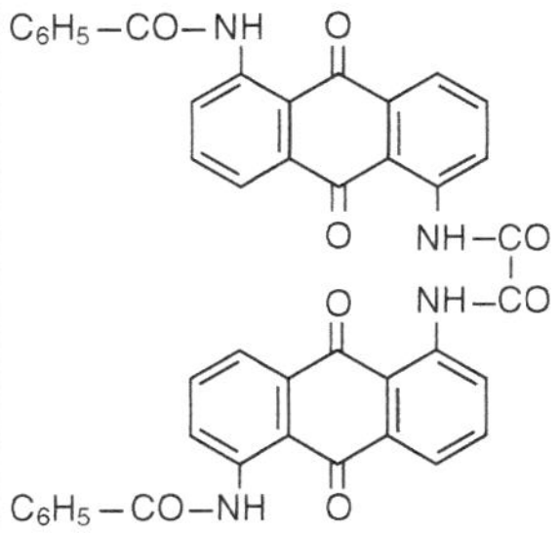	65,405
Vat yellow 20	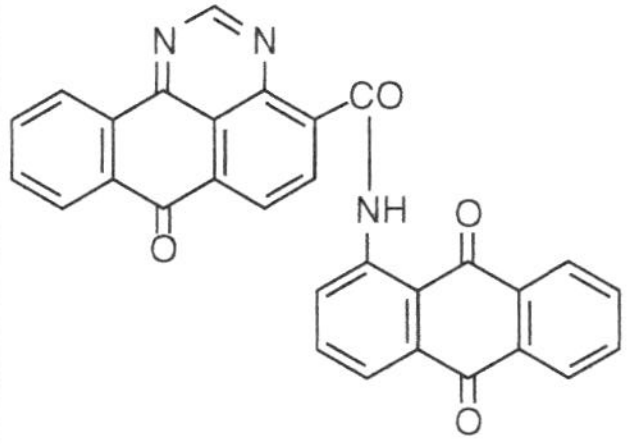	68,420
Vat yellow 28	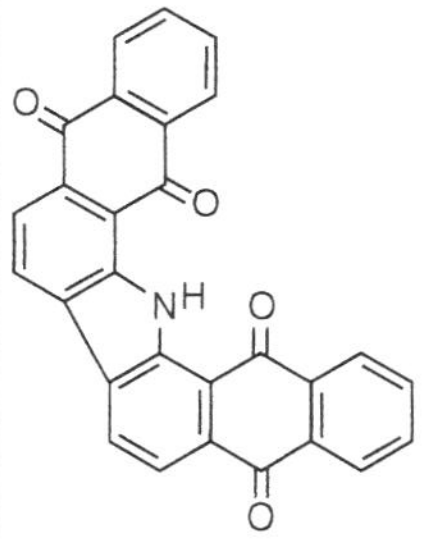	69,000

(continued)

Table 1.4 (continued)

Dyes	Molecular structure	C.I.
Vat orange 1	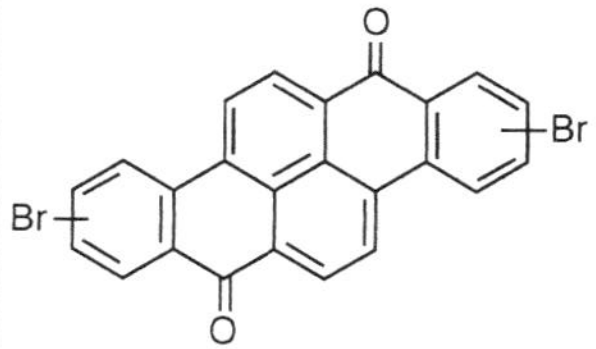	59,105
Vat orange 3	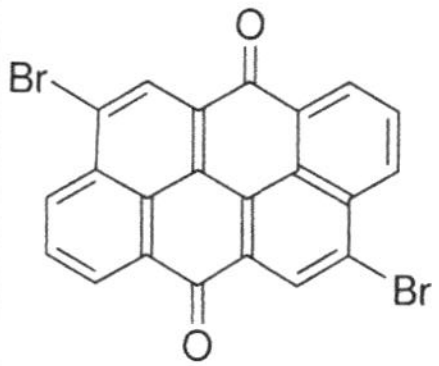	59,300
Vat orange 15	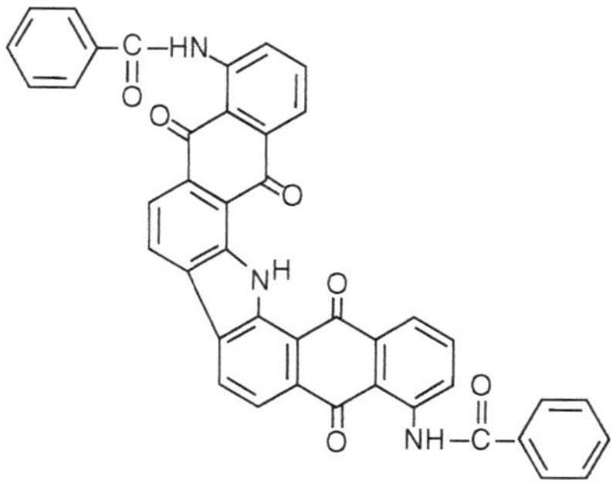	69,025

Vat orange 17	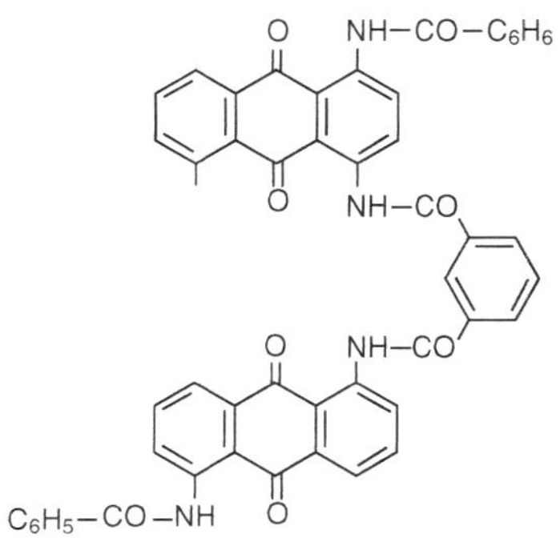	65,415
Vat green 1	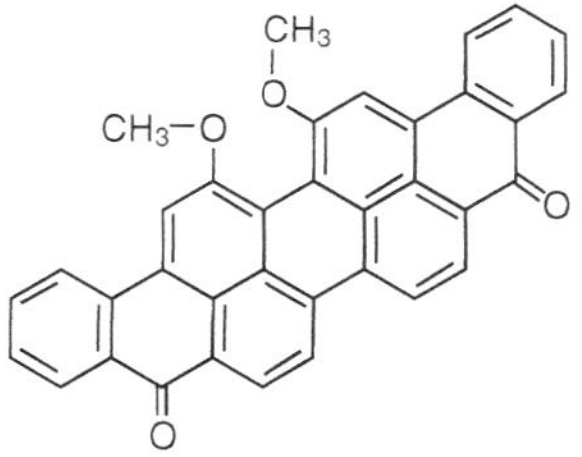	59,825

(continued)

Table 1.4 (continued)

Dyes	Molecular structure	C.I.
Vat green 8	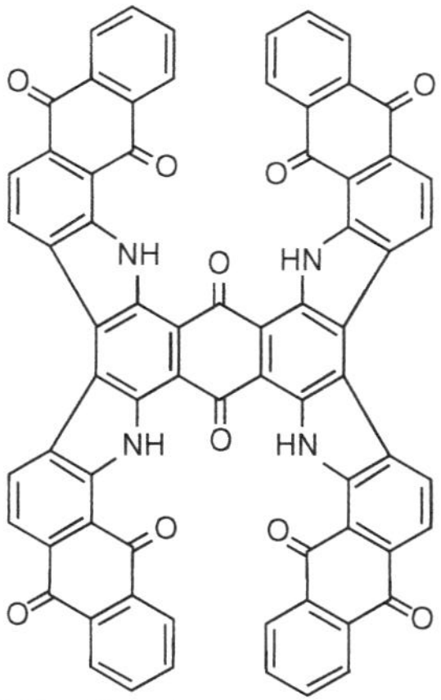	71,050
Vat green 11	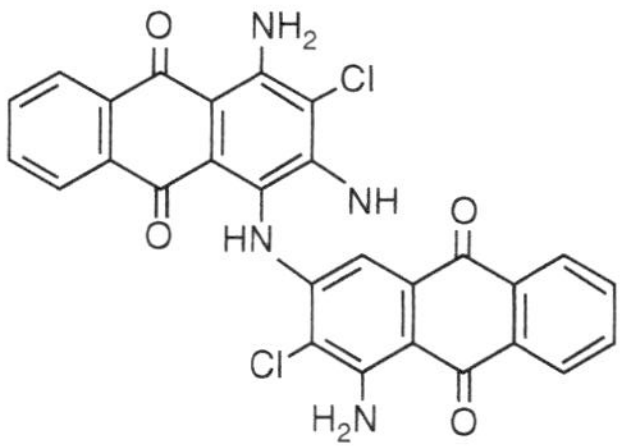	69,850

Vat green 12	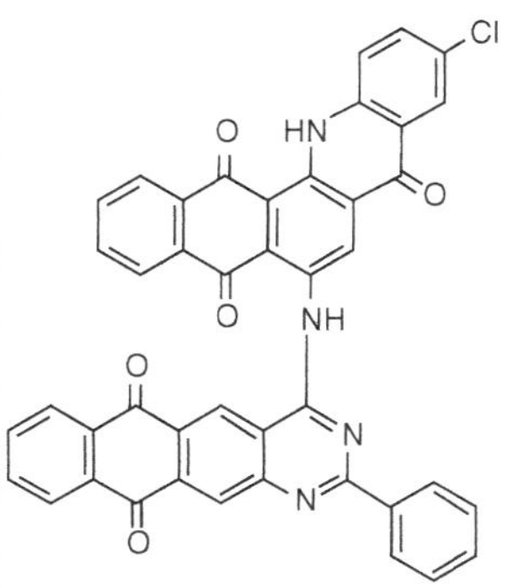	70,700
Vat red 10	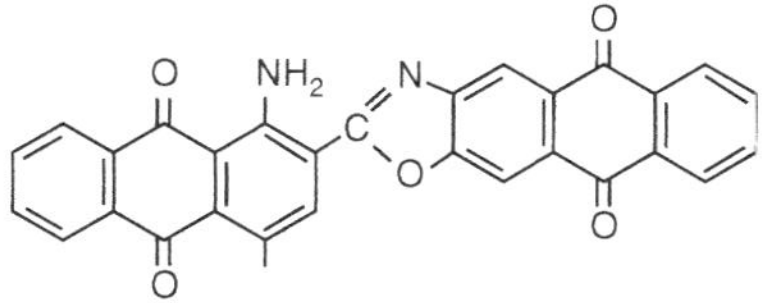	67,000
Vat red 28	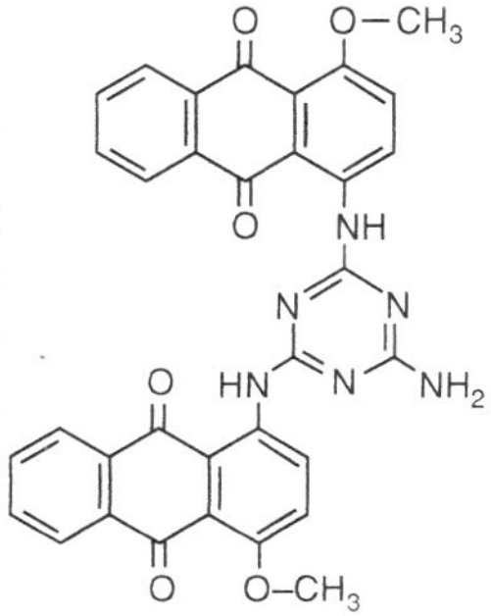	65,710

(continued)

Table 1.4 (continued)

Dyes	Molecular structure	C.I.
Vat blue 1 (indigo)	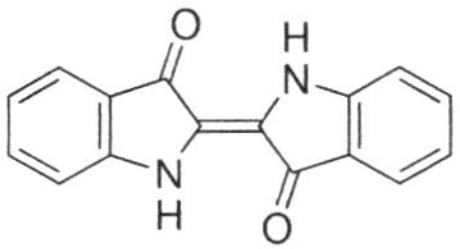	73,000
Vat blue 4	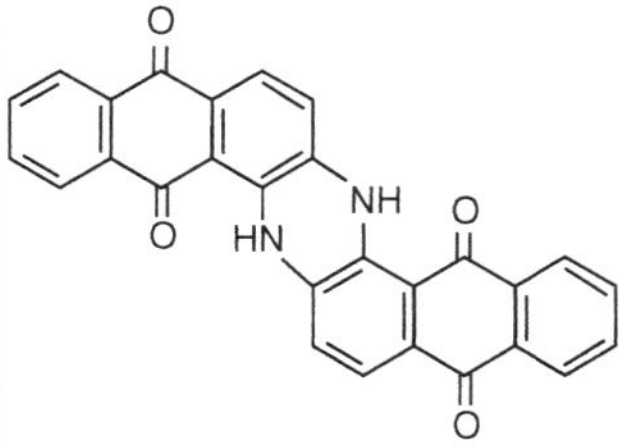	69,800
Vat blue 5	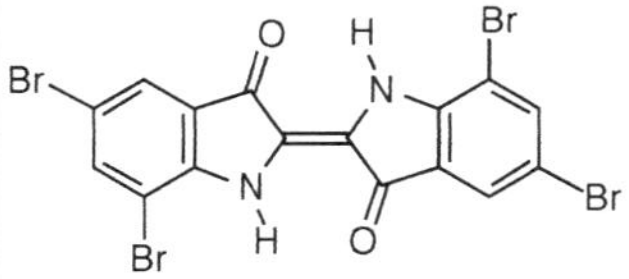	73,065
Vat blue 20	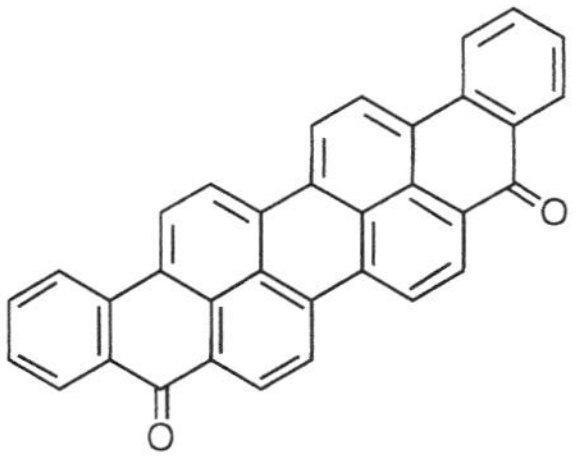	59,800

Vat blue 25	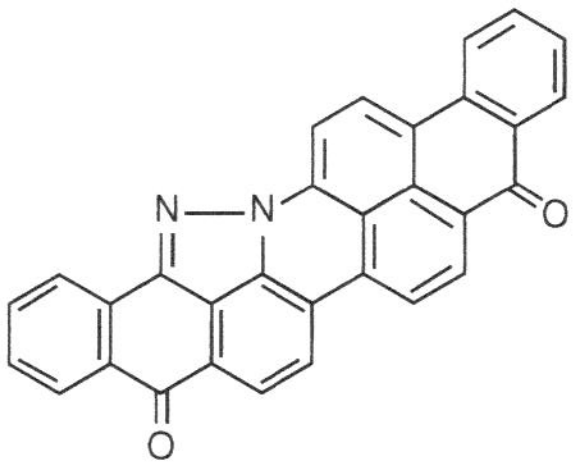	70,500
Vat blue 26	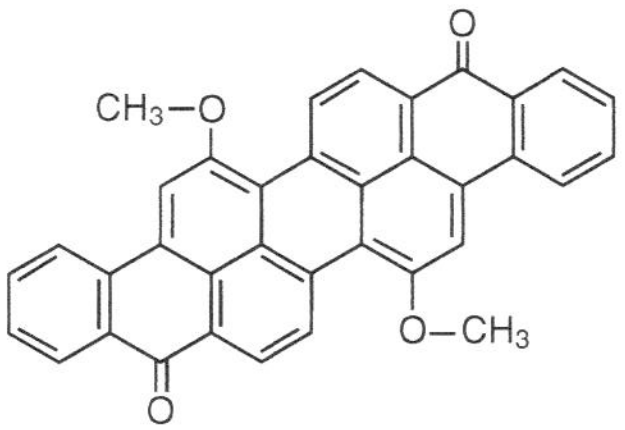	60,015
Vat blue 30	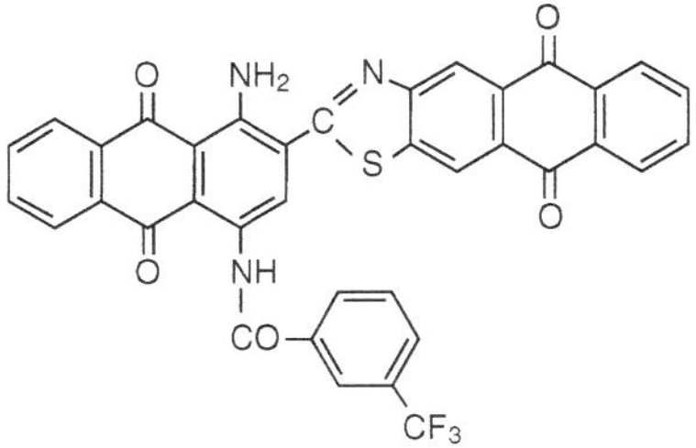	67,110

(continued)

Table 1.4 (continued)

Dyes	Molecular structure	C.I.
Vat blue 43	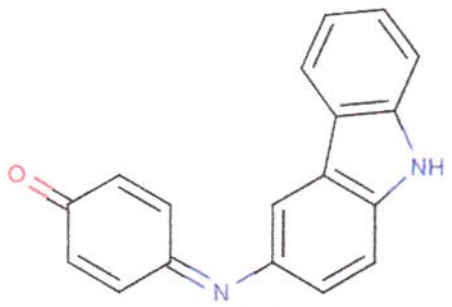	53,630
Vat blue 64	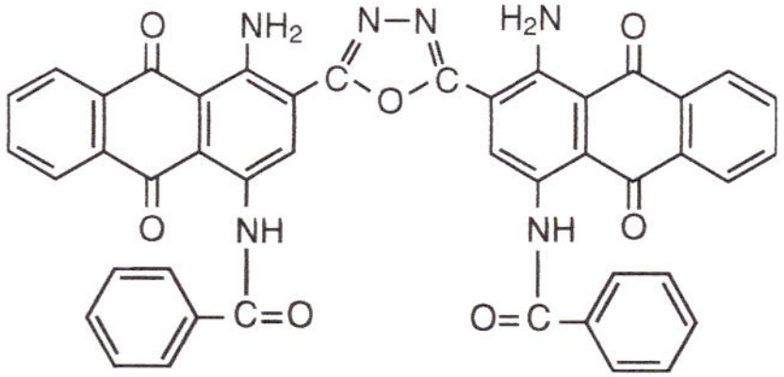	66,730
Vat violet 1	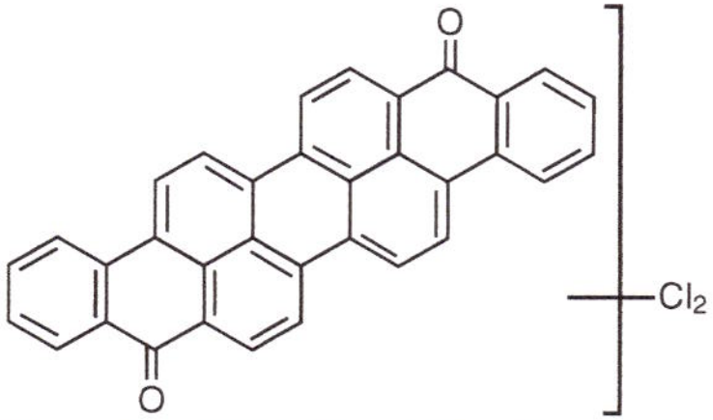	60,010
Vat violet 15	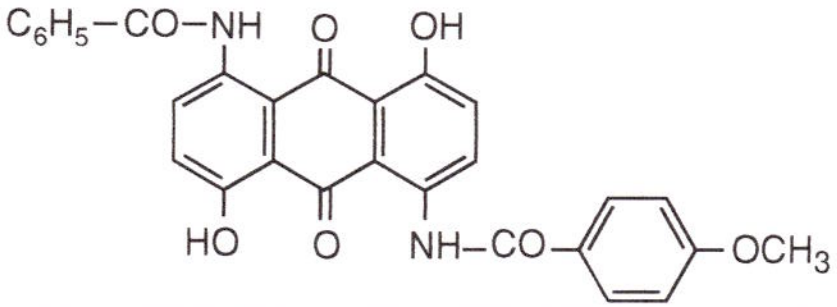	63,355

Vat violet 16	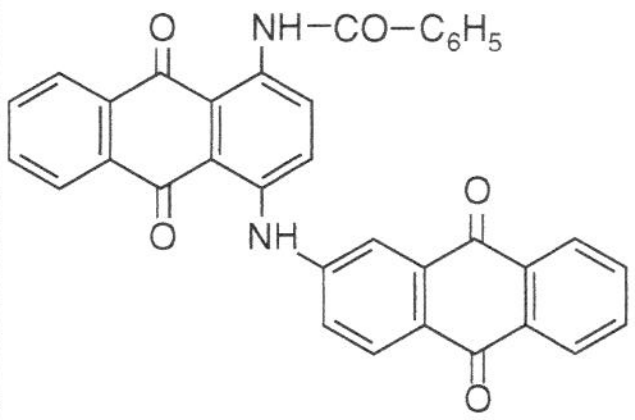	65,020
Vat brown 1	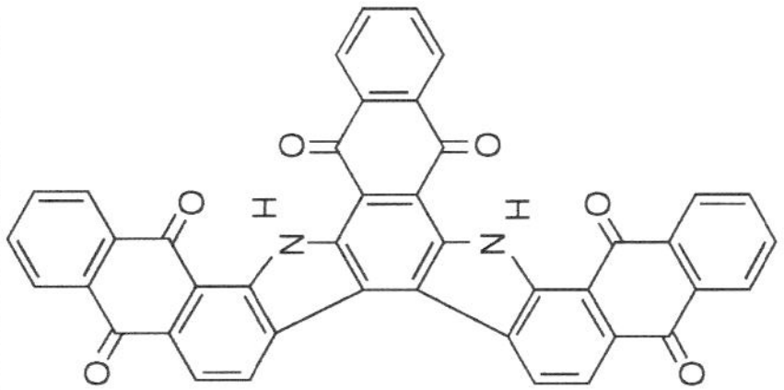	70,800
Vat brown 3	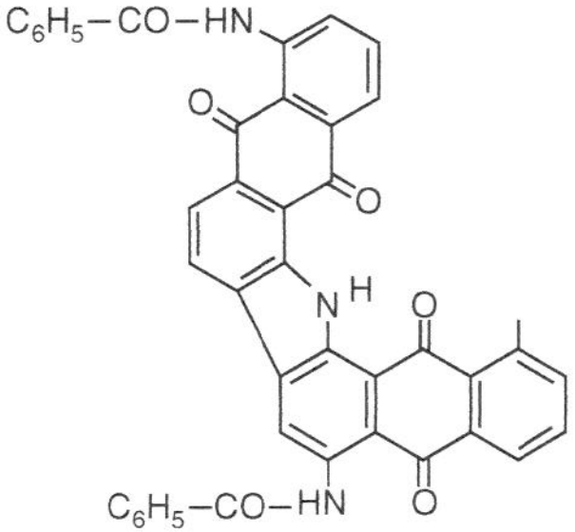	69,015

(continued)

Table 1.4 (continued)

Dyes	Molecular structure	C.I.
Vat black 25	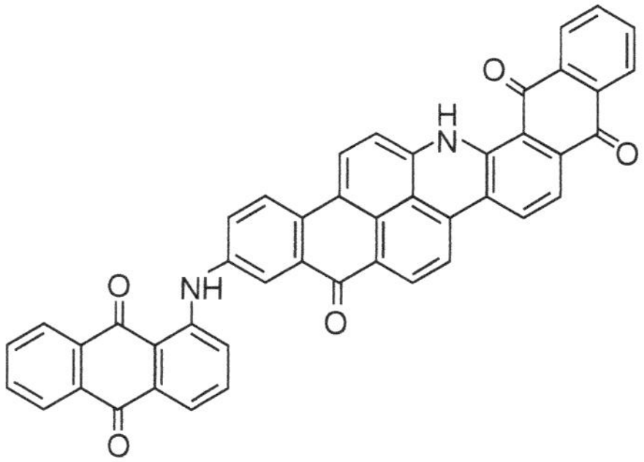	69,525
Vat black 27	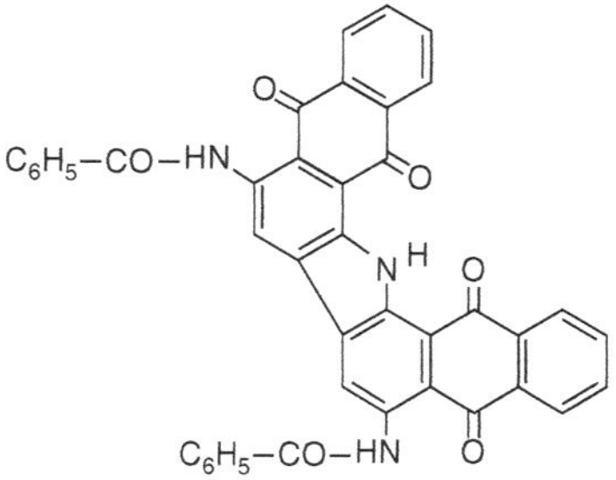	69,005

References

Al-Kdasi A, Idris A, Saed K, Guan CT (2004) Treatment of textile wastewater by advanced oxidation processes—a review. Global Nest J 6(3):222–230

Bae JS, Freeman HS (2007) Aquatic toxicity evaluation of new direct dyes to the Daphnia Magna. Dyes Pigments 73:81–85

Bafana A, Devi SS, Chakrabarti T (2011) Azo dyes: past, present and the future. Environ Rev 19:350–371

Banat IM, Nigam P, Singh D, Marchant R (1996) Microbial decolorization of textile-dye containing effluents: a review. Bioresour Technol 58:217–227

Baughman GL, Perenich TA (1988) Fate of dyes in aquatic systems: solubility and partitioning of some hydrophobic dyes and related compounds. Environ Toxicol Chem 7:183–199

Chen H (2006) Recent advances in azo dye degrading enzyme research. Current Protein-Peptide Sci 7:101–111

Choy KH, Mckay G, Porter JF (1999) Sorption of acid dyes from effluents using activated carbon. Resour Conserv Recycl 27:57–71

Christie RM (2007) Environmental aspects of textile dyeing. Boca Raton, Woodhead

Chu W (2001) Dye removal from textile dye waste water using recycled alum sludge. Wat Res 35:3147–3152

Chung KT, Cerniglia CE (1992) Mutagenicity of azo dyes: structure-activity relationships. Mutat Res 277:201–220

Combes RD, Havelandsmith RB (1982) A review of the genotoxicity of food, drug and cosmetic colors and other azo, triphenylmethane and xanthene dyes. Mutat Res 98:101–243

EWA (2005) Efficient use of water in the textile finishing industry, official publication of the European water association (EWA). Brussels

Georgion D, Aivazidis A, Hatiras J, Gimouhopoulos K (2003) Treatment of cotton textile waste water using lime and ferrous sulphate. Wat Res 37:2248–2250

Ghaly AE, Ananthashankar R, Alhattab M, Ramakrishnan VV (2014) Production, characterization and treatment of textile effluents: a critical review. J Chem Eng Proc Technol 5:1000182

Gupta VK, Mohan D, Sharma S, Sharma M (2000) Removal of basic dyes (rhodamine B and methylene blue) from aqueous solutions using bagasse fly ash. Sep Sci Technol 35:2097–2113

Haitan B, Teng TJT, Omar AKM (2000) Removal of dyes and industrial dye wastes by magnesium chloride. Wat Res 34:597–601

Hatch KL, Maibach HI (1999) Dyes as contact allergens: a comprehensive record. Text Chem Color 1:53–59

Hunger K (2003) Industrial dyes: chemistry, properties, applications. Wiley-VCH, Weinheim

Husain Q (2006) Potential applications of the oxidoreductive enzymes in the decolorization and detoxification of textile and other synthetic dyes from polluted water: a review. Crit Rev Biotechnol 26:201–221

Jeng-Shiou W, Hung LC, Chu KH, Suen SY (2008) Removal of cationic dye methyl violet 2B from water by cation exchange membranes. J Membr Sci 309:239–245

Kawamura S (2000) Integrated design and operation of water treatment facilities. Wiley, New York, pp 74–75

Kuo WG (1992) Decolorizing dye waste-water with fenton reagent. Water Res 26:881–886

Labanda J, Sabaté J, Llorens J (2009) Modeling of the dynamic adsorption of an anionic dye through ion-exchange membrane adsorber. J Membr Sci 340:234–240

Mathur N, Bhatnagar P (2007) Mutagenicity assessment of textile dyes from Sanganer (Rajasthan). J Environ Biol 28:123–126

Medvedev ZA, Crowne HM, Medvedeva MN (1988) Age related variations of hepato carcinogenic effect of azo dye (3′- MDAB) as linked to the level of hepatocyte polyploidization. Mech Age Devlop 46:159–174

Nohyneka GJ, Fautz R, Kieffer FB, Toutain H (2004) Toxicity and human health risk of hair dyes. Food Chem Toxicol 42:517–543

Novotny C, Dias N, Kapanen A, Malachova K, Vandrovcova M, Itavarra M, Lima N (2006) Comparative use of bacterial, algal and protozoan tests to study toxicity of azo and anthraquinone dyes. Chemosphere 63:1436–1442
Pagga U, Brown D (1986). The degradation of dyestuffs: part II behaviour of dyestuffs in aerobic biodegradation tests. Chemosphere 15(4):479–491
Panswed T, Wangchaisuwan M (1986) Mechanism of dye wastewater color removal by magnesium chloride. Wat Sci Technol 18:139–144
Percy AJ, Moore N, Chipman JK (1989) Formation of nuclear anomalies in rat intestine by benzidine and its biliary metabolites. Toxicology 57:217–223
Pinheiro HM, Touraud E, Thomas O (2004) Aromatic amines from azo dye reduction: status review with emphasis on direct UV spectrophotometric detection in textile industry wastewaters. Dyes Pigm 61:121–139
Puvaneswari N et al (2006) Toxicity assessment and microbial degradation of azo dyes. Indian J Exp Biol 44:618
Rai HS, Bhattacharyya MS, Singh J, Bansal TK, Vats P, Banerjee UC (2005) Removal of dyes from the effluent of textile and dyestuff manufacturing industry: a review of emerging techniques with reference to biological treatment. Crit Rev Env Sci Technol 35:219–238
Robinson T, McMullan G, Marchant R, Nigam P (2001) Remediation of dyes in textile effluent: a critical review on current treatment technologies with a proposed alternative. Biores Technol 77:247–255
Robinson T, Chandran B, Naidu S, Nigam P (2002) Studies on the removal of dyes from a synthetic textile effluent using barley husk in static-batch mode and in a continuous flow, packed-bed reactor. Biores Technol 85:43–49
Shaw CB, Carliell CM, Wheatley AD (2002) Anaerobic/aerobic treatment of colored effluents using sequencing batch reactors. Water Res 36:1993–2001
Slokar YM, Le Marechal AM (1997) Methods of decolorization of textile wastewaters. Dyes Pigm 37:335–356
Umbuzeiro GA, Freeman HS, Warren SH, Oliveira DP, Terao Y, Watanabe T, Claxton DL (2005) The contribution of azo dyes to the mutagenic activity of Cristais river. Chemosphere 60:55–64
Venkatamohan S, Rao N, Karthikeyan J (2002) Adsorptive removal of direct azo dye from aqueous phase onto coal based sorbents: a kinetic and mechanistic study. J Hazard Mater B90:189–204
Vogel EW (1982) Assessment of chemically induced genotoxic events, Prospectives and limitations. Universitaire Pers Leiden, Leiden
Weber E, Wolfe NL (1987) Kinetic studies of reduction of aromatic azo compounds in anaerobic sediment/water systems. Environ Toxicol Chem 6:911–920
Weisburger JH (2002) Comments on the history and importance of aromatic and heterocyclic amines in public health. Mutation Res-Fundam Mol Mech Muta 506-507:9–20
WTO (2012) World trade report 2012: trade and public policies. World Trade Organization, Geneva
Yaneva ZL, Georgieva NV (2012) Insights into Congo red adsorption on agro-industrial materials - spectral, equilibrium, kinetic, thermodynamic, dynamic and desorption studies. A review. Int Rev Chem Eng 4:127–146

Chapter 2
Adsorption of Dyes

Abstract Adsorption is one of the most commonly used, traditional separation technologies utilized for separation. Since it is an equilibrium-governed process, the process efficiency is excellent, but the throughput is relatively low. Nevertheless, because of its simplicity, this is one of the normally used technologies for dye removal from aqueous stream. Therefore, it is imperative to understand the modeling aspects of such adsorbent-based systems which is necessary for design and implementation of the technology. Additionally, the chapter describes the characteristics of the different commonly used adsorbents and its applicability.

Keywords Adsorbent • Chrysoidine • Eosin • Congo red • Activated charcoal

The word "adsorption" was formulated in 1881 by German physicist Heinrich Kayser to differentiate between the surface phenomena and intermolecular penetration. Adsorption can be divided into physical and chemical adsorption. Physical adsorption is controlled by the physical forces such as van der Waals forces, hydrophobicity, hydrogen bond, polarity, static interaction, dipole-dipole interaction, $\pi - \pi$ interaction, etc. When the species are adsorbed to the surface of the adsorbent by means of strong chemical interactions or bonding, it is referred to as chemisorption. The extent of adsorption depends on the nature of adsorbate such as molecular weight, molecular structure, molecular size, polarity, and solution concentration. It is also dependent on the surface properties of adsorbent such as particle size, porosity, surface area, surface charge, etc. The primary advantages of adsorption processes are:

1. Simple in design
2. Relatively safe and easy to operate
3. Inexpensive (compared to other separation processes)
4. Provides sludge-free cleaning operations (Gupta et al. 2000).

Selection of a suitable adsorbent is the primary concern for adopting adsorption in any process industries. The performance of the process is often limited by the equilibrium capacity of the adsorbent.

S. Mondal et al., *Advances in Dye Removal Technologies*, Green Chemistry and Sustainable Technology, DOI 10.1007/978-981-10-6293-3_2

2.1 Application of Adsorption in the Treatment of Process Wastewater

Various low-cost adsorbents for treatment of effluent containing heavy metals have been studied by various researchers in the past. Most of these adsorbents are prepared from the waste or by-products of other process plants or naturally occurring materials. A list of most commonly used low-cost adsorbents for heavy metal removal that is prepared from naturally occurring materials and processes are presented in Table 2.1.

2.2 Experimental Studies of Dye Adsorption

In the following sections, adsorption of chrysoidine, eosin, and Congo red by commercial activated carbon (CAC) has been presented.

2.2.1 Batch Adsorption

The batch adsorption is typically carried out in the solution phase containing dyes. The effects of agitation time and initial dye concentration on the percentage adsorption of dye by activated carbon at room temperature are shown in Figs. 2.1a, 2.1b, and 2.1c for chrysoidine, eosin, and Congo red. For all the cases, the percentage adsorption increases with agitation time for different initial dye concentration and attains equilibrium after some time.

Table 2.1 Low-cost high-capacity metal ion adsorbents

Metals	Adsorbent	Adsorption capacity (mg/g)	Reference
Zn^{2+}	Blast-furnace slag	103.3	83
	Powdered waste slag	168.0	147
Ni^{2+}	Red mud	160.0	112
Cu^{2+}	Blast-furnace slag	133.3	83
	Red mud	106.4	153
Cr^{6+}	Waste slurry	640.0	459
	Tea industry waste	455.0	165
Hg^{2+}	Waste slurry	560.0	159
Cd^{2+}	Fly ash	207.3	152
Pb^{2+}	Waste slurry	1030.0	159
V^{5+}	Waste metal sludge	24.8	97
As^{3+}, As^{5+}	Acid-activated laterite	24.5, 8.0	

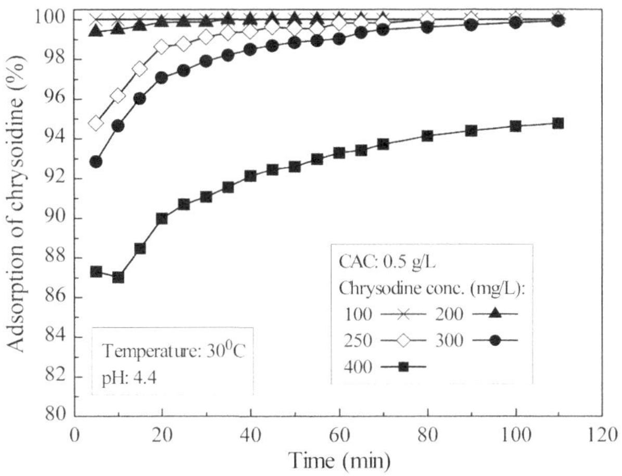

Fig. 2.1a Effects of agitation time and concentration of chrysoidine on percentage of adsorption (Reproduced from Purkait et al. (2004). With permission from Taylor & Francis Ltd)

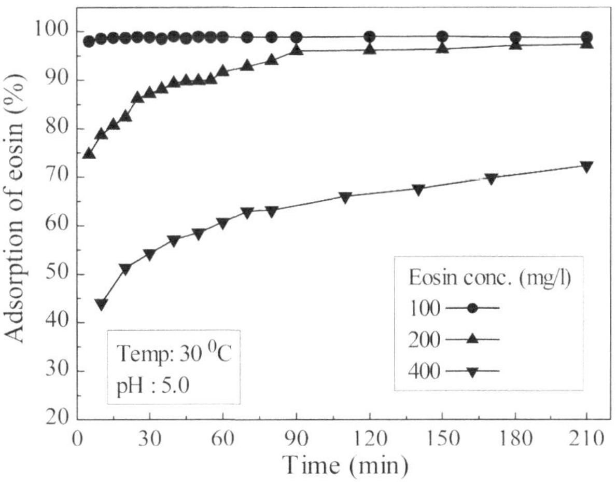

Fig. 2.1b Effects of agitation time and concentration of eosin on percentage of adsorption (Reproduced from Purkait et al. (2005). With permission from Elsevier)

From Fig. 2.1a, it may be observed that for all initial chrysoidine concentration, the percentage adsorption is found to be constant beyond 80 min. This indicates that equilibrium is attained at about 80 min for initial dye concentration in the range of 100–400 mg/L. It is also clear that the extent of adsorption depends on the initial dye concentration. For dye solution of lower initial concentration (up to 100 mg/L), the adsorption is very fast and almost 100% adsorption is achieved quickly. The dye adsorption at equilibrium decreases from 100% to about 94% as the dye concentration increases from 100 to 400 mg/L.

It is clear from Fig. 2.1b that up to an initial eosin concentration of 100 mg/L, more than 99% adsorption is achieved within 5 min. For an initial concentration of 200 mg/L, the percentage adsorption increases until 90 min and becomes constant thereafter. For a feed concentration of 400 mg/L, the percentage adsorption increases rapidly for about 90 min, and the increase becomes gradual thereafter.

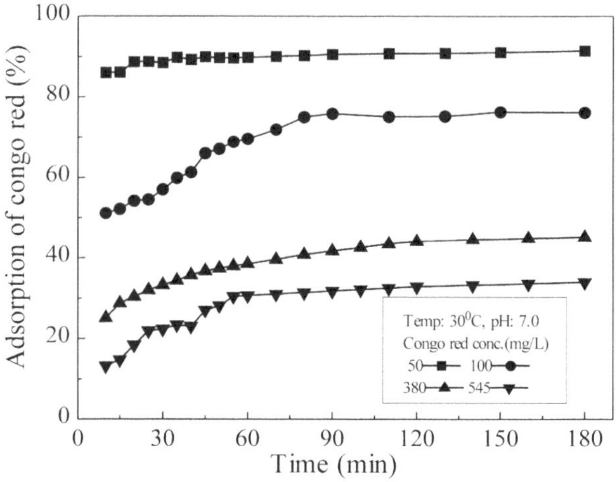

Fig. 2.1c Effects of agitation time and concentration of Congo red on percentage of adsorption (Reproduced from Purkait et al. (2007). With permission from Elsevier)

For 210 min of operation, the dye adsorption is 99.6% for an initial dye concentration of 100 mg/L but only 72.3% for 400 mg/L.

Figure 2.1c describes the variation of Congo red adsorption with time for different initial dye concentration. The percentage adsorption of Congo red is found to be constant beyond 50 min. This indicates that the equilibrium is attained within 50 min for the range of initial dye concentrations. It is also clear that the extent of adsorption depends on the initial dye concentration. For dye solution of lower initial concentration, the adsorption is very fast and 90% of adsorption is achieved quickly. The percentage dye adsorption at equilibrium decreases from 90% to 28% as the dye concentration increases from 50 to 545 mg/L.

The effects of adsorbent dose on the extent of chrysoidine adsorption are shown in Fig. 2.2 for initial dye concentrations of 700 mg/L. It is clear from the figure that percentage adsorption increases with time up to 80 min and also with adsorbent dose. Percentage adsorption increases from about 77 to 99% when the adsorbent dose increases from 0.75 to 1.40 g/L. This increase in percentage adsorption may be due to the fact that the number of available sites for adsorption increases with adsorbent dose.

The pH of the solution has significant influence in the rate of adsorption. The percentage dye adsorption at different pH is shown in Figs. 2.3a, 2.3b, and 2.3c for chrysoidine, eosin, and Congo red, respectively. Figure 2.3a describes the variation of chrysoidine adsorption at different pH for an initial dye concentration of 400 mg/L. The color of chrysoidine dye in aqueous medium is red (λ_{max}: 457 nm) in acidic pH but changes its color from red to yellow (λ_{max}: 442 nm) in basic pH. This is due to the presence of chromophore in the structure of chrysoidine. A chromophore is any structural feature (in this case, $-N=N-$) which produces light absorption in the ultraviolet region or color in the visible region. An auxochrome is any group (in this case $-NH_2$) which, although not a chromophore, leads to a red shift when attached to a chromophore. Thus, the combination of chromophore and auxochrome behaves as a new chromophore. Bathochromic effect (red shift) and

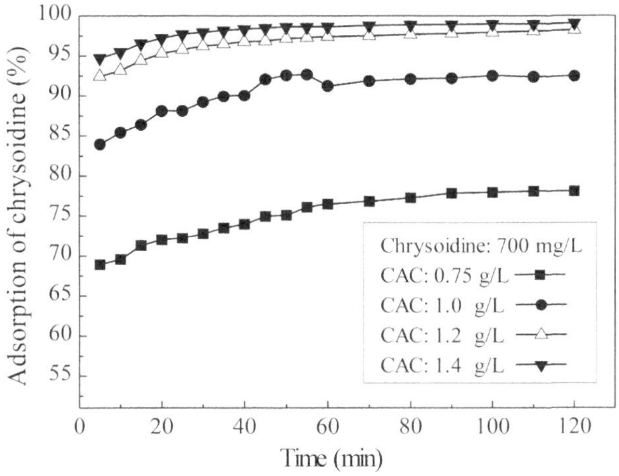

Fig. 2.2 Effects of agitation time and adsorbent dose on percentage adsorption (Reproduced from Purkait et al. (2004). With permission from Taylor & Francis Ltd)

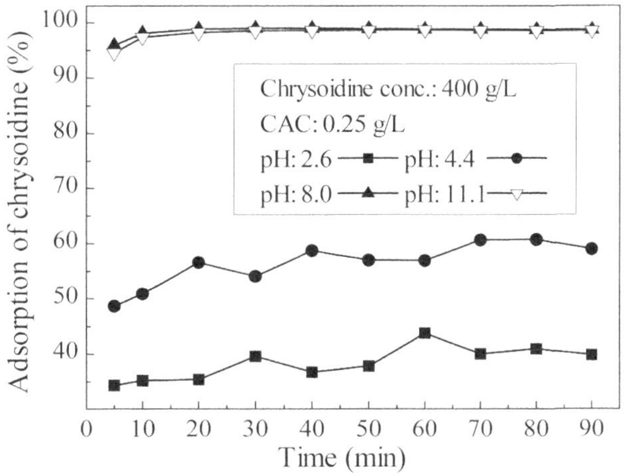

Fig. 2.3a Effect of initial pH on percentage of adsorption for 400 mg/L of feed chrysoidine (Reproduced from Purkait et al. (2004). With permission from Taylor & Francis Ltd)

hypsochromic effect (blue shift) are the shifting of the absorption band to the longer and shorter wavelengths (Finar 1973). Therefore, due to blue shift, chrysoidine changes its color in basic pH.

Adsorption followed by desorption technique is generally used to get the more concentrated form of the dye solution. One of the most common desorption technique is the pH treatment. But problem arises for the dyes which are highly pH sensitive, like chrysoidine as discussed in the previous paragraph.

Most of the activated carbon contains some oxygen complexes on the surface, e.g., (a) strongly carboxylic groups, (b) carbonyl groups, and (c) phenolic groups (Motoyuki 1990). These groups are nucleophilic in nature and potential adsorbing sites. In acidic pH, these active sites get blocked by hydrogen ion leading to reduction in adsorption. Hence, adsorption of chrysoidine on activated carbon is less in acidic pH. It is found from Fig. 2.3a that at pH 2.6, adsorption is nearly 62%

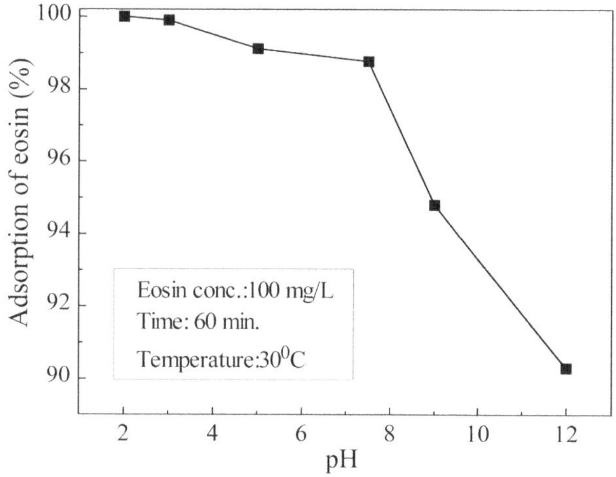

Fig. 2.3b Effect of pH on the percentage adsorption of feed eosin (Reproduced from Purkait et al. (2005). With permission from Elsevier)

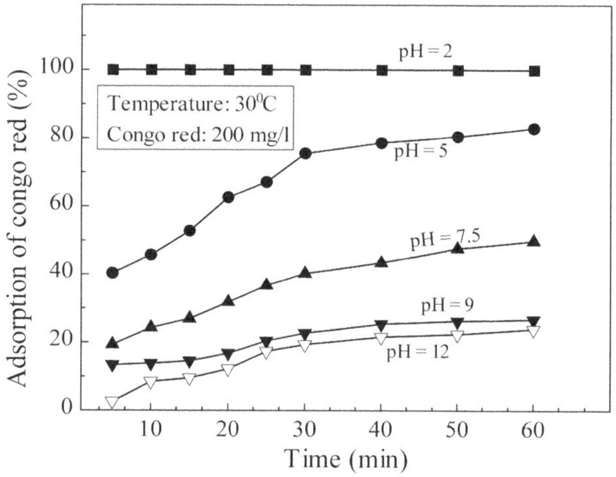

Fig. 2.3c Effect of initial pH on percentage of adsorption for 200 mg/L of feed Congo red (Reproduced from Purkait et al. (2007). With permission from Elsevier)

for the feed dye concentrations of 400 mg/L at the end of experiment. The percentage adsorption decreases from about 98 to 40%, when pH decreases from 11.1 to 2.6. From Fig. 2.3a, it may be observed that the adsorption of this dye is more at the basic pH.

The percentage of eosin adsorption at different pH levels are shown in Fig. 2.3b for an initial dye concentration of 100 mg/L. pH plays an important role on adsorption capacity by influencing the chemistry of both the dye molecule and the activated carbon in aqueous solution. Eosin is a dipolar molecule at low pH. Activated carbon contains oxygen complexes on its surface, e.g., strongly carboxylic groups, carbonyl groups, and phenolic groups (Motoyuki 1990). These groups are nucleophilic in nature. With decrease in pH of the dye solution, more dye molecules are protonated and get adsorbed on the surface of the activated carbon. It can be observed from Fig. 2.3b that at pH 2, adsorption is about 100% for an initial dye concentration of 100 mg/L. Percentage adsorption decreases with increase in pH. For the initial dye concentration of 100 mg/L, the removal is 91% for a pH of 12.

The percentage of Congo red adsorption at different pH has been shown in Fig. 2.3c for the initial dye concentrations of 200 mg/L. The initial pH of dye solution plays an important role particularly on the adsorption capacity by influencing the chemistry of both dye molecule and activated carbon in aqueous solution. Congo red is a dipolar molecular at lower pH and exists as anionic form at higher pH as shown in Fig. 2.3c. The sodium and potassium salt of anionic Congo red in aqueous medium is red in color in basic pH up to 10. Above the pH value of 10, the degree of red color changes from the original one. It has also been found that as the pH decreases, the color of Congo red solution changes from red to dark blue. Therefore, the pH of the medium needs to be maintained between 5 and 10 to treat Congo red. These variations of color with pH suggest that the extent and nature of ionic character of Congo red molecule depend on the pH of the medium. The variations in the extent of adsorption of Congo red on activated carbon with pH are due to the difference in ionic character of the dye molecule. With decrease in pH of dye solution, more dye molecules are protonated and chemisorbed on the nucleophilic sites of the surface of CAC. It is found from the figure that at pH 2, adsorption is about 100%. On the other hand, the percentage adsorption decreases with increase in pH of the dye solution. This is because at higher pH, dye molecules exist in anionic form, and due to interionic repulsion, less adsorption takes place. For the feed dye concentration of 200 mg/L, the percentage adsorption decreases to 25% at the end of the experiment when the pH is 12. From Fig. 2.3c, it may be observed that the adsorption of Congo red is maximum at the acidic pH. Therefore, when Congo red is present in the solution as red color, the operating pH for maximum adsorption should be kept at 5.

Effects of temperature on the extent of adsorption are shown in Figs. 2.4a, 2.4b, and 2.4c for chrysoidine, eosin, and Congo red, respectively. Adsorption experiments are carried out for aqueous solution of chrysoidine for two different concentrations (400 and 700 mg/L) at three different temperatures (30, 50, and 70 °C) and at a pH of 4.4. It has been observed that the adsorption capacity increases significantly with temperature as shown in Fig. 2.4a for the initial chrysoidine concentration of 400 mg/L. The percentage adsorption increases from about 94 to 99% for the feed dye concentration of 400 mg/L and about 80 to 87% for the feed dye concentration of 700 mg/L, at the end of experiment, when temperature is raised from 30 to 70 °C. This endothermic nature of adsorption is due to the positive ΔH^0 value as shown in Table 2.4a.

In order to observe the effect of temperature on the adsorption capacity, experiments are carried out for 100 mg/L eosin at three different temperatures (30, 40, and 50 °C) using 1.0 g of activated carbon per liter of the solution. It has been observed that with increase in temperature, adsorption capacity decreases as shown in Fig. 2.4b. This is due to the negative value of ΔH^0 value (refer to Table 2.4b).

Experiments are carried out to observe the effect of temperature on the extent of adsorption for Congo red of different initial concentration (50, 100, and 200 mg/L)

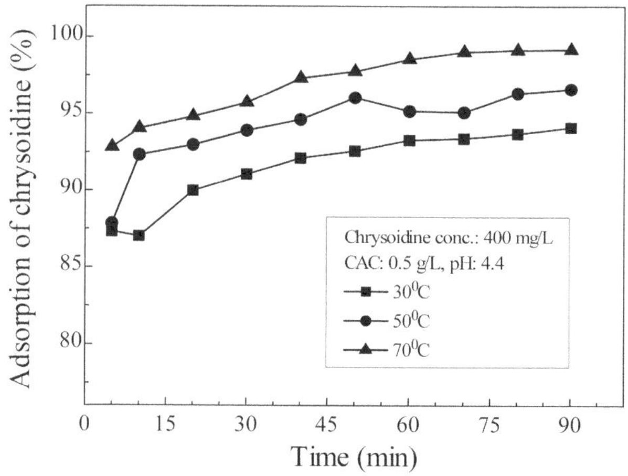

Fig. 2.4a Effect of temperature on adsorption capacity for 400 mg/L of feed chrysoidine (Reproduced from Purkait et al. (2004). With permission from Taylor & Francis Ltd)

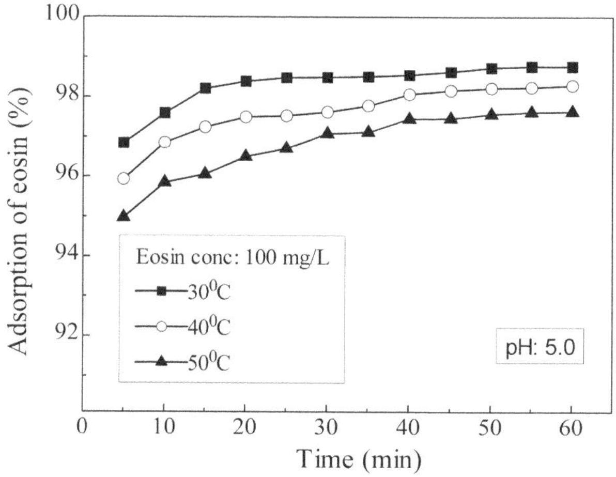

Fig. 2.4b Effect of temperature on adsorption capacity for 100 mg/L of feed eosin (Reproduced from Purkait et al. (2005). With permission from Elsevier)

in aqueous solution at three different temperatures (e.g., 30, 40, and 50 °C) and at neutral pH. It has been observed that the adsorption capacity increases significantly with temperature as shown in Fig. 2.4c for the initial Congo red concentration of 100 mg/L. This is because of positive ΔH^0 value as shown in Table 2.4c.

The thermodynamic parameters ΔG^0, $\Delta S,^0$ and ΔH^0 for the adsorption of chrysoidine, eosin, and Congo red have been determined by using the following equations (Khan et al. 1995):

$$\Delta G^0 = \Delta H^0 - T\Delta S^0 \tag{2.1}$$

$$\log(q_e/C_e) = \frac{\Delta S^0}{2.303R} + \frac{-\Delta H^0}{2.303RT} \tag{2.2}$$

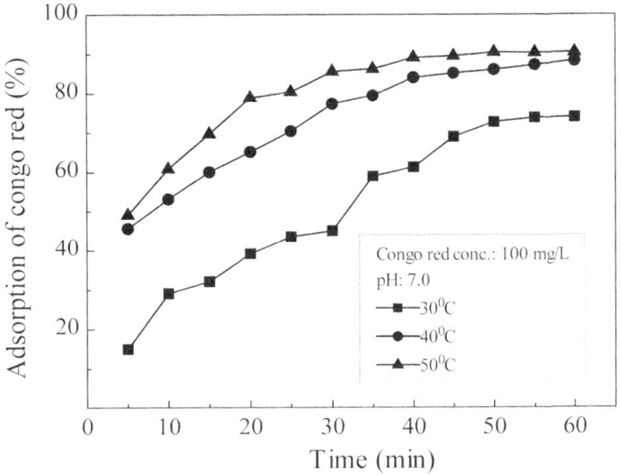

Fig. 2.4c Effect of temperature on adsorption capacity for 100 mg/L of feed Congo red (Reproduced from Purkait et al. (2007). With permission from Elsevier)

where q_e is the amount of dye adsorbed per unit mass of activated carbon (mg/g), C_e is equilibrium concentration (mg/L), and T is temperature in Kelvin. q_e/C_e is called the adsorption affinity. It may be noted here that the experimental data considered here for the calculation of the thermodynamic parameters, namely, ΔG^0,ΔH^0, and ΔS^0, are in the linear range of the equilibrium adsorption isotherm (e.g., for chrysoidine, $q_e$varies from 2.5 to 3.0 mmol/g and $C_e$varies from 0.012 to 0.2 mmol/L as shown in Fig. 2.5a). The values of Gibbs free energy (ΔG^0) have been calculated by knowing the enthalpy of adsorption (ΔH^0) and the entropy of adsorption (ΔS^0). ΔS^0 and ΔH^0 are obtained from a plot of $\log(q_e/C_e)$ versus $\frac{1}{T}$, from Eq. (2.2). Once these two parameters are obtained, ΔG^0 is determined from Eq. (2.1). The values ofΔG^0, ΔH^0, and $\Delta S^0$are listed in Tables 2.2a, 2.2b and 2.2c for chrysoidine, eosin, and Congo red, respectively.

Gibbs free energy (ΔG^0) for all the three dyes is negative (as shown in Tables 2.2a, 2.2b, 2.2c). This indicates that the adsorption process is spontaneous for all the three dyes. Adsorption of chrysoidine and Congo red is endothermic in nature (since ΔH^0 value is positive; refer to Tables 2.2a and 2.2c). On the other hand, eosin adsorption is exothermic in nature (as ΔH^0 value is negative; refer to Table 2.2b). The positive value of ΔS^0 for chrysoidine and Congo red (refer to Tables 2.2a and 2.2c) dictates that the adsorbed dye molecules on the activated carbon surface are organized in a more random fashion compared to those in the aqueous phase. Similar observations have been reported in the literature (Bhattacharyya and Sharma 2003). For eosin, the negative value of ΔS^0 (refer to Table 2.2b) suggests decreased randomness at the solid solution interface during adsorption (Manju et al. 1998).

Table 2.2a Thermodynamic parameters for adsorption of chrysoidine in activated charcoal at different temperature and dye concentrations

Adsorbent (g/L)	Chrysoidine (mg/L)	ΔH^0 (kJ/mol)	ΔS^0 (J/mol.K)	$-\Delta G^0$(kJ/mol) at temperature		
				303 K	323 K	343 K
0.5	400	43.6	171.4	8.3	11.8	15.2
1.0	700	48.9	180.9	5.9	9.6	13.2
Mean		46.3	176.2	7.1	10.7	14.2

Reproduced from Purkait et al. (2004). With permission from Taylor & Francis Ltd

Table 2.2b Thermodynamic parameters for adsorption of eosin in activated charcoal at different temperature and 100 mg/L of eosin

Eosin (mg/L)	ΔH^0 (kJ/mol)	ΔS^0 (J/mol.K)	$-\Delta G^0$ (kJ/mol) at temperature		
			303 K	313 K	323 K
100	27.0	52.6	11.0	10.5	10.0

Reproduced from Purkait et al. (2005). With permission from Elsevier

Table 2.2c Thermodynamic parameters for adsorption of Congo red in activated charcoal at different temperature and dye concentrations

Congo red (mg/L)	ΔH^0 (kJ/mol)	ΔS^0 (J/mol.K)	$-\Delta G^0$(kJ/mol) at temperature		
			303 K	313 K	323 K
50	21.5	86.8	4.8	5.7	6.6
100	51.1	178.1	2.9	4.6	6.4
200	10.1	34.5	0.3	0.67	1.0
Mean	27.6	99.8	2.7	3.7	4.7

2.2.1.1 Langmuir Adsorption Isotherm

Langmuir adsorption isotherm is applicable to explain the equilibrium data for many adsorption processes. The basic assumption of this process is the formation of monolayer of adsorbate on the outer surface of adsorbent, and after that no further adsorption takes place. The expression of the Langmuir model is given as follows (Ozacar and Sengil 2003):

$$q_e = \frac{QbC_e}{1 + bC_e} \tag{2.3}$$

A linear form of this expression is

$$\frac{1}{q_e} = \frac{1}{Q} + \frac{1}{Qb} \cdot \frac{1}{C_e} \tag{2.4}$$

where $q_e$is the amount of adsorbate adsorbed per unit weight of adsorbent (mg/g) and C_e is the equilibrium concentration of adsorbate (mg/L). The constant Q and

b are the Langmuir constants and are the significance of adsorption capacity (mg/g) and energy of adsorption (l/mg), respectively. Values of Q and b are calculated from the intercept and slope of the plot $1/q_e$ versus $1/C_e$.

2.2.1.2 Freundlich Adsorption Isotherm

This model is an indicative of the extent of heterogeneity of the surface of adsorbent and is given as follows:

$$q_e = K_F C_e^{1/n} \tag{2.5}$$

where K_F and n are Freundlich constants. A linear form of the Freundlich expression is as follows:

$$\log q_e = \log K_F + \frac{1}{n} \log C_e \tag{2.6}$$

The constants K_F and n are the Freundlich constants and are the significance of adsorption capacity and intensity of adsorption, respectively. Values of K_F and n are calculated from the intercept and slope of the plot $\log q_e$ versus $\log C_e$.

Adsorption isotherms of chrysoidine, eosin, and Congo red on activated carbon at 30 °C are shown in Figs. 2.5a, 2.5b, and 2.5c, respectively. The coefficients of these two isotherm models for the three dyes are given in Table 2.3. These data provide information on the amount of activated carbon required to adsorb a particular mass of dye under specified system conditions. Correlation coefficients are evaluated by fitting the experimental adsorption equilibrium data for three dyes separately using both Langmuir and Freundlich adsorption isotherms and are also shown in Table 2.3. It is found from the correlation coefficients (r^2) that adsorption

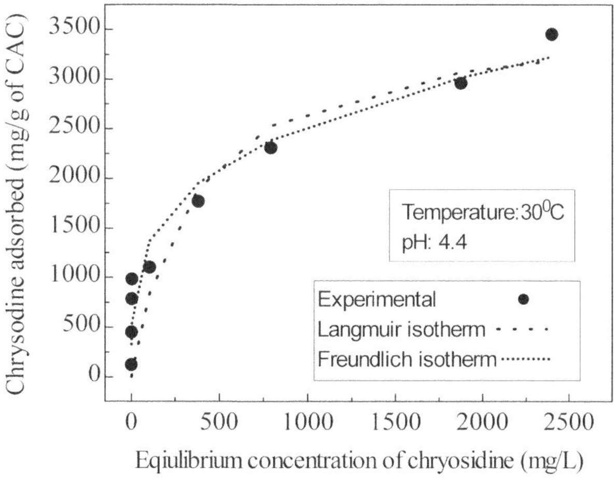

Fig. 2.5a Adsorption isotherms of chrysoidine on activated carbon (Reproduced from Purkait et al. (2004). With permission from Taylor & Francis Ltd)

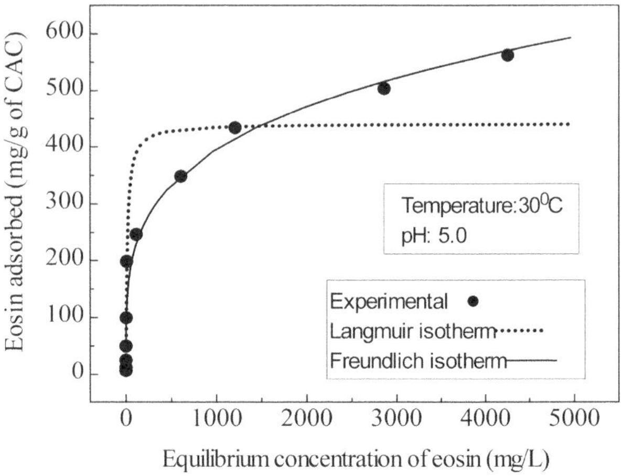

Fig. 2.5b Adsorption isotherms of eosin on activated carbon

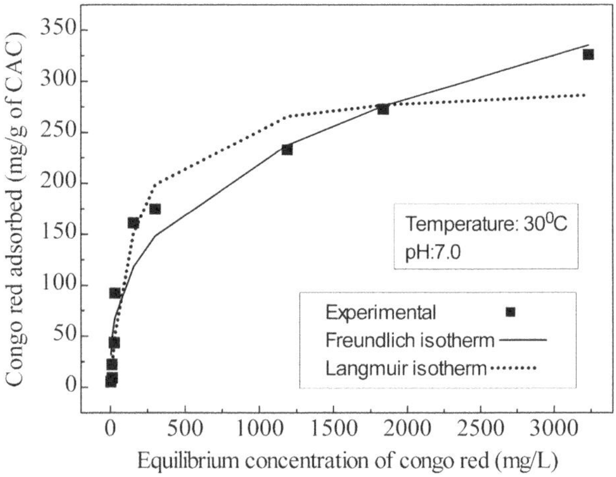

Fig. 2.5c Adsorption isotherms of Congo red on activated carbon (Reproduced from Purkait et al. (2007). With permission from Elsevier)

Table 2.3 Langmuir and Freundlich isotherm constants for adsorption of dyes on activated charcoal

	Langmuir isotherm constant			Freundlich isotherm constant		
Dye	Q (mg/g)	b (l/mg)	r^2	K_F (mg/g(L/mg)$^{1/n}$)	n	r^2
Chrysoidine	3652.0	2.84×10^{-3}	0.94	384.2	3.66	0.97
Eosin	571.4	4.35×10^{-3}	0.96	75.5	4.14	0.99
Congo red	300.0	6.50×10^{-3}	0.94	20.78	2.90	0.96

isotherm for the present three dye-activated charcoal systems is best explained by Freundlich equation.

2.2.1.3 Adsorption Kinetics

The kinetics of adsorption of chrysoidine, eosin, and Congo red on activated carbon have been described using both first- and pseudo-second-order model. The Lagergren's equation for first-order kinetics is as follows:

$$\log(q_e - q_t) = \log q_e - \frac{k_1 t}{2.303} \tag{2.7}$$

The expression for pseudo-second-order rate equation is given as (Ho et al. 1996)

$$\frac{t}{q_t} = \frac{1}{k_2 q_e^2} + \frac{t}{q_e} \tag{2.8}$$

where q_e and q_t are the amounts of dye adsorbed (mg/g) at equilibrium and at any time t and k_1 is the rate constant (min^{-1}). Figures 2.6a, 2.6b, and 2.6c show (t/q_t) versus t plot for pseudo-second-order kinetics for chrysoidine, eosin, and Congo red, respectively. In Eq. (2.8), k_2 (g/mg min) is the rate constant for the pseudo-second-order adsorption kinetics. The slope of the plot (t/q_t) versus t gives the value of q_e, and from the intercept, k_2 can be calculated. The values of k_1, k_2, and correlation coefficients (r^2), both in the first and pseudo-second-order kinetics, are presented in Tables 2.4a, 2.4b, and 2.4c for chrysoidine, eosin, and Congo red, respectively. It may be observed from Tables 2.4a, 2.4b, and 2.4c that the adsorption of chrysoidine, eosin, and Congo red on activated carbon follows pseudo-second-order kinetics more closely.

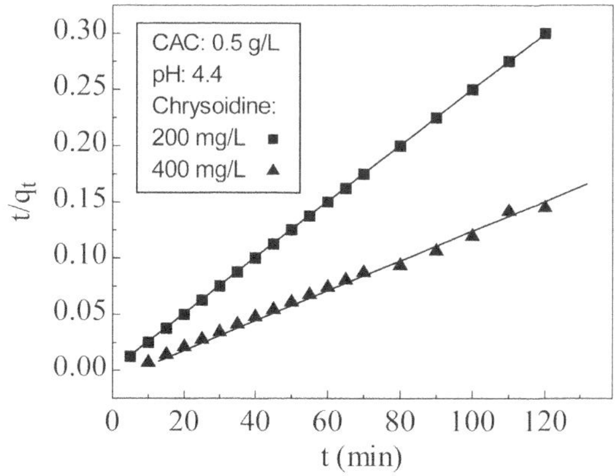

Fig. 2.6a Plot of the pseudo-second-order kinetic model for adsorption of chrysoidine on activated carbon (0.5 g/L). Feed chrysoidine: 200 and 400 mg/L (Reproduced from Purkait et al. (2004). With permission from Taylor & Francis Ltd)

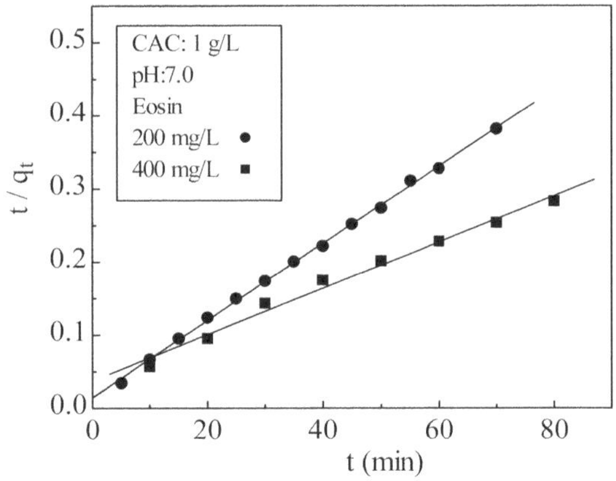

Fig. 2.6b Plot of the pseudo-second-order kinetic model for adsorption of eosin on activated carbon (1.0 g/L). Feed eosin: 200 and 400 mg/L (Reproduced from Purkait et al. (2005). With permission from Elsevier)

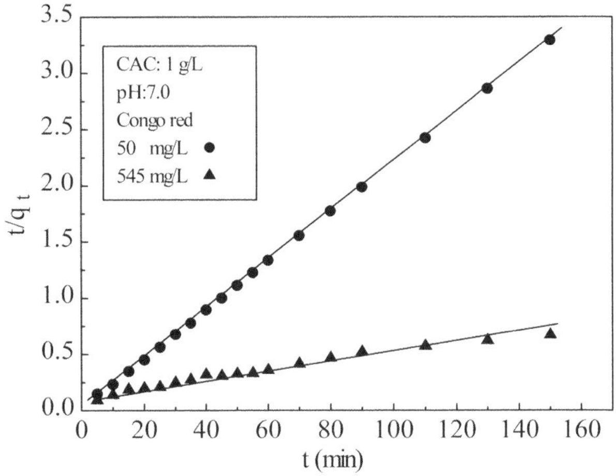

Fig. 2.6c Plot of the pseudo-second-order kinetic model for adsorption of Congo red on activated carbon (1.0 g/L). Feed Congo red: 50 and 545 mg/L

Table 2.4a Comparison of the first- and second-order adsorption rate constants, calculated and experimental q_e value for chrysoidine on activated charcoal

		First-order			Pseudo-second-order		
Chrysoidine (c_0) (mg/L)	q_e, exp t (mg/g)	q_e, *fit* (mg/g)	k_1 (1/min)	r^2	q_e, *fit* (mg/g)	k_2 (g/mg min)	r^2
Feed CAC: 0.25 g/L							
200	737.2	710.4	0.30	0.422	751.8	5.4×10^{-4}	0.998
250	831.9	770.2	0.46	0.202	833.3	3.5×10^{-4}	0.997
300	907.5	890.6	0.71	0.210	917.4	13.5×10^{-4}	0.999
400	972.0	921.9	0.44	0.395	980.4	4.3×10^{-4}	0.998
Feed CAC: 0.50 g/L							
200	400.0	399.4	1.08	0.454	400.0	7.4×10^{-1}	0.998
250	500.0	495.6	0.62	0.576	500.0	5.4×10^{-3}	0.999
300	599.5	589.9	0.55	0.547	602.4	2.3×10^{-3}	0.999
400	759.6	738.2	0.56	0.255	787.4	4.9×10^{-4}	0.998

Reproduced from Purkait et al. (2004). With permission from Taylor & Francis Ltd

Table 2.4b Comparison of the first- and second-order adsorption rate constants, calculated and experimental q_e value for eosin on activated charcoal

		First-order			Pseudo-second-order		
Eosin (c_0) (mg/L)	q_e, exp t (mg/g)	q_e, *fit* (mg/g)	k_1 (1/min)	r^2	q_e, *fit* (mg/g)	k_2 (g/mg min)	r^2
Feed CAC: 1.0 g/L							
200	175.0	152.3	6.7×10^{-2}	0.91	173.4	5.4×10^{-4}	0.998
400	286.0	232.2	4.4×10^{-2}	0.86	302.1	4.3×10^{-4}	0.998

Reproduced from Purkait et al. (2005). With permission from Elsevier

Table 2.4c Comparison of the first- and second-order adsorption rate constants, calculated and experimental q_e value for Congo red on activated charcoal

		First-order			Pseudo-second-order		
Congo red (c_0) (mg/L)	q_e, exp t (mg/g)	q_e, *fit* (mg/g)	k_1 (1/min)	r^2	q_e, *fit* (mg/g)	k_2 (g/mg min)	r^2
Feed CAC: 1.0 g/L							
50	45.7	2.2	1.8×10^{-2}	0.925	45.9	2×10^{-2}	0.999
100	76.2	75.0	4.1×10^{-2}	0.879	88.5	5.3×10^{-2}	0.987
380	171.1	119.0	2.9×10^{-2}	0.964	181.8	4.5×10^{-2}	0.997
545	183.1	144.8	3.0×10^{-2}	0.978	207.5	2.5×10^{-2}	0.995

Reproduced from Purkait et al. (2007). With permission from Elsevier

2.2.2 *Column Adsorption*

Column adsorptions studies are essential for design of industrial scale fixed-bed adsorber system. Figure 2.7 shows the breakthrough curves for different bed depths. It may be observed from Fig. 2.8 that the breakthrough time (duration for zero column outlet concentration) increases from 18 to 39 h, when the bed depth is increased from 4.5×10^{-2} m to 7.0×10^{-2} m, for the same flow rate of 0.18 L/hr. The shape and the gradient of the breakthrough curves for the two bed depths are almost identical.

2.3 Generalized Shrinking Core Model for Batch Adsorption Data

To develop a mathematical model that describes the adsorption dynamics, the following information are generally required:

1. A complete description of equilibrium behavior, i.e., the maximum level of adsorption attained in a sorbent/sorbate system as a function of the sorbate liquid-phase concentration

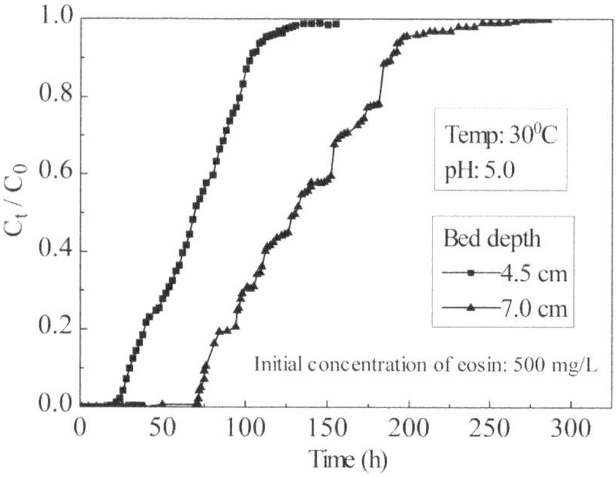

Fig. 2.7 Variation of the breakthrough curve with bed depth (Reproduced from Purkait et al. (2005). With permission from Elsevier)

2. Mathematical representation of associated rate of adsorption, which is controlled by the resistances within the sorbent particles

In adsorption, mainly two resistances prevail – the external liquid film resistance and the resistance in the adsorbent particle. The intraparticle diffusion resistance may be neglected for solutes that exhibit strong solid to liquid-phase equilibrium solute distribution, in the initial period of operation. However, even for such systems, the above assumption leads to errors that are substantial beyond the first few minutes if the agitation is high (Mathews and Weber 1976). So, both the resistances are important for kinetic study (Chatzopoulos et al. 1993; McKay 1984; Costa et al. 1987; Komiyama and Smith 1974; Liapis and Rippin 1977).

The external liquid film resistance is characterized by the external liquid film mass transfer coefficient (k_f). The mass transport within the adsorbent particles is assumed to be a pore diffusion (Dedrick and Beckman 1967; Weber and Rumer 1965; Furusawa and Smith 1973; McKay 1982) or homogeneous solid diffusion process (McKay 1982; Hand et al. 1983; Kapoor et al. 1989).

The pore diffusion model outlined in this paper is based on the unreacted shrinking core model (Yen 1968; Levenspiel 1972) with pseudo-steady-state approximation. This model has mostly been applied to gas-solid non-catalytic reactions, but a number of liquid-solid reactions also have been analyzed using this model (Neretnieks 1976; Spahn and Schlunder 1975). In the pore diffusion model, there is adsorption of the adsorbate into the pores with a cocurrent solute distributed all along the pore wall.

The assumptions made in this model are as follows:

(a) Pore diffusivity is independent of concentration.
(b) Adsorption isotherm is irreversible.
(c) Pseudo-steady-state approximation is valid.
(d) The driving force in both film and particle mass transfer is directly proportional.
(e) Adsorbent particles are spherical.

The major limitation of this model is that it is specific to the nature of isotherm. This means that the model available in literature is most suitable for Langmuir-type isotherm, i.e., formation of a monolayer of adsorbate on the adsorbent. Besides, this model is only applicable for higher initial adsorbate concentration in solution so that the batch process operating line intercepts the invariant zone of isotherm. For example, for Astrazone blue-silica system, the literature model is applicable for $C_0 >> 200$ mg/lit (McKay 1984). The present model, which is more generalized, overcomes the above limitations. The model proposed, here in, can be applied to wide ranges of initial adsorbate concentrations for all possible nature of isotherms. The system reported here is adsorption of Astrazone blue dye on Sorbsil Silica.

The equations considered for the kinetics of the adsorption process for spherical adsorbent particles for the present model are as follows:

The mass transfer from external liquid phase can be written as

$$N(t) = 4\pi R^2 K_f (C_t - C_{et}) \tag{2.9}$$

The diffusion of solute through the pores as per Fick's law can be written as

$$N(t) = \frac{4\pi D_p C_{et}}{\left[\frac{1}{R_f} - \frac{1}{R}\right]} \tag{2.10}$$

where D_p is the effective diffusivity in the porous adsorbent (Fogler 1997).

The mass balance on a spherical element of adsorbate particle can be written as

$$N(t) = -4\pi R_f{}^2 Y_{et} \rho \left[\frac{dR_f}{dt}\right] \tag{2.11}$$

The average concentration on adsorbent particle can be written as

$$\overline{Y_t} = Y_{et}\left[1 - \left(\frac{R_f}{R}\right)^3\right] \tag{2.12}$$

The differential mass balance over the system by equating the decrease in adsorbate concentration in the solution with the accumulation of the adsorbate in the adsorbent can be written as

$$N(t) = -V\left(\frac{dC_t}{dt}\right) = W\left(\frac{d\overline{Y_t}}{dt}\right) \tag{2.13}$$

The dimensionless terms used for simplification are as follows:

$$C_t^* = \frac{C_t}{C_0}, r = \frac{R_f}{R}, Bi = \frac{k_f R}{D_p}, \quad Ch = \frac{W}{VC_0}, C_{et}{}^* = \frac{C_{et}}{C_0} \text{ and } \tau = \frac{D_p t}{R^2}$$

Simplifying Eqs. (2.9) and (2.10)

$$C_{\mathrm{et}}{}^{*} = \frac{Bi(1-r)C_{\mathrm{t}}^{*}}{[r+Bi(1-r)]} = g_{1}\left(C_{\mathrm{t}}^{*}, r\right) \tag{2.14}$$

Now differentiating the above equation with respect to τ

$$\frac{dC_{\mathrm{et}}^{*}}{d\tau} = \frac{Bi(1-r)}{r+Bi(1-r)}\frac{dC_{\mathrm{t}}^{*}}{d\tau} - \frac{BiC_{\mathrm{t}}^{*}}{[r+Bi(1-r)]^{2}}\frac{dr}{d\tau} \tag{2.15}$$

From the equilibrium relationship

$$Y_{\mathrm{e}}(t) = g_{2}\left(C_{\mathrm{et}}^{*}\right) \tag{2.16}$$

where g_2 is any equilibrium isotherm relationship. Simplifying Eqs. (2.9) and (2.11)

$$\left(\frac{dr}{d\tau}\right) = \frac{-Bi\left(\frac{C_0}{\rho Y_{\mathrm{e}}}\right)\left(C_{\mathrm{t}}^{*} - C_{\mathrm{et}}{}^{*}\right)}{r^{2}} \tag{2.17}$$

Simplifying Eqs. (2.12) and (2.13)

$$\left(\frac{dC_{\mathrm{t}}^{*}}{d\tau}\right) + Ch\left(1-r^{3}\right)\left(\frac{dY_{\mathrm{et}}}{d\tau}\right) = 3Ch\cdot Y_{\mathrm{et}}r^{2}\left(\frac{dr}{d\tau}\right) \tag{2.18}$$

For Langmuir isotherm

$$Y_{\mathrm{et}} = \frac{Y_{\mathrm{s}}C_{\mathrm{et}}}{1+k_{0}C_{\mathrm{et}}} = \frac{Y_{\mathrm{s}}C_{0}C_{\mathrm{et}}^{*}}{1+k_{0}C_{0}C_{\mathrm{et}}^{*}} = \frac{Y_{\mathrm{es}}C_{\mathrm{et}}^{*}}{1+k_{0}^{*}C_{\mathrm{et}}^{*}} \tag{2.19}$$

where $Y_{\mathrm{es}} = Y_{\mathrm{s}}C_{0}$ and $k_{0}{}^{*} = k_{0}C_{0}$.

The time derivative of Eq. (2.19) becomes

$$\frac{dY_{\mathrm{et}}}{d\tau} = \frac{Y_{\mathrm{es}}}{\left(1+k_{0}^{*}C_{\mathrm{et}}^{*}\right)^{2}}\frac{dC_{\mathrm{et}}^{*}}{d\tau} \tag{2.20}$$

Combining Eqs. (2.14), (2.18), and (2.20) and after algebraic manipulation, the following expression is obtained (for Langmuir-type isotherm):

$$\left(\frac{dC_{\mathrm{t}}^{*}}{d\tau}\right) = \left(N/M\right)\left(\frac{dr}{d\tau}\right) \tag{2.21}$$

where $M = 1 + Ch\left(1-r^{3}\right)\dfrac{Y_{\mathrm{es}}Bi(1-r)}{\left(1+k_{0}^{*}C_{\mathrm{et}}^{*}\right)^{2}[r+(1-r)Bi]}$ and,

$N = 3ChY_{\mathrm{et}}r^{2} + \dfrac{ChY_{\mathrm{es}}Bi(1-r^{3})C_{\mathrm{t}}^{*}}{\left(1+k_{0}^{*}C_{\mathrm{et}}^{*}\right)^{2}[r+(1-r)Bi]^{2}}$. Using Eq. (2.14), Eq. (2.17) may be written as

$$\frac{dr}{d\tau} = \frac{-Bi\left(\frac{C_0}{\rho Y_e}\right)\left(C_t^* - C_{et}^*\right)}{r^2} = f_1\left(C_t^*, r\right) \tag{2.22}$$

Using Eqs. (2.14) and (2.22), Eq. (2.21) may be expressed as

$$\left(\frac{dC_t^*}{d\tau}\right) = \frac{N\left(C_t^*, r\right) f_1\left(C_t^*, r\right)}{M\left(C_t^*, r\right)} = f_2\left(C_t^*, r\right) \tag{2.23}$$

The initial conditions for Eqs. (2.22) and (2.23), $C_0 = 1.0$ and $r = 1.0$ at time, $\tau = 0.0$. Equations (2.22) and (2.23) can be solved to find the bulk concentration at any time "t" if we know all the process parameters. The two process parameters – the external mass transfer coefficient (k_f) and internal effective diffusivity (D_p) – are unknown to us. These two parameters are estimated by optimizing the experimental concentration profile as outlined in the next section.

2.3.1 Numerical Analysis

The above set of equations are numerically solved using fourth-order Runge-Kutta of step size (dτ) of the order 10^{-5} along with a nonlinear optimization technique (Levenberg-Marquardt) to estimate the two process parameters described above, so that the experiment kinetic profile (i.e., bulk concentration versus time) is matched. For this purpose, optimization subroutine UNLSF/DUNLSF from IMSL math library has been used.

The adsorption systems studied here encompass Radke-Prausnitz isotherm (Tables 2.5 and 2.6). The systems considered here are (1) Astrazone blue dye on silica, (2) para-nitrophenol on granular activated carbon from Lurgi, and (3) toluene on F300 activated carbon. The experimental data on kinetics and the isotherm constants have been reported in literature (McKay 1984; Costa et al. 1987; Chatzopoulos et al. 1993).

The adsorption of Astrazone blue on silica follows Langmuir isotherm (McKay 1984). The isotherm constants are $Y_s = 0.5$ lit/g and $K_0 = 0.016$ lit/mg, where Y_e in mg/g and C_e in mg/l. For $W = 17$ g, $V = 1.7$ l, $R = 0.3025$ mm, and $\rho = 2.2$ g/cc, the concentration decay data for $C_0 = 520$ mg/l has been used to determine the unknown process parameters using the above numerical procedure as shown in Fig. 2.8a. The estimated values of the parameters are as follows: $k_f = 130.0 \times 10^{-6}$ cm/s and $D_p = 16.16 \times 10^{-9}$ cm^2/s. These values of k_f and D_p are used to simulate the adsorption kinetics for different operating conditions. It is interesting to note that the estimated values of k_f and D_p are close to the values reported by Mckay (1984), i.e., $k_f = 80 \times 10^{-6}$ cm/s and $D_p = 18 \times 10^{-9}$ cm^2/s. The experimental observations and the model-simulated concentration profiles for different initial dye

Table 2.5 Radke-Prausnitz isotherm constants $\left[\left(\frac{1}{Y_e}\right) = \left(\frac{1}{AC_e}\right) + \left(\frac{1}{BC_e^{\delta}}\right)\right]$

Isotherm	T (°C)	A (L/g)	B (L/g)	δ
Radke-Prausnitz	10	958.91	2.523	0.195
	25	608.16	2.269	0.188
	40	315.37	2.078	0.196

Table 2.6 Model parameters using Radke-Prausnitz isotherm at various temperatures

Temp. (°C)	$k_f \times 10^5$ (m/s)	$D_p \times 10^8$ (m^2/s)
10	11.48	
25	19.10	6.0
40	27.86	

concentrations, masses of silica, and particle sizes of silica have been shown in Figs. 2.8b, 2.8c and 2.8d, respectively. From the above figures, it may be observed that beyond 120 min (2 h) of the process, the model underpredicts the bulk concentration profile. This may be due to the increase of the resistance inside the micropores which inhibits the process of adsorption. The present model can be used for multicomponent adsorption processes and also with concentration-dependent diffusivity. The model is useful to estimate k_f and D_p values, which are required for the design of fixed-bed adsorber.

2.4 Discussion of Mathematical Model Analysis

The adsorption experiments in the fixed-bed column are carried out to study the adsorption dynamics and quantify the breakthrough curve. One of the crucial aspects of design of adsorption columns for any separation process is the prediction of the breakthrough time. This is necessary to estimate the lifetime of the adsorption bed and its process efficiency. There have been several mathematical models developed in the past based on different assumption justifying the simplicity in the calculations.

2.4.1 Thomas Model (Thomas 1944)

Thomas solution is the most general and widely used equation for modeling performance of fixed-bed adsorption. The Thomas model assumes second-order reversible Langmuir kinetics of the adsorption-desorption process. Ideally the model is suitable for situations where the external and internal diffusion resistances are small. This is particularly true for adsorption scenarios in most liquid systems and therefore is most relevant for adsorption in aqueous environment. The expression describing the output concentration C_t/C_0 is given by

Fig. 2.8a Adsorption of Astrazone blue dye on silica

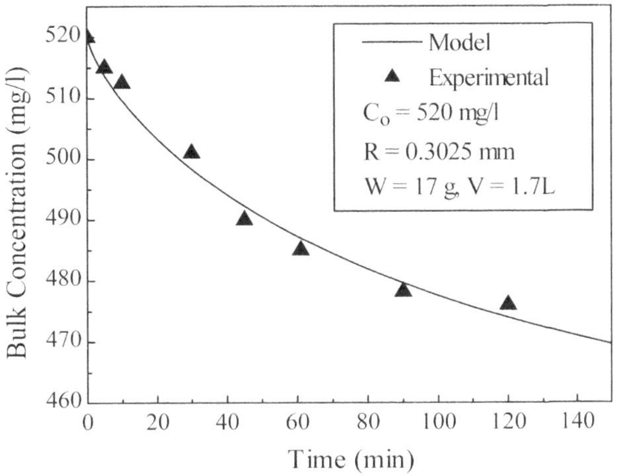

Fig. 2.8b Effect of initial adsorbate concentration. *Solid lines* are the model predictions and *symbols* are the experimental data

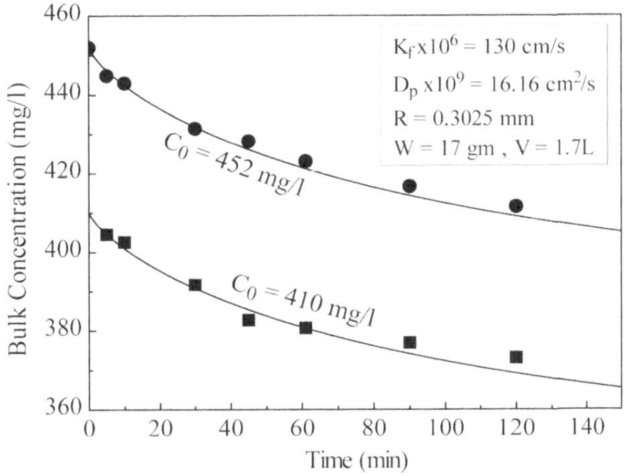

Fig. 2.8c Effect of the mass of adsorbent on concentration decay. *Solid lines* are the model predictions and *symbols* are the experimental data

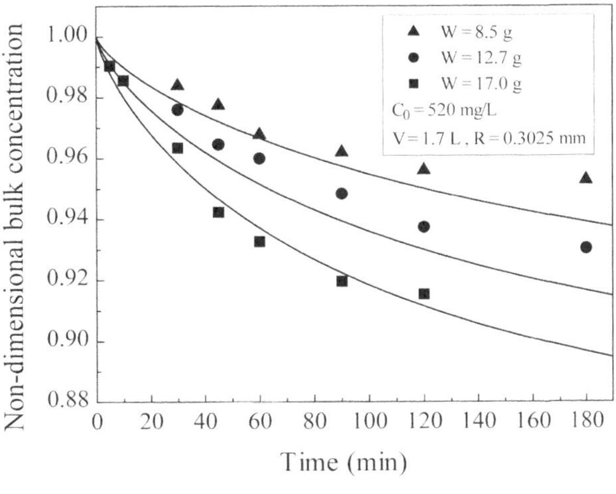

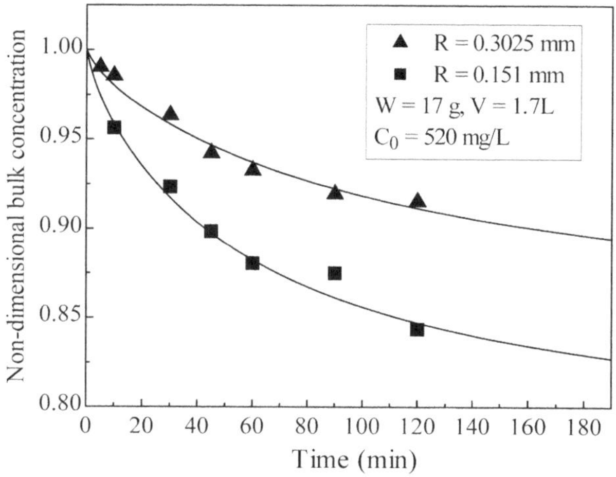

Fig. 2.8d Effect of silica particle size on concentration decay. *Solid lines* are the model predictions and *symbols* are the experimental data

$$\frac{c_t}{c_0} = \frac{1}{1 + \exp\left(k_{Th} q_e{}^{Z}/_{Q - k_{Th} c_0 t}\right)} \tag{2.24}$$

where k_{Th} is the model parameter obtained from nonlinear regression of the output concentration with time. The parameter q_e is the maximum adsorption capacity of the adsorbent.

2.4.2 *Adams-Bohart Model (Bohart and Adams 1920)*

The Adams-Bohart model considers that the adsorption rate is proportional to both the adsorbent leftover capacity and the concentration of the adsorbate species in the solution. The Adams-Bohart model is originally applied for prediction of adsorption behavior in gas-solid systems, but later on extended to liquid streams. It assumes that the adsorption rate is proportional to the residual capacity of the adsorbent and adsorbate concentration. Since the external mass transfer is not taken into account, it is particularly not suitable for describing the system at high flow rate and concentration. Theoretically, the model is applicable for predictions at early times, when $C/C_0 << 1$. The mathematical equation describing the output concentration is represented by Eq. (2.25):

$$\frac{c_t}{c_0} = \exp\left(k_{AB} c_0 t - k_{AB} q_e \frac{Z}{Q}\right) \tag{2.25}$$

where k_{AB} is the model parameter obtained from nonlinear regression of the experimental data.

2.4.3 *Yoon-Nelson Model (Yoon and Nelson 1984)*

The Yoon-Nelson model is based on the assumption that the probability of adsorption for each molecule decreases proportionately on the probabilities of the adsorbate adsorption and breakthrough. One of the features of the model is that the product of the parameters $K_{yn}\tau_{yn}$is constant for a particular adsorbent-adsorbate combination and independent on the operating conditions. This is a fairly simple model which does not require any knowledge of the adsorption capacity or type of the adsorbent:

$$\frac{c_t}{c_0 - c_t} = \exp(k_{\mathrm{YN}}t - \tau k_{\mathrm{YN}}) \tag{2.26}$$

where τ and k_{YN} are the model parameters obtained from nonlinear regression of the experimental data.

2.4.4 *Clark Model (Clark 1987)*

This model is based on the application of the mass transfer concept in combination with Freundlich equilibrium isotherm. The adsorption equilibrium isotherm satisfying Freundlich relationship can only be used for predicting the breakthrough profile of the adsorption column. The semiempirical relationship is presented in Eq. (2.27):

$$\frac{c_{\mathrm{t}}}{c_0} = \left(\frac{1}{1 + A\exp(-rt)}\right)^{1/(n-1)} \tag{2.27}$$

where A and r are the model constants obtained from nonlinear regression analysis. The constant 1/n is obtained from the Freundlich isotherm equation.

2.4.5 *Bed Depth/Service Time (BDST) Model (Goel et al. 2005)*

The BDST model is the linearized form of the Adams-Bohart model. The main consideration here is the assumption that the intraparticle diffusion and external mass transfer resistance is negligible and the adsorption kinetics is controlled by the surface chemical reaction between the adsorbate and adsorbent, which is generally uncommon in real systems. The popularity of the BDST is due to its simplicity in predicting breakthrough behavior owing to its rapid analysis. The expression predicting the breakthrough profile (C_t/C_0) is given by

$$\frac{c_0}{c_t} = 1 + K_{\mathrm{BDST}} \exp\left(\frac{q_e Z}{Q} - c_0 t\right) \tag{2.28}$$

where K_{BDST} is the model parameter determined by the nonlinear regression analysis of the experimental data. Although the BDST model provides a simple and comprehensive approach for evaluating sorption column test, its validity is limited and does not involve any sound understanding of the implicit transport mechanism (Bohart and Adams 1920; Poots et al. 1976a; Faust and Aly 1987). One of the major limitations of this model is the symmetry of the logistic function (S-shaped curve) around its midpoint $t = N_0 Z/C_0 U_0$ and $C = C_0/2$, which is not true for most breakthrough profiles. Therefore, a more detailed adsorption bed modeling based on the physical transport laws of pore diffusion is necessary for accuracy of the model prediction and scaling up of the process.

2.4.6 Pore Diffusion-Adsorption Model

The 1D single species convective-diffusive equation (Kunii and Levenspiel 1991) is described by Eq. (2.29):

$$\frac{\partial C}{\partial t} = D_L \frac{\partial^2 C}{\partial z^2} - v \frac{\partial C}{\partial z} - \left(\frac{3k_{\mathrm{f}}}{a_{\mathrm{p}}}\right)\left(\frac{1-\varepsilon}{\varepsilon}\right)\rho_{\mathrm{s}}(C - C_{\mathrm{e}}) \tag{2.29}$$

where the generation term accounted is dependent on the solid-fluid mass transfer rate and is linearly proportional to the concentration difference and C_{e} is the adsorbate concentration at the adsorbent-bulk interface. The solution of Eq. (2.29) provides information of the transient solute concentration at various bed depths. In deriving Eq. (2.29), by the material balance analysis, it is inherently assumed that all the interparticle void space in the bed is saturated and the fluid velocity is uniform and unhindered throughout. The initial and boundary conditions of Eq. (2.29) are

$$\text{at t} = 0, C = C_0 \text{ for } z = 0 \text{ and } C = 0 \text{ for } \quad 0 < z \le L \tag{2.30a}$$

$$\text{at } z = 0, \quad D_L \frac{\partial C}{\partial z} + V(C_0 - C) = 0 \tag{2.30b}$$

$$\text{and at } z = \text{L}, \quad \frac{\partial C}{\partial z} = 0 \tag{2.30c}$$

The intra-pellet adsorption is described by the pore diffusion transport model. Intraparticle mass transport is characterized by the pore diffusion coefficient D_{p}. The mass balance equation for the liquid phase (pore) in a spherical particle can be written as

$$\varepsilon_{\mathrm{p}}\frac{\partial C_{\mathrm{p}}}{\partial t}+(1-\varepsilon_{\mathrm{p}})\rho_{\mathrm{s}}\frac{\partial q}{\partial t}=D_{\mathrm{p}}\left(\frac{\partial^2 C_{\mathrm{p}}}{\partial r^2}+\frac{2}{r}\frac{\partial C_{\mathrm{p}}}{\partial r}\right) \tag{2.31}$$

where C_{p} is the contaminant concentration inside the particle and $\varepsilon_{\mathrm{p}}$is particle porosity. Assuming instantaneous equilibrium $\frac{\partial q}{\partial t}=\frac{\partial C_{\mathrm{p}}}{\partial t}\frac{\partial q}{\partial C_{\mathrm{p}}}$. Modifying Eq. (2.31), we get (Singha et al. 2012),

$$\frac{\partial C_{\mathrm{p}}}{\partial t}=\frac{1}{\left[1+(1-\varepsilon_{\mathrm{p}})\rho_s\frac{\partial q}{\partial C_{\mathrm{p}}}\right]}\left(\frac{D_{\mathrm{p}}}{\varepsilon_{\mathrm{p}}}\right)\left(\frac{\partial^2 C_{\mathrm{p}}}{\partial r^2}+\frac{2}{r}\frac{\partial C_{\mathrm{p}}}{\partial r}\right) \tag{2.32}$$

The initial condition ($t=0$) is given by $C_{\mathrm{p}}=0$ for $0<r<\mathrm{a_p}$.

The symmetry condition at the particle center ($r=0$) and continuity of the concentration on the external surface of the adsorbent bed are simultaneously expressed as

$$\text{at } r=0,\ \frac{\partial C_{\mathrm{p}}}{\partial r}=0 \tag{2.33a}$$

$$\text{and at } r=a_{\mathrm{p}},\ k_{\mathrm{f}}\left(C_{\mathrm{p}}-C_{\mathrm{e}}\right)=D_{\mathrm{p}}\varepsilon_{\mathrm{p}}\frac{\partial C_{\mathrm{p}}}{\partial r} \tag{2.33b}$$

2.5 Various Types of Adsorbents Used for Dye Adsorption

A summary of the various low-cost adsorbents for dye removal as studied by several researchers in the past is presented in Tables 2.7, 2.8 and 2.9. Natural materials or the wastes/by-products of industries or synthetically prepared materials, which cost less and can be used as such or after some minor treatment as adsorbents, are generally called low-cost adsorbents. Generally, the low-cost adsorbents are usually branded as substitutes for activated carbons because of their similar wide usage; however, in a clear sense, they are essentially substitutes for all available expensive adsorbents. These alternative low-cost adsorbents (Gupta et al. 2009) may be categorized in two ways (1) based on their availability, for, e.g., natural materials such as coal, wood, lignite, peat, etc., or agricultural/industrial/domestic wastes; or by-products such as sludge, slag, red mud, fly ash, etc., or synthesized products; and (2) depending on their nature, for, e.g., organic or inorganic. The adsorbents listed in Table 2.7, 2.8, and 2.9 provide useful information about the type and capacity of alternative adsorbents without going into too much detail of the preparation process.

Table 2.7 Adsorption capacities of commercial activated carbon and other alternative adsorbents for removal of acid dyes

Adsorbent	Adsorbate	Surface area (m^2/g)	Adsorption capacity (mg/g)	Concentration range (mg/L)	Source
GAC Filtrasorb 400	Acid blue 40	1100	57.5 mg/g	25–200	Ozacar and Sengil (2002)
Filtrasorb F 400	Acid blue 80	1200	112.3 mg/g	–	Choy et al. (2000)
Filtrasorb F 400	Acid red 114	1200	103.5 mg/g	–	Choy et al. (2000)
Filtrasorb F 400	Acid red 88	–	109 mg/g	–	Venkata Mohan et al. (1999)
Filtrasorb F 400	Acid yellow 117	1200	155.8 mg/g	–	Choy et al. (2000)
GAC Filtrasorb 400	Acid yellow 17	1100	133.3	25–200	Ozacar and Sengil (2002)
PAC	Acid brown 283	1026	22	30–250	Martin et al. (2003)
AC-charcoal	Acid blue	–	100.9	10–25	Choy et al. (1999)
	Acid yellow		128.8		
	Acid red 114		101		
AC rice husk	Acid blue	352	50	1–50	Mohamed (2004)
Blast furnace sludge	Acid blue 113	28	2.1	–	Jain et al. (2003c)
Bentonite	Acid blue 193	767	740.5	–	Ozcan et al. (2004)
Wood sawdust (raw)	Acid blue 25	–	5.92	–	Ho and McKay (1998a)
Treated cotton	Acid blue 25	–	589	–	Bouzaida and Rammah (2002)
Chitosan	Acid blue 25	–	77.4	–	Martel et al. (2001)
Hazelnut shell	Acid blue 25	–	60.2	50–500	Ferrero (2007)
Sawdust-walnut	Acid blue 25	–	37	50–500	Ferrero (2007)
Sawdust-cherry	Acid blue 25	–	32	50–500	Ferrero (2007)

(continued)

Table 2.7 (continued)

Adsorbent	Adsorbate	Surface area (m^2/g)	Adsorption capacity (mg/g)	Concentration range (mg/L)	Source
Sawdust-oak	Acid blue 25	–	27.8	50–500	Ferrero (2007)
Sawdust-pitch pine	Acid blue 25	–	26.2	50–500	Ferrero (2007)
AC-corncob	Acid blue 25	943	1060	–	Juang et al. (2002)
AC-bagasse	Acid blue 25	607	674	–	Juang et al. (2002)
AC-plum kernel	Acid blue 25	1162	904	–	Juang et al. (2002)
Cane pith	Acid blue 25	606.8	673.6	–	Juang et al. (2001)
Bagasse pith	Acid blue 25	–	17.5	10–1000	Chen et al. (2001)
Wood	Acid blue 25	3.8–6.4	7–11.6	–	Poots et al. (1976b)
Maize cob	Acid blue 25	–	41.4	0.05	El-Geundi and Aly (1992)
	Acid red 114		47.7		
Pine sawdust	Acid blue 256	–	280.3	–	Ozacar and Sengil (2005)
AC-pinewood	Acid blue 264	902	1176	–	Tseng et al. (2003)
Dead fungus *Aspergillus niger*	Acid blue 29	–	1.44–13.8	50	Fu and Viraraghavan (2001)
Living biomass *Aspergillus niger*	Acid blue 29	–	6.63	50	Fu and Viraraghavan (2001)
Modified fungal biomass – *Aspergillus niger*	Acid blue 29	–	17.6	46	Fu and Viraraghavan (2002b)
Calcined alunite	Acid blue 40	42.8	212.8	25–200	Ozacar and Sengil (2002)
Activated sewage sludge	Acid blue 74	390	60.0	100–1000	Otero et al. (2003b)
Pyrolyzed sewage sludge	Acid blue 74	80	30.8	100–1000	Otero et al. (2003b)
AC-bagasse	Acid blue 80	1433	391	20–1050	Valix et al. (2004)

(continued)

Table 2.7 (continued)

Adsorbent	Adsorbate	Surface area (m^2/g)	Adsorption capacity (mg/g)	Concentration range (mg/L)	Source
Activated clay	Acid blue 9	–	57.8	–	Ho et al. (2001)
Soy meal hull	Acid blue 92	0.76	114.9	50–150	Arami et al. (2006)
Banana pith	Acid brilliant blue	–	4.4	–	Namasivayam et al. (1998)
Coir pith	Acid brilliant blue	–	16.7	–	Namasivayam et al. (2001)
Leather industry waste	Acid brown	–	2.84–6.24	50–125	Sekaran et al. (1995)
Chitosan	Acid green 25	–	645.1	–	Wong et al. (2004)
	Acid orange 10		922.9		
Banana peel	Acid orange 52	20.6–23.5	21	10–120	Annadurai et al. (2002)
Orange peel	Acid orange 52	20.6–23.5	20.5	10–120	Annadurai et al. (2002)
Sewage sludge	Acid red 1	–	35–73	10–1000	Seredych and Bandosz (2007)
Bagasse pith (raw)	Acid red 114	–	20	10–1000	Chen et al. (2001)
AC from gingelly seed shell	Acid red 114	229.6	102	–	Thinakaran et al. (2008)
AC from cottonseed shell	Acid red 114	124.3	153.8	–	Thinakaran et al. (2008)
AC from pongam seed shell	Acid red 114	324.8	204.1	–	Thinakaran et al. (2008)
Soy meal hull	Acid red 14	0.76	109.9	50–150	Arami et al. (2006)
Hen feathers	Acid red 51	–	129.1	8.79–52.7	Gupta et al. (2006)
Charfines (raw)	Acid red 88	–	33.3	–	Venkata Mohan et al. (1999)
Lignite coal (raw)	Acid red 88	–	30.9	–	Venkata Mohan et al. (1999)
Bituminous coal	Acid red 88	–	26.1	–	Venkata Mohan et al. (1999)

(continued)

Table 2.7 (continued)

Adsorbent	Adsorbate	Surface area (m^2/g)	Adsorption capacity (mg/g)	Concentration range (mg/L)	Source
Coir pith (raw)	Acid violet	–	1.65	–	Namasivayam et al. (2001)
Orange peel (raw)	Acid violet 17	19.9	–	–	Sivaraj et al. (2001)
Pine sawdust (raw)	Acid yellow 132	–	398.8	–	Ozacar and Sengil (2005)
Calcined alunite	Acid yellow 17	42.8	151.5	25–200	Ozacar and Sengil (2005)
Sawdust carbon	Acid yellow 36	516.3	183.8	–	Malik (2003)
Rice husk carbon	Acid yellow 36	272.5	86.9	–	Malik (2003)
Blast furnace sludge	Acid yellow 36	28	1.4	–	Jain et al. (2003c)
Treated cotton	Acid yellow 99	–	448	–	Bouzaida and Rammah (2002)
Blast furnace sludge	Ethyl orange	28	1.3	–	Jain et al. (2003c)
Fly ash	Metomega chrome orange	–	0.743	10	Gupta and Shukla (1996)
Wollastonite	Metomega chrome orange	–	0.7	10	Gupta and Shukla (1996)
Kaolinite	Metomega chrome orange	–	0.65	10	Gupta and Shukla (1996)
Coal	Metomega chrome orange	–	0.77	10	Gupta and Shukla (1996)
Activated bentonite	Sella fast brown H	–	360.5	–	Espantaleon et al. (2003)

Reproduced from Gupta and Suhas (2009). With permission from Elsevier

Table 2.8 Adsorption capacities of commercial activated carbon and other alternative adsorbents for removal of basic dyes

Adsorbent	Adsorbate	Surface area of adsorbent (m_2/g)	Adsorption capacity (mg/g)	Concentration range (mg/L)	Source
Commercial AC (E. Merck India)	Basic blue 9 (methylene blue)		980.3	100–400	Kannan and Sundaram (2001)
CAC granular Wako (Wako Pure Chemicals)	Basic blue 9	1150	260	–	Okada et al. (2003)
CAC fiber FE400 (Toho Rayon Co.)	Basic blue 9	1010	170	–	Okada et al. (2003)
CAC felt KF1500 (Toyobo Co.)	Basic blue 9	1480	300	–	Okada et al. (2003)
Activated carbon	Basic blue 3	–	648.6	50–600	Nassar and Magdy (1997)
GAC (Miloje Zakic)	Basic dye Maxilon Goldgelb GL EC	–	159	20–200	Meshko et al. (2001)
GAC (Miloje Zakic)	Basic dye Maxilon Schwarz FBL-01	–	309.2	50–500	Meshko et al. (2001)
CAC Merck	Basic green 4 (malachite green)	765	222.22	–	Malik et al. (2007)
Chemviron F-400	Basic red 22	–	720	50–1000	Allen et al. (2003)
Activated carbon	Basic red 22	–	790	50–600	Nassar and Magdy (1997)
PAC Chemviron GW	Basic red 46	1026	106	30–250	Martin et al. (2003)
Chemviron F-400	Basic yellow 21	–	860	50–1000	Allen et al. (2003)
Activated carbon	Basic yellow 21	–	600	50–600	Nassar and Magdy (1997)
Activated sludge biomass	Basic blue 3	–	36.5	–	Chu and Chen (2002a)
Palm-fruit bunch (raw)	Basic blue 3	–	92	50–600	Nassar and Magdy (1997)
Activated sludge biomass	Basic blue 47	–	157.5	–	Chu and Chen (2002a)

Activated sludge biomass	Basic blue 54	–	86.6	–	Chu and Chen (2002a)
Carbonaceous adsorbent	Basic blue 6 (MeldolÃ¡s blue)	380	170	–	Jain et al. (2003b)
Blast furnace (BF) sludge, BF dust, BF slag	Basic blue 6 (MeldolÃ¡s blue)	28, 13, 4	67, 34, 3.7	–	Jain et al. (2003b)
AC-pinewood	Basic blue 69	902	1119	–	Tseng et al. (2003)
Bagasse pith raw	Basic blue 69	–	152	10–1000	Chen et al. (2001)
Wood sawdust (raw)	Basic blue 69	–	71.9	–	Ho and McKay (1998a)
Peat	Basic blue 69	–	195	–	Ho and McKay (1998b)
Wood	Basic blue 69 (Astrazone blue)	–	100.1	–	Poots et al. (1978)
Peat	Basic blue 69 (Astrazone blue)	–	40–910	200	McKay et al. (1981)
Hardwood sawdust	Basic blue 69 (Astrazone blue)	–	82.2–105.7	200	Asfour et al. (1985)
Activated clay	Basic blue 69, basic red 22	–	585, 488.4	–	El-Guendi et al. 1995
Bagasse pith	Basic blue 69, basic red 22	–	157.4, 76.6	200	McKay et al. (1997)
AC-pinewood	Basic blue 9	902	556	–	Tseng et al. (2003)
AC-waste newspaper	Basic blue 9	1740	390	–	Okada et al. (2003)
Bentonite	Basic blue 9 (methylene blue)	28	1667	100–1000	Ozacar and Sengil (2006)
Coal	Basic blue 9	–	250	10–1000	McKay et al. (1999)
Bark	Basic blue 9	–	914	10–1000	McKay et al. (1999)
Rice husk	Basic blue 9	–	312	10–1000	McKay et al. (1999)
Cotton waste	Basic blue 9	–	277	10–1000	McKay et al. (1999)
Hair	Basic blue 9	–	158	10–1000	McKay et al. (1999)

(continued)

Table 2.8 (continued)

Adsorbent	Adsorbate	Surface area of adsorbent (m_2/g)	Adsorption capacity (mg/g)	Concentration range (mg/L)	Source
Sewage sludge	Basic blue 9	–	114.94	–	Otero et al. (2003a)
Bamboo dust carbon	Basic blue 9	–	143.2	100–400	Kannan and Sundaram (2001)
Coconut shell carbon	Basic blue 9	–	277.9	100–400	Kannan and Sundaram (2001)
Groundnut shell carbon	Basic blue 9	–	164.9	100–400	Kannan and Sundaram (2001)
Rice husk carbon	Basic blue 9	–	343.5	100–400	Kannan and Sundaram (2001)
Straw carbon	Basic blue 9	–	472.1	100–400	Kannan and Sundaram (2001)
Raw date pits	Basic blue 9	–	80.3	20–400	Banat et al. (2003)
AC-apricot shell	Basic blue 9	783	4.11	–	Aygun et al. (2003)
AC-hazelnut shell	Basic blue 9	793	8.82	–	Aygun et al. (2003)
AC-walnut shell	Basic blue 9	774	3.53	–	Aygun et al. (2003)
Fly ash-Slovakia	Basic blue 9	3.26	1.47	16–64	Janos et al. (2003)
Fly ash-Czech Republic	Basic blue 9	5.47	6.04	16–64	Janos et al. (2003)
Fe(III)/Cr(III) hydroxide	Basic blue 9	–	22.8	–	Namasivayam and Sumithra (2005)
Banana peel (raw)	Basic blue 9	20.6–23.5	20.8	10–120	Annadurai et al. (2002)
Orange peel (raw)	Basic blue 9	20.6–23.5	18.6	–	Annadurai et al. (2002)
Clay	Basic blue 9	71	300	–	Bagane and Guiza (2000)
Diatomite	Basic blue 9	27.8	198	100–400	Al-Ghouti et al. (2003)
Diatomite	Basic blue 9	33	134	41–600	Shawabkeh and Tutunji (2003)

Clay	Basic blue 9	30	6.3	–	Gurses et al. (2004)
Activated sludge	Basic blue 9	–	256.41	–	Gulnaz et al. (2004)
Spirodela polyrrhiza biomass	Basic blue 9	–	144.93	–	Waranusantigul et al. (2003)
Dead fungus *Aspergillus niger*	Basic blue 9	–	10.49–18.54	50	Fu and Viraraghavan (2000)
Living biomass *Aspergillus niger*	Basic blue 9	–	1.17	50	Fu and Viraraghavan (2000)
Neem sawdust	Basic blue 9	–	3.622	12	Khattri and Singh (2000)
Yellow passion fruit	Basic blue 9	30	44.7	–	Pavan et al. (2008)
Guava leaf powder	Basic blue 9	–	295	100–800	Ponnusami et al. (2008)
Beer brewery waste	Basic blue 9	4.5	4.92	–	Tsai et al. (2008)
Jackfruit peel	Basic blue 9 (methylene blue)	–	285.713	35–400	Hameed (2009a)
Spent tea leaves	Basic blue 9 (methylene blue)	–	300.052	30–390	Hameed (2009b)
Sugarcane dust	Basic blue 9	–	3.745	12	Khattri and Singh (1999)
Carbonaceous adsorbent	Basic blue 9 (methylene blue)	380	92	–	Jain et al. (2003a)
Blast furnace (BF) sludge, BF dust, BF slag	Basic blue 9 (methylene blue)	28, 13, 4	6.4, 3.3, 2.1	–	Jain et al. (2003a)
Diatomite	Basic blue 9 (methylene blue)	27.8	198	100–400	Al-Ghouti et al. (2003)
Cedar sawdust, crushed brick	Basic blue 9 (methylene blue)	–	142.36, 96.41	–	Hamdaoui (2006)
Fly ash (treated with H_2SO_4)	Basic blue 9 (methylene blue)	6.236	0.67	8.5–85	Lin et al. (2008)

(continued)

Table 2.8 (continued)

Adsorbent	Adsorbate	Surface area of adsorbent (m_2/g)	Adsorption capacity (mg/g)	Concentration range (mg/L)	Source
Fly ash, zeolite, unburned carbon	Basic blue 9 (methylene blue)	15.6, 16.0, 224	6.4, 14.4, 80	0.32–3.2	Wang et al. (2005)
PET carbon	Basic blue 9 (methylene blue)	–	33.4	–	Zhang and Itoh (2003)
Hazelnut shell	Basic blue 9	–	76.9	50–1000, 50–500	Ferrero (2007)
Sawdust-walnut	Basic blue 9	–	59.17	50–1000, 50–500	Ferrero (2007)
Sawdust-cherry	Basic blue 9	–	39.84	50–1000, 50–500	Ferrero (2007)
Sawdust-oak	Basic blue 9	–	29.94	50–1000, 50–500	Ferrero (2007)
Sawdust-pitch pine	Basic blue 9	–	27.78	50–1000, 50–500	Ferrero (2007)
Sunflower stalk	Basic blue 9 (methylene blue), basic red 9	1.2054	205, 317	100–2000, 100–2000	Sun and Xu (1997)
Beech sawdust untreated	Basic blue 9 (methylene blue), red basic 22	–	9.78, 20.2	–	Batzias and Sidiras (2004)
Zeolite	Basic dye Maxilon Goldgelb GL EC	–	14.91	20–200	Meshko et al. (2001)
Zeolite	Basic dye Maxilon Schwarz FBL-01	–	55.86	50–500	Meshko et al. (2001)
Sawdust carbon	Basic green 4	–	74.5	50–250	Garg et al. (2003)
Neem sawdust	Basic green 4	–	3.42	12	Khattri and Singh (2000)
AC from pine sawdust	Basic green 4	1390	370.37	50–2000	Akmil-Basar et al. (2005)
Oil palm trunk fiber	Basic green 4 (malachite green)	–	149.35	25–300	Hameed and el-Khaiary (2008)
AC-groundnut shell	Basic green 4 (malachite green)	1114	222.22	–	Malik et al. (2007)

Waste material from paper industry, pine bark	Basic green 4 (malachite green)	–	–	100	Mendez et al. (2007)
Carbonaceous material	Basic green 4 (malachite green)	629	75.08	36.4–364	Gupta et al. (1997)
Sugarcane dust	Basic green 4 (malachite green)	–	3.999	12	Khattri and Singh (1999)
Carbonaceous adsorbent	Basic orange 2 (Chrysoidine G)	380	75	–	Jain et al. (2003b)
Blast furnace (BF) sludge, BF dust, BF slag	Basic orange 2 (Chrysoidine G)	28, 13, 4	10.1, 5.4, 1.9	–	Jain et al. (2003b)
Tree fern	Basic red 13	–	408	–	Ho et al. (2005)
Activated sludge	Basic red 18	–	285.71	–	Gulnaz et al. (2004)
Activated sludge biomass	Basic red 18	–	133.9	–	Chu and Chen (2002a)
Activated clay	Basic red 18 (C.I. 11,085)	–	157	–	Ho et al. (2001)
Coal	Basic red 2 (C.I. 50,240)	–	120	10–1000	McKay et al. (1999)
Bark	Basic red 2 (C.I. 50,240)	–	1119	10–1000	McKay et al. (1999)
Rice husk	Basic red 2 (C.I. 50,240)	–	838	10–1000	McKay et al. (1999)
Cotton waste	Basic red 2 (C.I. 50,240)	–	875	10–1000	McKay et al. (1999)
Human hair	Basic red 2 (C.I. 50,240)	–	190	10–1000	McKay et al. (1999)
Bentonite	Basic red 2 (C.I. 50,240)	47.73	274	50–450	Hu et al. (2006)
AC-plum kernel	Basic red 22	1162	710	–	Juang et al. (2002)
Sugar-industry-mud	Basic red 22	–	519	50–2000	Magdy and Daifullah (1998)
Kudzu	Basic red 22	–	210	50–1000	Allen et al. (2003)
Bagasse pith raw	Basic red 22	–	75	10–1000	Chen et al. (2001)
Palm-fruit bunch RAW	Basic red 22	–	180	50–600	Nassar and Magdy (1997)
AC-bagasse	Basic red 22 (C.I. 11,055)	607	942	–	Juang et al. (2002)

(continued)

Table 2.8 (continued)

Adsorbent	Adsorbate	Surface area of adsorbent (m_2/g)	Adsorption capacity (mg/g)	Concentration range (mg/L)	Source
Cane pith	Basic red 22 (C.I. 11,055)	606.8	941.7	–	Juang et al. (2001)
AC-corncob	Basic red 22 (C.I. 11,055)	943	790	–	Juang et al. (2002)
Activated sludge biomass	Basic red 29	–	113.2	–	Chu and Chen (2002a)
AC sludge based	Basic red 46	253	188	30–250	Martin et al. (2003)
Neem sawdust	Basic violet 10	–	2.355	12	Khattri and Singh (2000)
Fly ash-Slovakia	Basic violet 10	3.26	1.91	24–95.8	Janos et al. (2003)
Fly ash-Czech Republic	Basic violet 10	5.47	5.5	24–95.8	Janos et al. (2003)
Banana peel	Basic violet 10	20.6–23.5	20.6	10–120	Annadurai et al. (2002)
Orange peel	Basic violet 10	20.6–23.5	14.3	10–120	Annadurai et al. (2002)
Coir pith carbonized	Basic violet 10 (Rhodamine B)	259	2.56	–	Namasivayam et al. (2001)
Coir pith (raw)	Basic violet 10 (Rhodamine B)	–	203.25	–	Namasivayam et al. (2001)
Sugarcane dust	Basic violet 10 (Rhodamine B)	–	3.24	12	Khattri and Singh (1999)
Sewage sludge	Basic violet 14 (basic fuchsin)	–	70–127	10–1000	Seredych and Bandosz (2007)
Neem sawdust	Basic violet 3	–	3.789	12	Khattri and Singh (2000)
Activated sludge biomass	Basic violet 3	–	113.6	–	Chu and Chen (2002a)
Carbonaceous adsorbent	Basic violet 3 (crystal violet)	380	161	–	Jain et al. (2003b)
Blast furnace (BF) sludge, BF dust, BF slag	Basic violet 3 (crystal violet)	28, 13, 4	25, 11, 3	–	Jain et al. (2003b)
Activated sewage sludge	Basic violet 3 (crystal violet)	390	270.88	100–1000	Otero et al. (2003b)
Pyrolyzed sewage sludge	Basic violet 3 (crystal violet)	80	184.68	100–1000	Otero et al. (2003b)

Sugarcane dust	Basic violet 3 (crystal violet)	–	3.798	12	Khattri and Singh (1999)
Palm-fruit bunch (raw)	Basic yellow 21	–	327	50–600	Nassar and Magdy (1997)
Kudzu	Basic yellow 21	–	160	50–1000	Allen et al. (2003)
Activated sludge biomass	Basic yellow 24	–	105.6	–	Chu and Chen (2002b)

Reproduced from Gupta and Suhas (2009). With permission from Elsevier

Table 2.9 Adsorption capacities of commercial activated carbon and other alternative adsorbents for removal of dyes (apart from acid or basic dyes)

Adsorbent	Adsorbate	Surface area of adsorbent (m^2/g)	Adsorption capacity (mg/g)	Concentration range (mg/L)	Source
PAC Chemviron GW	Direct black 168	1026	18.7	30–250	Martin et al. (2003)
Filtrasorb 400 (Calgon Corporation)	Direct brown 1 (C.I. 30,110)	–	7.69	–	Venkata Mohan et al. (2002)
GAC Filtrasorb 400 (Calgon Corporation)	Direct red 28 (C.I. 22,120)	–	13.8	–	Fu and Viraraghavan (2002a)
PAC from Filtrasorb 400 (Calgon Corporation)	Direct red 28 (C.I. 22,120)	–	16.81	–	Fu and Viraraghavan (2002a)
PAC Chemviron GW	Direct red 89	1026	8.4	30–250	Martin et al. (2003)
CAC Aldrich	Reactive red X6BN Sandoz	–	163	25–1000	Oliveira et al. (2007)
Filtrasorb 400 (Chemviron Carbon UK)	Remazol golden yellow, Remazol red, Remazol black B	1100	1111, 400, 434	50–1000	Al-Degs et al. (2000)
AC sludge based	Direct black 168	253	28.9	30–250	Martin et al. (2003)
AC-orange peel	Direct blue 86	–	33.78	25–125	Nemr et al. (2009)
Charfines, lignite coal, bituminous coal	Direct brown (C.I. 30,110)	–	6.4, 4.1, 2.04	50	Venkata Mohan et al. (2002)
Banana pith	Direct red	–	5.92	–	Namasivayam et al. (1998)
Fe(III)/Cr(III) hydroxide	Direct red 12B	–	5	–	Namasivayam and Sumithra (2005)
Biogas residual slurry	Direct red 12B	–	3.46	–	Namasivayam and Yamuna (1995)
Orange peel	Direct red 23, direct red 80	–	10.72, 21.05	–	Arami et al. (2005)
Coir pith	Direct red 28	–	6.72	–	Namasivayam and Kavitha (2002)

Rice hull ash	Direct red 28	236.4	171	–	Chou et al. (2001)
Red mud	Direct red 28	–	4.05	–	Namasivayam and Arasi (1997)
Dead fungus *Aspergillus niger*	Direct red 28	–	14.16	–	Fu and Viraraghavan (2002a)
Banana peel	Direct red 28 (Congo red)	20.6–23.5	18.2	10–120	Annadurai et al. (2002)
Orange peel	Direct red 28 (Congo red)	20.6–23.5	14	10–120	Annadurai et al. (2002)
Activated red mud	Direct red 28 (Congo red)	–	7.08	10–90	Tor and Cengeloglu (2006)
Chitosan	Direct red 28 (Congo red)	–	81.23	–	Wang and Wang (2007)
Sunflower stalk	Direct red 28 (Congo red), direct blue	1.2054	37.78, 26.84	50–1000, 50–1000	Sun and Xu (1997)
Crude sewage sludge	Direct red 79	5.28	19.6	–	Dhaouadi and M'Henni (2008)
Mixture almond shells	Direct red 80	10.5	22.422	50–150	Doulati Ardejani et al. (2008)
Soy meal hull	Direct red 80, direct red 81	0.7623	178.57, 120.48	50–150	Arami et al. (2006)
Chitosan bead (chemically cross-linked)	Direct red 81	–	2383	–	Chiou et al. (2004)
AC sludge based	Direct red 89	253	49.2	30–250	Martin et al. (2003)
Powdered activated sludge	Direct yellow 12	–	98	–	Kargi and Ozmihci (2004)
Palm oil ash	Disperse blue, disperse red	–	49.5, 61.35	–	Hasnain Isa et al. (2007)
AC from biomass *Euphorbia rigida*	Disperse orange 25	741–21	118.93	50–125	Gercel et al. (2008)

(continued)

Table 2.9 (continued)

Adsorbent	Adsorbate	Surface area of adsorbent (m^2/g)	Adsorption capacity (mg/g)	Concentration range (mg/L)	Source
Modified fungal biomass (*Aspergillus niger*)	Disperse red 1 (C.I. 11,110)	–	5.59	13.5	Fu and Viraraghavan (2002b)
Modified sepiolite	Reactive black 5	50.5	120.5	–	Ozdemir et al. (2004)
Modified zeolite	Reactive black 5	11.8	60.5	–	Ozdemir et al. (2004)
Rhizopus arrhizus biomass	Reactive black 5	–	588.2	20–800	Aksu and Tezer (2000)
Sunflower seed shells, mandarin peelings	Reactive black 5	–	–	50	Osma et al. (2007)
High lime fly ash	Reactive black 5	5.35	7.184	5–100	Eren and Acar (2007)
Biomass *Chlorella vulgaris*	Reactive black B	–	555.6	20–800	Aksu and Tezer (2005)
Untreated alunite	Reactive blue 114 (C.I. 21,620)	–	2.92	–	Ozacar and Sengil (2003)
Calcined alunite	Reactive blue 114 (C.I. 21,620)	66	170.7	–	Ozacar and Sengil (2003)
Metal hydroxide sludge	Reactive blue 19	–	275	–	Santos et al. (2008)
Biomass *Rhizopus arrhizus*	Reactive blue 19 (C.I. 61,200)	–	90	0–500	O'Mahony et al. (2002)
Activated sludge	Reactive blue 2 (C.I. 61,211)	–	250	–	Aksu (2001)
Chitosan bead (chemically cross-linked)	Reactive blue 2, reactive red 2, reactive yellow 2, reactive yellow 86	–	2498, 2422, 2436, 1911	–	Chiou et al. (2004)
Charred dolomite	Reactive dye Levafix brilliant red E-4BA	36	950	100–2000	Walker et al. (2003)
Squid pens	Reactive green 12, direct green 26	8.82	39.8, 4.83	–	Figueiredo et al. (2000)

Sepia pens	Reactive green 12, direct green 26	4.11	3.46, 56.0	–	Figueiredo et al. (2000)
Anodonta shell	Reactive green 12, direct green 26	1.42	0.436, 11.3	–	Figueiredo et al. (2000)
Biomass *Rhizopus arrhizus*	Reactive orange 16 (C.I. 17,757)	–	190	0–500	O'Mahony et al. (2002)
Metal hydroxide sludge	Reactive red 120	–	48.31	10–200	Netpradit et al. (2003)
Untreated alunite	Reactive red 124 (C.I. 17,780)	–	2.85	–	Ozacar and Sengil (2003)
Calcined alunite	Reactive red 124 (C.I. 17,780)	66	153	–	Ozacar and Sengil (2003)
Metal hydroxide sludge	Reactive red 141	–	56.18	10–200	Netpradit et al. (2003)
Cross-linked chitosan bead	Reactive red 189	–	1936	–	Chiou and li (2002)
Non-cross-linked chitosan bead	Reactive red 189	–	1189	–	Chiou and li (2002)
Metal hydroxide sludge	Reactive red 2	–	62.5	10–200	Netpradit et al. (2003)
Chitosan bead (crab)	Reactive red 222	–	1106	–	Wu et al. (2000)
Chitosan bead (lobster)	Reactive red 222	–	1037	–	Wu et al. (2000)
Chitosan bead (shrimp)	Reactive red 222	–	1026	–	Wu et al. (2000)
Chitosan flake (shrimp)	Reactive red 222	–	494	–	Wu et al. (2000)
Chitosan flake (crab)	Reactive red 222	–	293	–	Wu et al. (2000)
Modified sepiolite	Reactive red 239	50.5	108.8	–	Ozdemir et al. (2004)
Modified zeolite	Reactive red 239	11.8	111.1	–	Ozdemir et al. (2004)
Biomass *Rhizopus arrhizus*	Reactive red 4 (C.I. 18,105)	–	150	0–500	O'Mahony et al. (2002)

(continued)

Table 2.9 (continued)

Adsorbent	Adsorbate	Surface area of adsorbent (m^2/g)	Adsorption capacity (mg/g)	Concentration range (mg/L)	Source
Chromium-containing leather waste	Reactive red X6BN Sandoz	–	48	25–1000	Oliveira et al. (2007)
Modified sepiolite	Reactive yellow 176	50.5	169.1	–	Ozdemir et al. (2004)
Modified zeolite	Reactive yellow 176	11.8	88.5	–	Ozdemir et al. (2004)
Activated sludge	Reactive yellow 2 (CI 18972)	–	333.3	–	Aksu (2001)
Treated cotton	Reactive yellow 23	–	302	–	Bouzaida and Rammah (2002)
Untreated alunite	Reactive yellow 64 (C.I. 29,025)	–	5	–	Ozacar and Sengil (2003)
Calcined alunite	Reactive yellow 64 (C.I. 29,025)	66	236	–	Ozacar and Sengil (2003)
Eucalyptus bark	Remazol BB	–	90	500	Morais et al. (1999)
Yeast (*Saccharomyces cerevisiae*)	Remazol black B	–	88.5	10–400	Aksu (2003)
Yeast (*Saccharomyces cerevisiae*)	Remazol blue	–	84.6	10–400	Aksu (2003)
Yeast-*C. lipolytica*	Remazol blue	–	250	100–400	Aksu and Donmez (2003)
Yeast-*C. tropicalis*	Remazol blue	–	182	100–400	Aksu and Donmez (2003)
Yeast-*Candida* sp.	Remazol blue	–	167	100–400	Aksu and Donmez (2003)
Yeast-*C. quilliermendii*	Remazol blue	–	154	100–400	Aksu and Donmez (2003)
Yeast-*C. utilis*	Remazol blue	–	114	100–400	Aksu and Donmez (2003)

Biomass *Chlorella vulgaris*	Remazol golden-yellow RNL	–	71.9	10–200	Aksu and Tezer (2005)
Yeast (*Saccharomyces cerevisiae*)	Remazol red RB	–	48.8	10–400	Aksu (2003)
Biomass *Chlorella vulgaris*	Remazol red RR	–	196.1	20–800	Aksu and Tezer (2005)
Crude sewage sludge	Vat blue 4	5.28	248.3	–	Dhaouadi and M'Henni (2008)

Reproduced from Gupta and Suhas (2009). With permission from Elsevier

References

Akmil-Basar C, Onal Y, Kilicer T, Eren D (2005) Adsorptions of high concentration malachite green by two activated carbons having different porous structures. J Hazard Mater 127:73–80
Aksu Z (2001) Biosorption of reactive dyes by dried activated sludge: equilibrium and kinetic modelling. Biochem Eng J 7:79–84
Aksu Z (2003) Reactive dye bioaccumulation by Saccharomyces cerevisiae. Process Biochem 38:1437–1444
Aksu Z, Donmez G (2003) A comparative study on the biosorption characteristics of some yeasts for Remazol blue reactive dye. Chemosphere 50:1075–1083
Aksu Z, Tezer S (2000) Equilibrium and kinetic modelling of biosorption of Remazol black B by Rhizopus arrhizus in a batch system: effect of temperature. Process Biochem 36:431–439
Aksu Z, Tezer S (2005) Biosorption of reactive dyes on the green alga Chlorella Vulgaris. Process Biochem 40:1347–1361
Al-Degs Y, Khraisheh MAM, Allen SJ, Ahmad MN (2000) Effect of carbon surface chemistry on the removal of reactive dyes from textile effluent. Water Res 34:927–935
Al-Ghouti MA, Khraisheh MAM, Allen SJ, Ahmad MN (2003) The removal of dyes from textile wastewater: a study of the physical characteristics and adsorption mechanisms of diatomaceous earth. J Environ Manag 69:229–238
Allen SJ, Gan Q, Matthews R, Johnson PA (2003) Comparison of optimized isotherm models for basic dye adsorption by kudzu. Bioresour Technol 88:143–152
Annadurai G, Juang RS, Lee DJ (2002) Use of cellulose-based wastes for adsorption of dyes from aqueous solutions. J Hazard Mater 92:263–274
Arami M, Limaee NY, Mahmoodi NM, Tabrizi NS (2006) Equilibrium and kinetics studies for the adsorption of direct and acid dyes from aqueous solution by soy meal hull. J Hazard Mater 135:171–179
Asfour HM, Fadali OA, Nassar MM, El-Geundi MS (1985) Equilibrium studies on adsorption of basic dyes on hardwood. J Chem Technol Biotechnol 35A:21–27
Aygun A, Yenisoy-Karakas S, Duman I (2003) Production of granular activated carbon from fruit stones and nutshells and evaluation of their physical, chemical and adsorption properties. Microporous Mesoporous Mater 66:189–195
Bagane M, Guiza S (2000) Elimination d'un colorant des effluents de l'industrie textile par adsorption. Ann Chim Sci Mater 25:615–625
Banat F, Al-Asheh S, Al-Makhadmeh L (2003) Evaluation of the use of raw and activated date pits as potential adsorbents for dye containing waters. Process Biochem 39:193–202
Batzias FA, Sidiras DK (2004) Dye adsorption by calcium chloride treated beech sawdust in batch and fixed-bed systems. J Hazard Mater 114:167–174
Bhattacharyya KG, Sarma A (2003) Adsorption characteristics of the dye, Brilliant Green, on Neem leaf powder. Dyes Pigm 57(3):211–222
Bohart G, Adams EQ (1920) Some aspects of the behavior of charcoal with respect to chlorine. J Am Chem Soc 42:523–544
Bouzaida I, Rammah MB (2002) Adsorption of acid dyes on treated cotton in a continuous system. Mater Sci Eng C 21:151–155
Chatzopoulos D, Verma A, Irvine RL (1993) Activated carbon adsorption and desorption of toluene in the aqueous phase. AICHE J 39:2027–2041
Chen BN, Hui CW, McKay G (2001) Film-pore diffusion modeling and contact time optimization for the adsorption of dyestuffs on pith. Chem Eng J 84:77–94
Chiou MS, Li HY (2002) Equilibrium and kinetic modeling of adsorption of reactive dye on cross-linked chitosan beads. J Hazard Mater 93:233–248
Chiou MS, Ho PY, Li HY (2004) Adsorption of anionic dyes in acid solutions using chemically cross-linked chitosan beads. Dyes Pigments 60:69–84
Choy KKH, McKay G, Porter JF (1999) Sorption of acid dyes from effluents using activated carbon. Resour Conserv Recycl 27:57–71

Choy KKH, Porter JF, McKay G (2000) Langmuir isotherm models applied to the multicomponent sorption of acid dyes from effluent onto activated carbon. J Chem Eng Data 45:575–584
Chu HC, Chen KM (2002a) Reuse of activated sludge biomass: I. Removal of basic dyes from wastewater by biomass. Process Biochem 37:595–600
Chu HC, Chen KM (2002b) Reuse of activated sludge biomass: II. The rate processes for the adsorption of basic dyes on biomass. Process Biochem 37:1129–1134
Clark RM (1987) Evaluating the cost and performance of field-scale granular activated carbon systems. Environ Sci Technol 21:573–580
Costa E, Calleja G, Marijuan L (1987) Adsorption of phenol and p-nitrophenol on activated carbon: determination of effective diffusion coefficient. Adsorp Sci Technol 4:59–77
Dedrick RL, Beckmann RB (1967) Kinetics of adsorption by activated carbon from dilute aqueous solution. AIChe Symp Ser 63:68–78
Dhaouadi H, M'Henni F (2008) Textile mill effluent decolorization using crude dehydrated sewage sludge. Chem Eng J 138:111–119
Doulati Ardejani F, Badii K, Limaee NY, Shafaei SZ, Mirhabibi AR (2008) Adsorption of direct red 80 dye from aqueous solution onto almond shells: effect of pH, initial concentration and shell type. J Hazard Mater 151:730–737
El-Guendi MS, Ismail HM, Attyia KME (1995) Activated clay as an adsorbent for cationic dyestuffs. Adsorpt Sci Technol 12:109–117
Eren Z, Acar FN (2007) Equilibrium and kinetic mechanism for reactive black 5 sorption onto high lime Soma fly ash. J Hazard Mater 143:226–232
Espantaleon AG, Nieto JA, Fernandez M, Marsal A (2003) Use of activated clays in the removal of dyes and surfactants from tannery waste waters. Appl Clay Sci 24:105–111
Faust SD, Aly OM (1987) Adsorption processes for water treatment. Butterworth Publishers, USA
Ferrero F (2007) Dye removal by low cost adsorbents: hazelnut shells in comparison with wood sawdust. J Hazard Mater 142:144–152
Figueiredo SA, Boaventura RA, Loureiro JM (2000) Color removal with natural adsorbents: modeling, simulation and experimental. Sep Purif Technol 20:129–141
Finar IL (1973) Addison Wesley Longman ltd., organic chemistry, Vol-1: the fundamental principles
Fogler HS (1997) Elements of chemical reaction engineering. Prentice Hall (India) Ltd, New Delhi
Fu YZ, Viraraghavan T (2000) Removal of a dye from an aqueous solution by the fungus Aspergillus niger. Water Qual Res J Can 35:95–111
Fu YZ, Viraraghavan T (2001) Removal of CI acid blue 29 from an aqueous solution by Aspergillus niger. Am Assoc Text Chem Color Rev 1:36–40
Fu Y, Viraraghavan T (2002a) Removal of Congo Red from an aqueous solution by fungus Aspergillus niger. Adv Environ Res 7:239–247
Fu YZ, Viraraghavan T (2002b) Dye biosorption sites in Aspergillus niger. Bioresour Technol 82:139–145
Furusawa T, Smith JM (1973) Fluid-particle and intraparticle mass transport rates in slurries. Ind Eng Chem Fundamen 12:197–203
Garg VK, Gupta R, Bala Yadav A, Kumar R (2003) Dye removal from aqueous solution by adsorption on treated sawdust. Bioresour Technol 89:121–124
Gercel O, Gercel HF, Koparal AS, Ogutveren UB (2008) Removal of disperse dye from aqueous solution by novel adsorbent prepared from biomass plant material. J Hazard Mater 160:668–674
Goel J, Kadirvelu K, Rajagopal C, Garg VK (2005) Removal of lead(II) by adsorption using treated granular activated carbon: batch and column studies. J Hazard Mater 125:211–220
Gulnaz O, Kaya A, Matyar F, Arikan B (2004) Sorption of basic dyes from aqueous solution by activated sludge. J Hazard Mater 108:183–188
Gupta GS, Shukla SP (1996) An inexpensive adsorption technique for the treatment of carpet effluents by low cost materials. Adsorp Sci Technol 13:15–26

Gupta VK, Suhas (2009) Application of low cost adsorbents for dye removal – a review. J Environ Manage 90:2313–2342
Gupta VK, Srivastava SK, Mohan D (1997) Equilibrium uptake, sorption dynamics, process optimization, and column operations for the removal and recovery of malachite green from wastewater using activated carbon and activated slag. Ind Eng Chem Res 36:2207–2218
Gupta VK, Mohan D, Sharma S, Sharma M (2000) Removal of basic dyes (Rhodamine B and Methylene Blue) from aqueous solutions using bagasse fly ash. Sep Sci Technol 35:2097–2113
Gupta VK, Mittal A, Kurup L, Mittal J (2006) Adsorption of a hazardous dye, erythrosine, over hen feathers. J Colloid Interface Sci 304:52–57
Gupta VK, Carrott PJM, Carrott R, Suhas MML (2009) Low cost adsorbents: growing approach to wastewater treatment – a review. Crit Rev Env Sci Technol 39:783–842
Gurses A, Karaca S, Dogar C, Bayrak R, Acikyildiz M, Yalcin M (2004) Determination of adsorptive properties of clay/water system: methylene blue sorption. J Colloid Interface Sci 269:310–314
Hamdaoui O (2006) Batch study of liquid-phase adsorption of methylene blue using cedar sawdust and crushed brick. J Hazard Mater 135:264–273
Hameed BH (2009a) Removal of cationic dye from aqueous solution using jackfruit peel as non-conventional low-cost adsorbent. J Hazard Mater 162:344–350
Hameed BH (2009b) Spent tea leaves: a new non-conventional and low-cost adsorbent for removal of basic dye from aqueous solutions. J Hazard Mater 161:753–759
Hameed BH, El-Khaiary MI (2008) Batch removal of malachite green from aqueous solutions by adsorption on oil palm trunk fibre: equilibrium isotherms and kinetic studies. J Hazard Mater 154:237–244
Hand DW, Crittenden JE, Thacker WE (1983) User oriented batch reactor solutions to the homogeneous surface diffusion model. J Environ Eng 109:82–101
Hasnain Isa M, Siew Lang L, Asaari FAH, Aziz HA, Azam Ramli N, Dhas JPA (2007) Low cost removal of disperse dyes from aqueous solution using palm ash. Dyes Pigments 74:446–453
Ho YS, McKay G (1998a) Kinetic models for the sorption of dye from aqueous solution by wood. Process Saf Environ Prot 76:183–191
Ho YS, McKay G (1998b) Sorption of dye from aqueous solution by peat. Chem Eng J 70:115–124
Ho YS, Wase DAJ, Forster CF (1996) Kinetic studies of competitive heavy metal adsorption by sphagnum moss peat. Environ Technol 17:71–77
Ho YS, Chiang TH, Hsueh YM (2005) Removal of basic dye from aqueous solution using tree fern as a biosorbent. Process Biochem 40:119–124
Hu QH, Qiao SZ, Haghseresht F, Wilson MA, Lu GQ (2006) Adsorption study for removal of basic red dye using bentonite. Ind Eng Chem Res 45:733–738
Jain AK, Gupta VK, Bhatnagar A, Jain S, Suhas (2003a) A comparative assessment of adsorbents prepared from industrial wastes for the removal of cationic dye. J Indian Chem Soc 80:267–270
Jain AK, Gupta VK, Bhatnagar A, Suhas (2003b) A comparative study of adsorbents prepared from industrial wastes for removal of dyes. Sep Sci Technol 38:463–481
Jain AK, Gupta VK, Bhatnagar A, Suhas (2003c) Utilization of industrial waste products as adsorbents for the removal of dyes. J Hazard Mater 101:31–42
Janos P, Buchtova H, Ryznarova M (2003) Sorption of dyes from aqueous solutions onto fly ash. Water Res 37:4938–4944
Juang RS, Tseng RL, Wu FC (2001) Role of microporosity of activated carbons on their adsorption abilities for phenols and dyes. Adsorption 7:65–72
Juang RS, Wu FC, Tseng RL (2002) Characterization and use of activated carbons prepared from bagasses for liquid-phase adsorption. Colloids Surf A Physicochem Eng Asp 201:191–199
Kannan N, Sundaram MM (2001) Kinetics and mechanism of removal of methylene blue by adsorption on various carbons – a comparative study. Dyes Pigments 51:25–40
Kapoor A, Yang RT, Wong C (1989) Surface diffusion. Catal Rev Sci Eng 31:129–214

Kargi F, Ozmihci S (2004) Biosorption performance of powdered activated sludge for removal of different dyestuffs. Enzym Microb Technol 35:267–271

Khan SA, Rehma R, Khan MM (1995) Adsorption of Cr(III), Cr(VI) and Ag(I) on Bentonite. Waste Manag 15:271–282

Khattri SD, Singh MK (1999) Colour removal from dye wastewater using sugar cane dust as an adsorbent. Adsorpt Sci Technol 17:269–282

Khattri SD, Singh MK (2000) Colour removal from synthetic dye wastewater using a bioadsorbent. Water Air Soil Pollut 120:283–294

Komiyama H, Smith J (1974) Surface diffusion in liquid filled pores. AICHE J 20:1110–1117

Kunii D, Levenspiel O (1991) Fluidization engineering. Butterworth-Heinemann, Oxford

Levenspiel O (1972) Chemical reaction engineering. Wiley, New York

Liapis AI, Rippin DWT (1977) A general model for the simulation of multicomponent adsorption from a finite batch. Chem Eng Sci 23:619–629

Lin JX, Zhan SL, Fang MH, Qian XQ, Yang H (2008) Adsorption of basic dye from aqueous solution onto fly ash. J Environ Manag 87:193–200

Magdy YH, Daifullah AAM (1998) Adsorption of a basic dye from aqueous solutions onto sugar-industry-mud in two modes of operations. Waste Manag 18:219–226

Malik PK (2003) Use of activated carbons prepared from sawdust and rice-husk for adsorption of acid dyes: a case study of acid yellow 36. Dyes Pigments 56:239–249

Malik R, Ramteke DS, Wate SR (2007) Adsorption of malachite green on groundnut shell waste based powdered activated carbon. Waste Manag 27:1129–1138

Manju GN, Raji C, Anirudhan TS (1998) Evaluation of coconut husk carbon for the removal of arsenic from water. Water Res 32:3062–3070

Martel B, Devassine M, Crini G, Weltrowski M, Bourdonneau M, Morcellet M (2001) Preparation and sorption properties of a beta-cyclodextrin-linked chitosan derivative. J Polym Sci Part A: Polym Chem 39:169–176

Martin MJ, Artola A, Balaguer MD, Rigola M (2003) Activated carbons developed from surplus sewage sludge for the removal of dyes from dilute aqueous solutions. Chem Eng J 94:231–239

Mathews P, Weber WJ Jr (1976) Effects of external mass transfer and intraparticle diffusion on adsorption rates in slurry reactors. AIChe Symp Ser 73

McKay G (1984) Analytical solution using a pore diffusion model for a pseudoirreversible isotherm for the adsorption of basic dye on silica. AIChE J 30(4):692–697

McKay G, Allen SJ, McConvey IF, Otterburn MS (1981) Transport processes in the sorption of colored ions by peat particles. J Colloid Interface Sci 80:323–339

McKay G, El-Geundi M, Nassar MM (1997) Equilibrium studies for the adsorption of dyes on bagasse pith. Adsorpt Sci Technol 15:251–270

McKay G, Porter JF, Prasad GR (1999) The removal of dye colours from aqueous solutions by adsorption on low-cost materials. Water Air Soil Pollut 114:423–438

Mendez A, Fernandez F, Gasco G (2007) Removal of malachite green using carbon-based adsorbents. Desalination 206:147–153

Meshko V, Markovska L, Mincheva M, Rodrigues AE (2001) Adsorption of basic dyes on granular activated carbon and natural zeolite. Water Res 35:3357–3366

Mohamed MM (2004) Acid dye removal: comparison of surfactant-modified mesoporous FSM-16 with activated carbon derived from rice husk. J Colloid Interface Sci 272:28–34

Morais LC, Freitas OM, Goncalves EP, Vasconcelos LT, Gonzalez Beca CG (1999) Reactive dyes removal from wastewaters by adsorption on eucalyptus bark: variables that define the process. Water Res 33:979–988

Motoyuki S (1990) Adsorption engineering. Elsevier Science Publishers, Tokyo

Namasivayam C, Arasi D (1997) Removal of Congo Red from wastewater by adsorption onto waste red mud. Chemosphere 34:401–417

Namasivayam C, Kavitha D (2002) Removal of Congo Red from water by adsorption onto activated carbon prepared from coir pith, an agricultural solid waste. Dyes Pigments 54:47–58

Namasivayam C, Sumithra S (2005) Removal of direct red 12B and methylene blue from water by adsorption onto Fe (III)/Cr(III) hydroxide, an industrial solid waste. J Environ Manag 74:207–215

Namasivayam C, Yamuna RT (1995) Adsorption of direct red 12B by biogas residual slurry equilibrium and rate processes. Environ Pollut 89:1–7

Namasivayam C, Prabha D, Kumutha M (1998) Removal of direct red and acid brilliant blue by adsorption on to banana pith. Bioresour Technol 64:77–79

Namasivayam C, Dinesh Kumar M, Selvi K, Ashruffunissa Begum R, Vanathi T, Yamuna RT (2001) 'Waste' coir pith–a potential biomass for the treatment of dyeing wastewaters. Biomass Bioenergy 21:477–483

Nassar MM, Magdy YH (1997) Removal of different basic dyes from aqueous solutions by adsorption on palm-fruit bunch particles. Chem Eng J 66:223–226

Nemr AE, Abdelwahab O, El-Sikaily A, Khaled A (2009) Removal of direct blue- 86 from aqueous solution by new activated carbon developed from orange peel. J Hazard Mater 161:102–110

Neretnieks I (1976) Adsorption in finite batch and counter flow with systems having a nonlinear isotherm. Chem Eng Sci 31:107–114

Netpradit S, Thiravetyan P, Towprayoon S (2003) Application of 'waste' metal hydroxide sludge for adsorption of azo reactive dyes. Water Res 37:763–772

O'Mahony T, Guibal E, Tobin JM (2002) Reactive dye biosorption by Rhizopus arrhizus biomass. Enzym Microb Technol 31:456–463

Okada K, Yamamoto N, Kameshima Y, Yasumori A (2003) Adsorption properties of activated carbon from waste newspaper prepared by chemical and physical activation. J Colloid Interface Sci 262:194–199

Oliveira LCA, Goncalves M, Oliveira DQL, Guerreiro MC, Guilherme LRG, Dallago RM (2007) Solid waste from leather industry as adsorbent of organic dyes in aqueous-medium. J Hazard Mater 141:344–347

Osma JF, Saravia V, Toca-Herrera JL, Couto SR (2007) Sunflower seed shells: a novel and effective low-cost adsorbent for the removal of the diazo dye Reactive Black 5 from aqueous solutions. J Hazard Mater 147:900–905

Otero M, Rozada F, Calvo LF, Garcia AI, Moran A (2003a) Kinetic and equilibrium modelling of the methylene blue removal from solution by adsorbent materials produced from sewage sludges. Biochem Eng J 15:59–68

Otero M, Rozada F, Calvo LF, Garcıá AI, Morań A (2003b) Elimination of organic water pollutants using adsorbents obtained from sewage sludge. Dyes Pigments 57:55–65

Ozacar M, Sengil IA (2002) Adsorption of acid dyes from aqueous solutions by calcined alunite and granular activated carbon. Adsorption 8:301–308

Ozacar M, Sengil IA (2003) Adsorption of reactive dyes on calcined alunite from aqueous solutions. J Hazard Mater B98:211–224

Ozacar M, Sengil IA (2005) Adsorption of metal complex dyes from aqueous solutions by pine sawdust. Bioresour Technol 96:791–795

Ozacar M, Sengil IA (2006) A two stage batch adsorber design for methylene blue removal to minimize contact time. J Environ Manag 80:372–379

Ozcan AS, Erdem B, Ozcan A (2004) Adsorption of acid blue 193 from aqueous solutions onto Na-bentonite and DTMA-bentonite. J Colloid Interface Sci 280:44–54

Ozdemir O, Armagan B, Turan M, Celik MS (2004) Comparison of the adsorption characteristics of azo-reactive dyes on mezoporous minerals. Dyes Pigments 62:49–60

Pavan FA, Lima EC, Dias SLP, Mazzocato AC (2008) Methylene blue biosorption from aqueous solutions by yellow passion fruit waste. J Hazard Mater 150:703–712

Ponnusami V, Vikram S, Srivastava SN (2008) Guava (Psidium Guajava) leaf powder: novel adsorbent for removal of methylene blue from aqueous solutions. J Hazard Mater 152:276–286

Poots VJP, McKay G, Healy JJ (1976a) The removal of acid dye from effluent using natural adsorbents—I peat. Water Res 10:1061–1066

Poots VJP, McKay G, Healy JJ (1976b) The removal of acid dye from effluent using natural adsorbents – II wood. Water Res 10:1067–1070

Poots VJP, McKay G, Healy JJ (1978) Removal of basic dye from effluent using wood as an adsorbent. J Water Pollut Control Fed 50:926

Purkait MK, Gusain DS, Dasgupta S, De S (2004) Adsorption behavior of Chrysoidine dye on activated Charcoal and its regeneration characteristics by using different surfactants. Sep Sci Technol 39:2419–2440

Purkait MK, Dasgupta S, De S (2005) Adsorption of eosin dye on activated carbon and its surfactant based desorption. J Environ Manage 76:135–142

Purkait MK, Maiti A, Dasgupta S, De S (2007) Removal of congo red using activated carbon and its regeneration. J Hazard Mater 145:287–295

Santos SCR, Vilar VJP, Boaventura RAR (2008) Waste metal hydroxide sludge as adsorbent for a reactive dye. J Hazard Mater 153:999–1008

Sekaran G, Shanmughasundaram KA, Mariaappan M, Raghavan KV (1995) Utilisation of a solid waste generated in leather industry for removal of dye in aqueous solution. Indian J Chem Technol 2:311–316

Seredych M, Bandosz TJ (2007) Removal of cationic and ionic dyes on industrial-municipal sludge based composite adsorbents. Ind Eng Chem Res 46:1786–1793

Shawabkeh RA, Tutunji MF (2003) Experimental study and modeling of basic dye sorption by diatomaceous clay. Appl Clay Sci 24:111–120

Singha S, Sarkar U, Mondal S, Saha S (2012) Transient behavior of a packed column of Eichhornia Crassipes stem for the removal of hexavalent chromium. Desalination 297:48–58

Sivaraj R, Namasivayam C, Kadirvelu K (2001) Orange peel as an adsorbent in the removal of acid violet 17 (acid dye) from aqueous solutions. Waste Manag 21:105–110

Spahn H, Schlunder EU (1975) the scale-up of activated carbon columns for water purification, based on results from batch tests—I: theoretical and experimental determination of adsorption rates of single organic solutes in batch tests. Chem Eng Sci 30:529–537

Sun G, Xu X (1997) Sunflower stalk as adsorbents for color removal from textile wastewater. Ind Eng Chem Res 36:808–812

Thinakaran N, Panneerselvam P, Baskaralingam P, Elango D, Sivanesan S (2008) Equilibrium and kinetic studies on the removal of acid red 114 from aqueous solutions using activated carbons prepared from seed shells. J Hazard Mater 158:142–150

Thomas HC (1944) Heterogeneous ion exchange in a flowing system. J Am Chem Soc 66:1664–1466

Tor A, Cengeloglu Y (2006) Removal of congo red from aqueous solution by adsorption onto acid activated red mud. J Hazard Mater 138:409–415

Tsai WT, Hsu HC, Su TY, Lin KY, Lin CM (2008) Removal of basic dye (methylene blue) from wastewaters utilizing beer brewery waste. J Hazard Mater 154:73–78

Tseng RL, Wu FC, Juang RS (2003) Liquid-phase adsorption of dyes and phenols using pinewood-based activated carbons. Carbon 41:487–495

Valix M, Cheung WH, McKay G (2004) Preparation of activated carbon using low temperature carbonisation and physical activation of high ash raw bagasse for acid dye adsorption. Chemosphere 56:493–501

Venkata Mohan S, Sailaja P, Srimurali M, Karthikeyan J (1999) Colour removal of monoazo acid dye from aqueous solution by adsorption and chemical coagulation. Environ Eng Policy 1:149–154

Walker GM, Hansen L, Hanna JA, Allen SJ (2003) Kinetics of a reactive dye adsorption onto dolomitic sorbents. Water Res 37:2081–2089

Wang L, Wang A (2007) Adsorption characteristics of Congo red onto the chitosan/montmorillonite nanocomposite. J Hazard Mater 147:979–985

Wang S, Li L, Wu H, Zhu ZH (2005) Unburned carbon as a low-cost adsorbent for treatment of methylene blue-containing wastewater. J Colloid Interface Sci 292:336–343

Waranusantigul P, Pokethitiyook P, Kruatrachue M, Upatham ES (2003) Kinetics of basic dye (methylene blue) biosorption by giant duckweed (Spirodela polyrrhiza). Environ Pollut 125:385–392
Weber WJ Jr, Rumer RR Jr (1965) Intraparticle transport of sulfonated alkylbenzenes in a porous solid: diffusion and non-linear adsorption. Water Resour Res 1:361–373
Wong YC, Szeto YS, Cheung WH, McKay G (2004) Adsorption of acid dyes on chitosan–equilibrium isotherm analyses. Process Biochem 39:695–704
Wu FC, Tseng RL, Juang RS (2000) Comparative adsorption of metal and dye on flake- and bead-types of chitosans prepared from fishery wastes. J Hazard Mater 73:63–75
Yoon YH, Nelson JH (1984) Application of gas adsorption kinetics. I. A theoretical model for respirator cartridge service time. Am Ind Hyg Assoc J 45:509–516
Zhang FS, Itoh H (2003) Adsorbents made from waste ashes and post-consumer PET and their potential utilization in wastewater treatment. J Hazard Mater 101:323–337

Chapter 3
Adsorption of Dyes from Actual Effluent

Abstract Adsorption of color components from the effluent in practice is a challenge as often the actual efficiency of the adsorption process is masked by the competitive adsorption of multiple species present in an effluent. The chapter illustrates the adsorption phenomena using one of the commonly used adsorbents detailed in Chap. 2 for removal of toxic dyes from an effluent. Further analyses were done for the kinetics and isotherm for multicomponent adsorption.

Keywords Adsorption • Isotherm • Equilibrium • Kinetics • Effluent

Among the various available technologies for treatment of dye effluent, adsorption is one of the effective, scalable, and economic physicochemical processes. The process of adsorption is considered to be most effective for the bulk removal of dyes from aqueous solution and is used prior to any polishing final filtration step (membrane separation). The development of adsorbent, equilibrium, and kinetic studies together with effect of various process parameters is described in this chapter.

The first step of adsorption in any industrial process is to search for a suitable adsorbent, possessing adequate adsorbing capacity and reasonable cost. In the present study, sawdust (Ajmal et al. 1998; Garg et al. 2003) is used to prepare the adsorbent. Aqueous solution of crystal violet and methylene blue (single component systems) is selected for the detailed adsorption study to determine the kinetics and the equilibria processes involved. The effects of various process parameters, viz., pH of the solution, agitation, particle size of the adsorbent, temperature, initial dye concentration, and the adsorbent loading, are observed.

Various isotherms are used to fit the equilibrium data. A two-resistance mass transfer model developed from unreacted shrinking core mechanism (Levenspiel 1972) is used for the prediction of concentration profile in batch adsorption study for both the single component as well as the two-component systems. The model is fitted with a few selective experimental data for the estimation of its parameters, i.e., the external mass transfer coefficient (k_f) and the internal effective diffusivity (D_p). Using these parameter values, the concentration profiles at various other operating conditions are predicted.

S. Mondal et al., *Advances in Dye Removal Technologies*, Green Chemistry and Sustainable Technology, DOI 10.1007/978-981-10-6293-3_3

Table 3.1 Characterization of the effluent from the textile plant

$C_{0,1}$ (reactive black) M_w 923 Da (mg/l)	$C_{0,2}$ (reactive red) M_w 854 Da (mg/l)	pH	Conductivity $\times 10^{-3}$ (mΩ/cm)	TS (mg/l)	Equivalent salt content in terms of equivalent NaCl (g/l)	COD (mg/l)
173	118	6.86	116.3	1.05×10^5	56	6312

3.1 Characterization of the Textile Effluent

The effluent is characterized in terms of dye concentrations, chemical oxygen demand (COD), total solid content (TS), conductivity, pH, and salt content and is presented in Table 3.1.

The concentration of the dyes in the aqueous solution is measured by a Genesys2 spectrophotometer (Thermo Spectronic, USA) at wavelengths of 535 and 599 nm for reactive red and reactive black, respectively. For dye mixture, the method of Vogel (1970) is used for the determination of concentration of each dye in the effluent.

The chemical oxygen demand (COD) is the measure of oxygen consumed during the oxidation of the oxidizable organic matter by a strong oxidizing agent. The samples (original feed, feed after adsorption, and permeate streams after nanofiltration) are refluxed with potassium dichromate and sulfuric acid in the presence of mercuric sulfate (to neutralize the effect of chlorides) and silver sulfate (catalyst).

3.2 Characterization of the Adsorbent

Different properties of the adsorbent (charred saw dust or CSD) are determined by standard methods (IS: 877). A commercial granular activated carbon GAC-1240, supplied by NORIT, Netherlands, is used for comparison of the performance of the prepared adsorbent. Table 3.2 shows some physical properties measured for the prepared adsorbent and that of the commercial carbon (GAC). Surface area of the adsorbents is measured by BET method.

The adsorbent developed from saw dust (CSD) is characterized by scanning electron microscope (SEM) model JSM-5800, from JEOL. The SEM photograph of CSD and GAC is compared in Figs. 3.1 and 3.2. The presence of irregular and porous structure in the SEM photograph indicates high surface area of CSD. A quantitative analysis of the adsorbent is done in the SEM, which shows that the adsorbent (CSD) contains 93.01% carbon, 5.04% phosphorus, and negligible amount of silicon, calcium, and aluminum.

Table 3.2 Physical properties of the adsorbents

Properties	CSD	GAC
Solid density (kg/m^3)	1056	440
Moisture content (%)	1.80	2.0
Ash content (%)	14.68	5.8
BET surface area (m^2/g of adsorbent)	559	1100

Fig. 3.1 Image of charred saw dust in scanning electron microscope

Fig. 3.2 Image of granular activated carbon in scanning electron microscope

3.3 Adsorption Equilibrium Studies

The equilibrium study consists of the single component adsorption with crystal violet, methylene blue, reactive black, and reactive red and the two-component adsorption with the mixture of reactive black and reactive red. The common isotherms are Langmuir and Freundlich, and they are already presented in Chap. 2 (Eqs. 2.3 and 2.5).

$$\text{Radke and Prausnitz isotherm,}\quad \frac{1}{Y_{\mathrm{e}}} = \frac{1}{A_{\mathrm{R}} C_{\mathrm{e}}} + \frac{1}{B_{\mathrm{R}} C_{\mathrm{e}}^{\delta_R}} \tag{3.1}$$

where, A_{R}, B_{R}, and δ_{R} are the three isotherm constants.

$$\text{Redlich and Peterson isotherm,}\quad Y_{\mathrm{e}} = \frac{K_{\mathrm{R}} C_{\mathrm{e}}}{1 + K_{\mathrm{P}} C_{\mathrm{e}}^{\iota}} \tag{3.2}$$

$$\text{Tempkin isotherm,}\ Y_{\mathrm{e}} = \frac{R_{\mathrm{g}} T}{b_{\mathrm{T}}} \ln A_{\mathrm{T}} + \frac{R_{\mathrm{g}} T}{b_{\mathrm{T}}} \ln C_{\mathrm{e}} \tag{3.3}$$

$$\text{Fritz-Schlunder isotherm,}\ Y_{\mathrm{e}} = \frac{K_{F_1} C_{\mathrm{e}}^{\phi_1}}{1 + K_{F_2} C_{\mathrm{e}}^{\phi_2}} \tag{3.4}$$

For adsorption isotherm models based on more than two parameters, the reader is suggested to refer the literature work (Hamdaouia and Naffrechoux 2007).

Figure 3.3 shows a comparison of the equilibrium adsorption capacities between the two adsorbents: charred saw dust (CSD) and the available granular activated carbon (GAC) using the two dyes crystal violet and methylene blue. The equilibrium study is performed at room temperature (298 K) with smaller particle size of the adsorbent (0.044 mm). From the figure it is observed that at 298 K, the equilibrium adsorption capacity (Y_{e}) of crystal violet is higher for CSD than the granular activated carbon (GAC). For methylene blue, although at lower equilibrium concentrations, Y_e is less for CSD, but at higher equilibrium dye concentrations, CSD shows reasonable adsorption capacity compared to GAC.

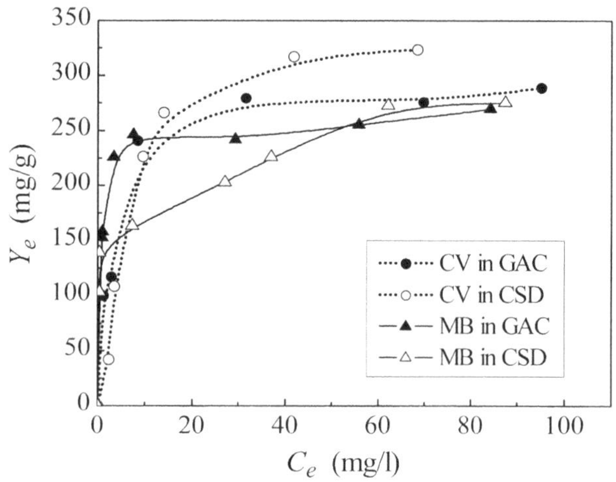

Fig. 3.3 Equilibrium adsorption of CV and MB on CSD and comparison with GAC ($T = 298$ K; $d_p = 0.044$ mm)

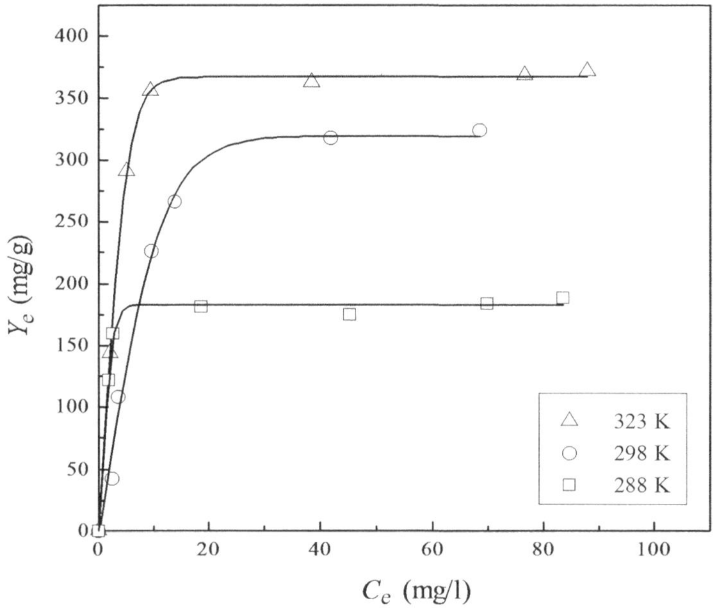

Fig. 3.4 Equilibrium adsorption of crystal violet on CSD at different temperatures (Reproduced from Chakraborty et al. (2005) with permission from Elsevier)

3.3.1 Effect of Temperature on the Equilibrium Study of Crystal Violet and Methylene Blue

Figures 3.4 and 3.5 show the equilibrium adsorption of crystal violet and methylene blue, respectively, on CSD at three different temperatures, viz., 288, 298, and 323 K, for smaller particle size (0.044 mm) of adsorbent. From the figures, it is observed that adsorption is favored at higher temperatures for both the dyes. In Fig. 3.4, the maximum adsorption capacity of crystal violet increases from 188 mg/g at 288 K to 379 mg/g at 323 K. For methylene blue (Fig. 3.5), the maximum adsorption capacity increases from 228 mg/g at 288 K to 327 mg/g at 323 K. Similar trend is also observed by some other researchers for aqueous-phase adsorption (Namasivayam, and Yamuna 1995).

The rise of adsorption with temperature may be due to the chemisorption of the dye molecules on the adsorbent surface through the formation of a complex surface compound (McKay 1996). Another possible reason may be due to the enlargement of the pore sizes of adsorbent particles at elevated temperatures (Namasivayam, and Yamuna 1995).

Thermodynamic parameters such as change in free energy (ΔG^0), enthalpy (ΔH^0), and entropy (ΔS^0) are determined in the temperature range of 288 K–323 K, at an initial dye concentration of 150 mg/l, adsorbent dose of 0.1 g/l, and 0.044 mm particle

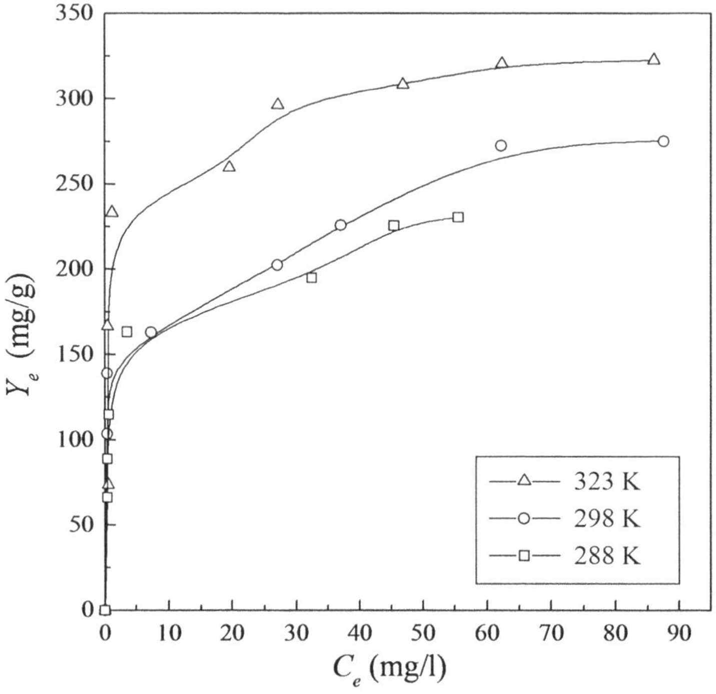

Fig. 3.5 Equilibrium adsorption of methylene blue on CSD at different temperatures

size of the adsorbent. The change in the free energy (ΔG^0) is estimated using the following equation.

$$\Delta G^0 = -R_g T \ln K_c \tag{3.5}$$

ΔH^0 and ΔS^0 are estimated from the slope and intercept of the Van't Hoff plot of log K_c vs. $1/T$.

$$\log K_c = \frac{\Delta S^0}{2.303R_g} - \frac{\Delta H^0}{2.303R_g T} \tag{3.6}$$

Where T is the temperature in Kelvin, R_g is the gas constant, and K_c is the adsorption equilibrium constant. The values of ΔG^0, ΔH^0, and ΔS^0 are shown in Tables 3.3 and 3.4. Positive values of ΔH^0 indicate that the adsorption process for both the dyes is endothermic in nature. The negative values of ΔG^0 indicate the spontaneous nature of adsorption in the temperature range studied here. The positive values of ΔS^0 show the increased randomness at the solid-solution interface during the adsorption of dye on charred saw dust.

Table 3.3 Thermodynamic parameters for the adsorption of crystal violet by CSD

T (K)	K_c (kJ/mol)	ΔG^0 (kJ/mol)	ΔH^0 (J/mol K)	ΔS^0
288	1.569	−1.078		
298	9.493	−5.576	59.166	212.487
323	28.65	−9.010		

Reproduced from Chakraborty et al. (2006) with permission from American Chemical Society

Table 3.4 Thermodynamic parameters for the adsorption of methylene blue by CSD

T (K)	K_c (kJ/mol)	ΔG^0 (kJ/mol)	ΔH^0 (J/mol K)	ΔS^0
288	2.533	−2.225		
298	3.048	−2.761	21.833	83.129
323	6.634	−5.081		

3.3.2 *Effect of Adsorbent Particle Sizes on the Equilibrium Study of Crystal Violet and Methylene Blue*

The effect of particle size on equilibria is observed with particle size varying from 0.044 to 0.226 mm, at 298 K. It is observed in Figs. 3.6 and 3.7 that the maximum adsorption capacity increases with decrease in the particle size. Smaller particles provide more surface area per unit mass, resulting in enhanced adsorption. Using a higher particle size (0.226 mm), the maximum adsorption capacity for crystal violet is about 195 mg/g, which increases to 341 mg/g with a decrease in the particle size to 0.044 mm. Similarly for methylene blue, the maximum adsorption capacity increases from about 133 mg/g to 275 mg/g with the decrease in the adsorbent particle size from 0.226 mm to 0.044 mm (Fig. 3.7).

3.3.3 *Equilibrium Isotherm Studies for Crystal Violet and Methylene Blue*

Adsorption equilibrium data are fitted to various isotherms, viz., Langmuir, Freundlich, Redlich-Peterson, Fritz-Schlunder, Radke-Prausnitz, and Temkin isotherm. The values of the equilibrium constants and the correlation coefficients (CR) obtained for various isotherms are shown in Tables 3.5, 3.6, 3.7, 3.8, 3.9, 3.10, 3.11, and 3.12. Figures 3.8 and 3.9 show the Langmuir plots for crystal violet and methylene blue, respectively, at various temperatures and particle sizes of the adsorbent. The values of the parameters (Y_o and k_0) are obtained from the slopes, and intercepts of the linearized plots of C_e/Y_e vs. C_e. Y_o (mg/g) is the maximum adsorption capacity of the adsorbent to form a complete monolayer on the surface, at the equilibrium dye concentration C_e, and k_0 is the Langmuir constant (l/mg). It is observed from the results that the equilibrium data for all temperatures and particle

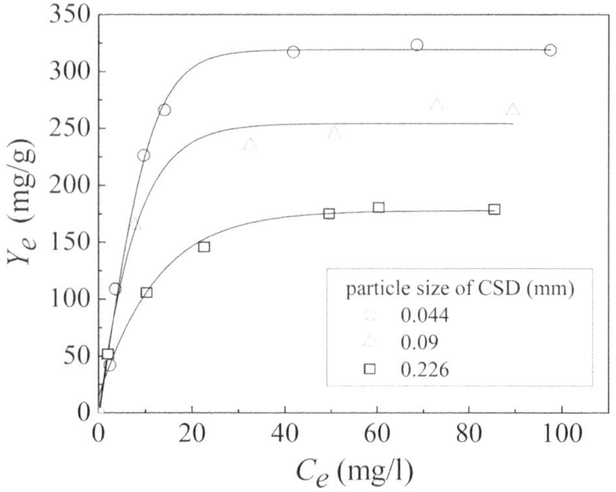

Fig. 3.6 Equilibrium adsorption of crystal violet on CSD at different particle sizes (Reproduced from Chakraborty et al. (2005) with permission from Elsevier)

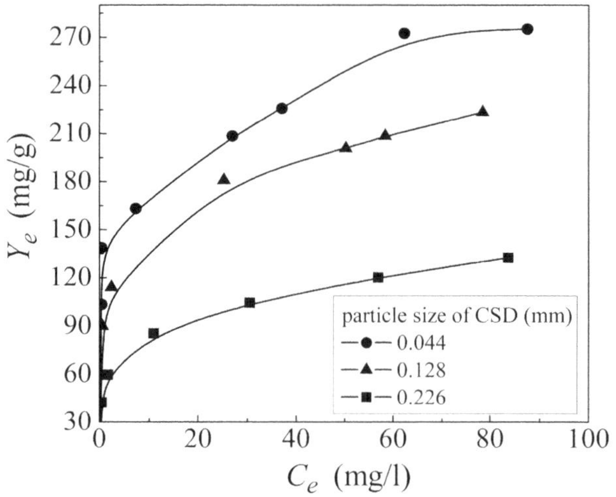

Fig. 3.7 Equilibrium adsorption of methylene blue on CSD at different particle sizes

Table 3.5 Langmuir constants for CV – CSD systems

T (K)	d_p (mm)	Y_o (mg/g)	K_o (l/mg)	CR
323	0.044	378.80	0.536	0.999
298	0.044	341.30	0.197	0.999
288	0.044	187.97	0.943	0.999
298	0.090	289.86	0.151	0.998
298	0.226	195.69	0.148	0.998

Reproduced from Chakraborty et al. (2006) with permission from American Chemical Society

Table 3.6 Langmuir constants for MB – CSD systems

T (K)	d_p (mm)	Y_o (mg/g)	K_o (l/mg)	CR
323	0.044	326.79	0.503	0.998
298	0.044	280.89	0.237	0.992
288	0.044	228.31	0.767	0.996
298	0.128	223.71	0.320	0.997
298	0.226	133.87	0.239	0.995

Table 3.7 Freundlich Constants for CV – CSD systems

T (K)	d_p (mm)	Y_f (l/g)	n	CR
323	0.044	172.71	5.147	0.801
298	0.044	51.955	2.156	0.868
288	0.044	131.57	11.983	0.849
298	0.090	50.99	2.432	0.926
298	0.226	45.019	2.948	0.982

Table 3.8 Freundlich Constants for MB – CSD systems

T (K)	d_p (mm)	Y_f (l/g)	n	CR
323	0.044	230.22	13.02	0.998
298	0.044	102.76	4.500	0.981
288	0.044	138.92	8.427	0.948
298	0.128	96.714	5.263	0.999
298	0.226	53.341	4.960	0.999

Table 3.9 Redlich-Peterson constants for CV – CSD system

T (K)	d_p (mm)	K_R (l/g)	K_P (mg/l)$^{-\nu}$	ν	CR
323	0.044	102.73	0.133	1.162	0.984
298	0.044	33.44	0.028	1.271	0.990
288	0.044	202.03	0.969	1.027	0.932
298	0.090	33.71	0.059	1.149	0.997
298	0.226	39.16	0.316	0.896	0.993

Table 3.10 Redlich-Peterson constants for MB – CSD system

T (K)	d_p (mm)	K_R (l/g)	K_P (mg/l)$^{-\nu}$	ν	CR
323	0.044	641.14	2.549	0.946	0.905
298	0.044	6.87×10^8	5.3×10^6	0.837	0.963
288	0.044	473.39	2.922	0.918	0.984
298	0.128	2.21×10^4	223.12	0.815	0.999
298	0.226	3.75×10^4	71.43	0.793	0.999

sizes fit well to Langmuir isotherm (Tables 3.5 and 3.6). The values of correlation coefficients (CR) are higher than 0.99 in all the cases.

The dimensionless separation factor R_L (McKay 1982), which is an essential characteristic of the Langmuir isotherm, is calculated from the relation $R_L = \frac{1}{1+K_oC_o}$.

Table 3.11 Fritz-Schlunder constants for CV – CSD system

T (K)	d_p (mm)	K_{F1} (mg/g)(mg/l)$-\varphi_1$	K_{F2} (mg/l)$-\varphi_2$	φ_1	φ_2	CR
323	0.044	55.213	0.14	1.925	1.94	0.998
298	0.044	13.318	0.043	1.838	1.825	0.997
288	0.044	15.00	0.089	5.33	5.309	0.988
298	0.09	23.972	0.078	1.359	1.386	0.999
298	0.226	35.664	0.042	0.597	0.842	0.997

Table 3.12 Fritz-Schlunder constants for MB – CSD system

T (K)	d_p (mm)	K_{F1} (mg/g)(mg/l)$-\varphi_1$	K_{F2} (mg/l)$-\varphi_2$	φ_1	φ_2	CR
323	0.044	1259.27	5.439	2.007	1.935	0.92
298	0.044	207.17	0.602	0.163	9.47×10^{-7}	0.963
288	0.044	9593.22	70.15	4.216	4.092	0.992
298	0.128	8487.76	85.19	0.10	0.086	0.999
298	0.226	19.4e5	36.9e3	0.912	0.706	0.998

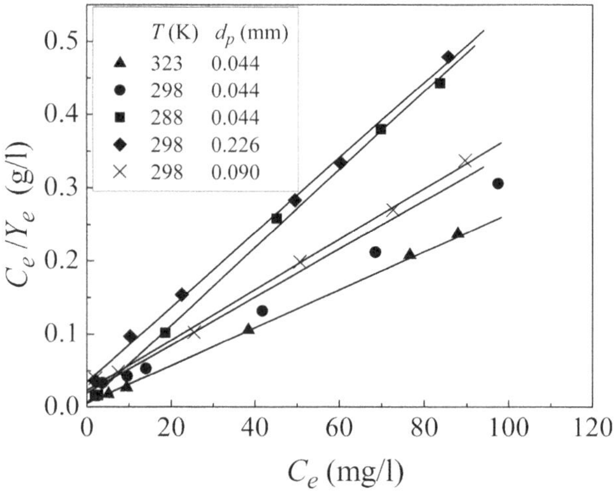

Fig. 3.8 Linearized Langmuir plot at different operating conditions for crystal violet

For both the dyes, the factor, R_L, is between 0 and 1 at any operating condition, indicating a favorable adsorption.

The values of the Freundlich isotherm constants, i.e., Y_f and n, are presented in Tables 3.7 and 3.8. Here, the exponent n is larger than 2 and 4 for crystal violet and methylene blue, respectively, which indicate favorable adsorption condition for both the dyes. The correlation coefficient values (CR) for crystal violet are not high and indicate that the Freundlich isotherm does not fit well. However for methylene blue, it shows a good fitting.

Tables 3.9 and 3.10 show the Redlich-Peterson isotherm constants for crystal violet and methylene blue, respectively. From the tables, it is observed that the correlation coefficients are not high at all the temperatures for both the dyes. The

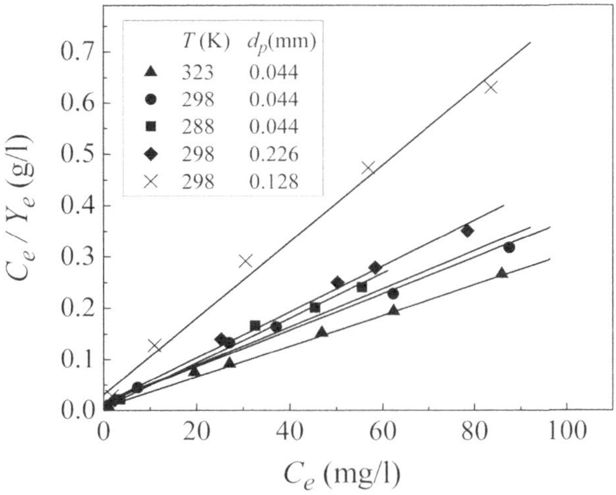

Fig. 3.9 Linearized Langmuir plot at different operating conditions for methylene blue

Fritz-Schlunder isotherm constants are represented in Tables 3.11 and 3.12. It is observed that the Fritz-Schlunder isotherm also shows a very good fitting for crystal violet (correlation coefficient CR is higher than 0.99 in almost all the cases), and for methylene blue except at high temperatures, the fitting is good. For the negative values of the empirical parameters of Radke-Prausnitz isotherm (Tables 3.13 and 3.14), it may be rejected.

Tables 3.15 and 3.16 show the Temkin isotherm constants. It is observed that the Temkin isotherm does not fit well. It can be concluded that the values of correlation coefficients (CR) for different temperature and particle sizes are the highest for Langmuir isotherm (0.99 Tables 3.5 and 3.6) for both the dyes. Hence the Langmuir isotherm is chosen for the process.

3.3.4 Equilibrium Study with the Reactive Dyes

Figure 3.10 shows the equilibrium adsorption of the reactive black (dye 1) and reactive red (dye 2), where the solid symbols represent single component systems and the hollow symbols for mixture. The adsorbent, charred saw dust (CSD) shows a good adsorption capacity for the single component systems of both the dyes. From the figure, it is observed that the equilibrium adsorption of both the dyes decreases in the mixture. This is probably due to the mutual interaction effects.

Table 3.13 Radke-Prausnitz constants for CV – CSD system

T (K)	d_p (mm)	A_R (l/g)	B_R (l/g)	δ_R	CR
323	0.044	1.55×10^{11}	−204.67	0.145	0.860
298	0.044	2.32×10^{11}	−95.45	0.290	0.888
288	0.044	2.27×10^{11}	−133.97	0.078	0.888
298	0.090	-5.2×10^{11}	−83.85	0.276	0.924
298	0.226	-1.0×10^{11}	−52.73	0.294	0.973

Table 3.14 Radke-Prausnitz constants for MB – CSD system

T (K)	d_p (mm)	A_R (l/g)	B_R (l/g)	δ_R	CR
323	0.044	2.68×10^{11}	−168.75	0.154	0.894
298	0.044	2.47×10^{11}	−129.50	0.163	0.963
288	0.044	3.59×10^{11}	−110.38	0.183	0.967
298	0.128	3.18×10^{11}	−98.47	0.185	0.999
298	0.226	3.33×10^{11}	−51.72	0.210	0.999

Table 3.15 Temkin constants for CV – CSD systems

T (K)	d_p (mm)	A_T (l/g)	b_T (J/mol)	CR
323	0.044	37.504	55.107	0.840
298	0.044	1.357	33.563	0.945
288	0.044	2.49×10^4	184.02	0.869
298	0.090	1.798	43.739	0.968
298	0.226	2.211	68.885	0.991

Table 3.16 Temkin constants for MB – CSD systems

T (K)	d_p (mm)	A_T (l/g)	b_T (J/mol)	CR
323	0.044	113.82	74.96	0.903
298	0.044	163.04	91.99	0.938
288	0.044	54.95	84.55	0.977
298	0.128	39.443	92.26	0.993
298	0.226	22.696	148.55	0.989

3.3.5 Single Component Systems

Adsorption equilibrium data for the aqueous single component system of the reactive dyes are fitted with various isotherms. Tables 3.17, 3.18, 3.19, and 3.20 show the isotherm constants and the correlation coefficient values for the different isotherms. It is observed that both the Langmuir and the Fritz-Schlunder isotherms fit good for reactive red, as the average value of correlation coefficient, (CR) is higher than 0.98. For reactive black, all the three isotherms, viz., the Langmuir, Redlich-Peterson, and the Fritz-Schlunder isotherm fit good (CR is higher than 0.99). The maximum adsorption capacity of reactive red is about 424 mg/g and that of reactive black is about 542 mg/g at 298 K (from Langmuir isotherm).

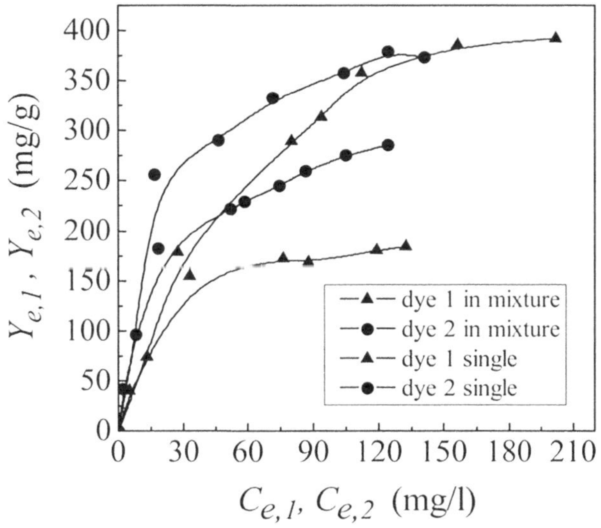

Fig. 3.10 Equilibrium adsorption of the reactive dyes in single component systems and in mixture (Reproduced from Chakraborty et al. (2006) with permission from American Chemical Society)

Table 3.17 Langmuir constants for reactive dye systems

T (K)	d_p (mm)	A_T (l/g)	b_T (J/mol)	CR	T (K)
Reactive black	298	0.044	541.96	0.016	0.993
Reactive red	298	0.044	423.68	0.059	0.985

Table 3.18 Freundlich constants for reactive dye systems

Dye	T (K)	d_p (mm)	Y_f (l/g)	n	CR
Reactive black	298	0.044	36.58	0.463	0.971
Reactive red	298	0.044	84.47	0.311	0.932

Table 3.19 Redlich-Peterson constants for reactive dye systems

Dye	T (K)	d_p (mm)	Y_f (l/g)	n	CR	Dye
Reactive black	298	0.044	7.75	0.008	1.10	0.99
Reactive red	298	0.044	23.6	0.047	1.04	0.96

Table 3.20 Fritz-Schlunder constants for reactive dye systems

Dye	T (K)	d_p (mm)	K_{F1} (mg/g)(mg/l)$-\varphi_1$	K_{F2} (mg/l)$-\varphi_2$	φ_1	φ_2	CR
Reactive black	298	0.044	3.806	0.015	1.319	1.218	0.994
Reactive red	298	0.044	0.929	0.005	2.588	2.458	0.985

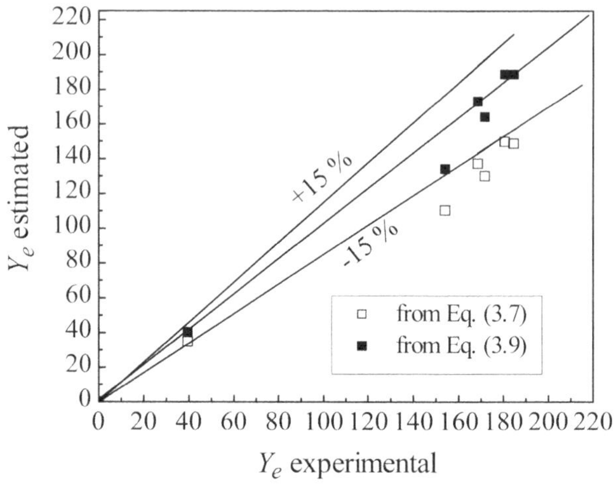

Fig. 3.11 Comparison between the experimental and estimated Y_e values for reactive black (Reproduced from Chakraborty et al. (2006) with permission from American Chemical Society)

3.3.6 Two-Component System

For the two-component system of the reactive dyes, the extended Langmuir isotherms, Eqs. (3.7) and (3.8), are used.

$$Y_{e,1} = \frac{Y_{o,1}K_{o,1}C_{e,1}}{1 + K_{o,1}C_{e,1} + K_{o,2}C_{e,2}} \tag{3.7}$$

$$Y_{e,2} = \frac{Y_{o,2}K_{o,2}C_{e,2}}{1 + K_{o,1}C_{e,1} + K_{o,2}C_{e,2}} \tag{3.8}$$

The bisolute Langmuir isotherm model assumes that all the adsorption sites are equally available to all adsorbed species and no interaction between the adsorbed species. The constants of these two isotherms are from the single component Langmuir isotherms, reported in Table 3.17.

The comparison between the experimental and calculated values of Y_e from the bisolute Langmuir isotherm is shown in Figs. 3.11 and 3.12 for reactive black and reactive red, respectively. It is observed that although for reactive red, the calculated values (hollow circles) lie within $\pm15\%$ of the experimental values (Fig. 3.12); however for reactive black, there is wide variation between the experimental and the calculated Y_e values (hollow squares in Fig. 3.11). Hence the extended Langmuir equation fails to predict the equilibrium adsorption capacity of reactive black. This is due to the difference in the equilibrium adsorption capacity (Y_o) value of the two dyes (Leyva-Ramos et al. 2001). As mentioned earlier, Y_o for reactive red is 424 mg/g and that for reactive black is 542 mg/g (1.28 times greater than previous).

To encounter this deviation, the modified bisolute Langmuir isotherm (Ho and McKay 1999) is used. The model equations are represented in Eqs. (3.9 and 3.10), where $\eta_{i,j}$ is the interaction factor of component *i* for the adsorption of component *j*.

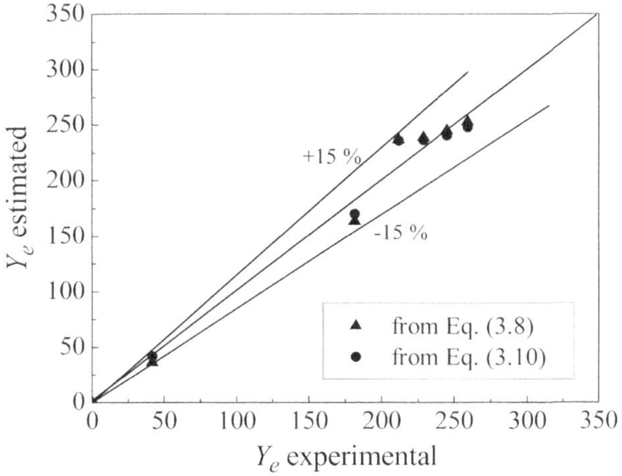

Fig. 3.12 Comparison between the experimental and estimated Y_e values for reactive red (Reproduced from Chakraborty et al. (2006) with permission from American Chemical Society)

$$Y_{e,1} = \frac{Y_{o,1}K_{o,1}\left(\frac{C_{e,1}}{\eta_{1,1}}\right)}{1 + K_{o,1}\left(\frac{C_{e,1}}{\eta_{1,1}}\right) + K_{o,2}\left(\frac{C_{e,2}}{\eta_{2,1}}\right)} \tag{3.9}$$

$$Y_{e,2} = \frac{Y_{o,2}K_{o,2}\left(\frac{C_{e,2}}{\eta_{2,2}}\right)}{1 + K_{o,1}\left(\frac{C_{e,1}}{\eta_{1,2}}\right) + K_{o,2}\left(\frac{C_{e,2}}{\eta_{2,2}}\right)} \tag{3.10}$$

The values of $\eta_{i,j}$ are found out by minimizing the sum of square of errors between the experimental and the estimated Y_e data, using an optimization algorithm. The values are as follows: $\eta_{1,1} = 0.881$; $\eta_{1,2} = 0.711$; $\eta_{2,1} = 1.319$; $\eta_{2,2} = 0.826$.

In Figs. 3.11 and 3.12, the solid symbols represent the Y_e values estimated from the modified bisolute Langmuir isotherm. It is evident from Fig. 3.11 that there is a remarkable improvement on the prediction for reactive black; here the calculated values lie within ±15% (solid squares). In Fig. 3.12, there is only slight improvement in the predictions compared to the previous approach (see solid and hollow circles).

3.4 Adsorption Kinetic Studies

Adsorption kinetic study is performed to observe the effect of various process parameters using the synthetic solutions of crystal violet and methylene blue. The rate of adsorption using the present adsorbent is compared with that of the GAC. The effect of various process parameters, viz., pH, stirrer speed, adsorbent particle size, initial dye concentration, temperature, etc., is observed in the concentration

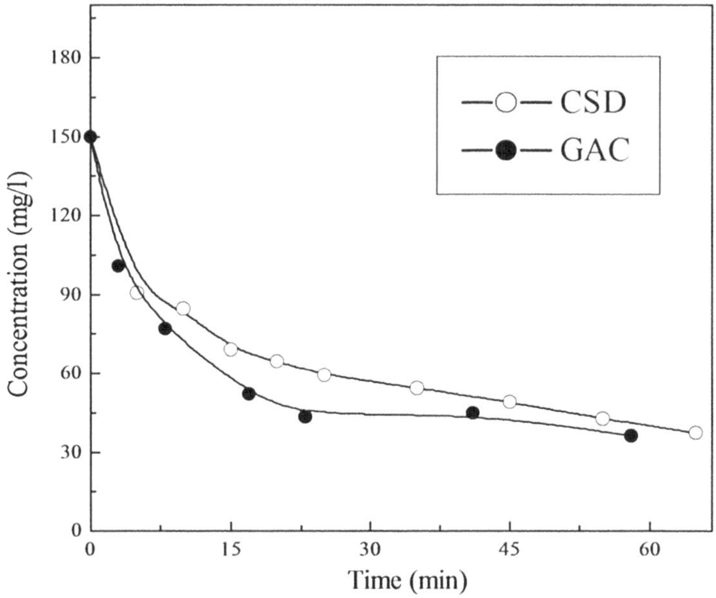

Fig. 3.13 Concentration decay of CV using CSD and comparison with GAC (C_o = 150 mg/l, d_p = 0.044 mm, T = 298 K, M_a = 0.5 g)

decay of the dyes with time. The concentration profiles are predicted by solving the model equations developed for the single component system.

3.4.1 Comparison of the Dye Removal Rate of CSD and GAC

The dye removal rate of charred saw dust (CSD) is compared with that of GAC in Figs 3.13 and 3.14, where the concentration decay of crystal violet and methylene blue are shown, respectively. From Fig. 3.13, it is observed that using 0.5 g/l CSD, the concentration of crystal violet drops from 150 mg/l to about 43 mg/l after a process time of 55 min, whereas for GAC, concentration reduces up to about 38 mg/l. Hence, GAC shows slightly better adsorbing capacity in comparison to CSD. In Fig. 3.14, the concentration decay curve of methylene blue for the two adsorbents is very close to each other. After 55 min of experiment, concentration of methylene blue reduces from 150 mg/l to about 41 mg/l, for both the adsorbents. It may be concluded that the present adsorbent (CSD) has a reasonable removal rate in comparison to the activated carbon (GAC) for both the dyes.

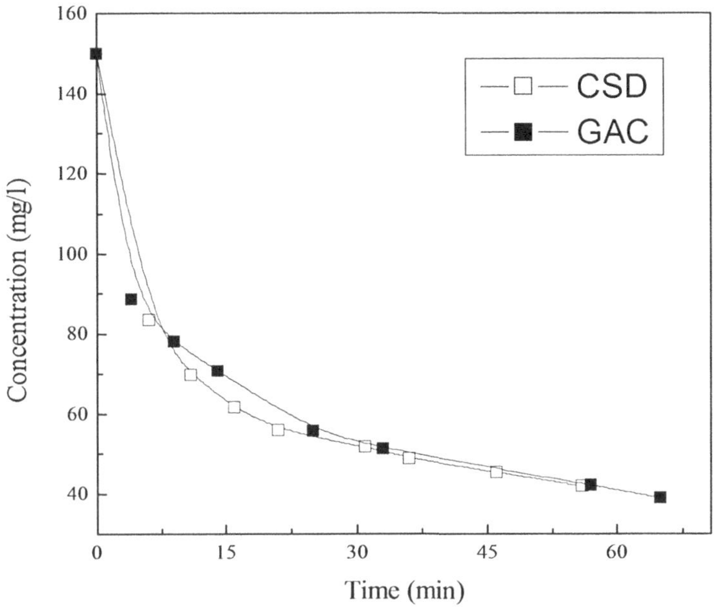

Fig. 3.14 Concentration decay of MB using CSD and comparison with GAC (C_o = 150 mg/l, d_p = 0.044 mm, T = 298 K, M_a = 0.5 g)

3.4.2 Effect of pH on the Rate of Adsorption

The effect of the pH on the adsorption is observed by varying the solution pH from 2 to 10. Figure 3.15 shows the dye removal (%) after 1 hour with variation in the solution pH of crystal violet and methylene blue. From the figure, it is observed that the adsorption is favored at basic pH for both the dyes. However, the adsorption of methylene blue is more sensitive with change in the pH in the range of 2 to 7, in comparison with crystal violet. At solution pH of 2.4, the removal of methylene blue is about 51%, which increases to about 72% with the increase in pH to 6.99. The removal of dye increases to 75% at pH of 10.2. For crystal violet, the removal of dye varies from 71% to about 83%, with an increase in the pH of 2.5 to 10.8, respectively, which is very small. Both the dyes are cationic in nature. In acidic medium (pH < 7), the excess H^+ ions reduces the amount of the negatively charged adsorbent sites and enhance the positively charged sites. This does not favor the adsorption of cationic dyes (Al-Degs et al. 2001). Similarly, an increase in the pH (alkaline) increases the negatively charged surface sites thereby enhancing the adsorption of the cationic dye.

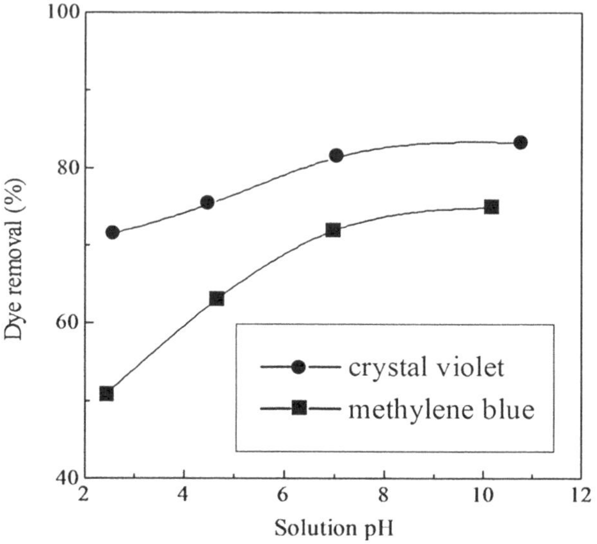

Fig. 3.15 Effect of pH on the adsorption of the crystal violet and methylene blue ($C_o = 150$ mg/l, $d_p = 0.044$ mm, $T = 298$ K, $M_a = 0.5$ g)

3.4.3 Effect of Stirrer Speed in the Adsorption

Figures 3.16 and 3.17 show the effect of stirrer speed in the adsorption of crystal violet and methylene blue, respectively. Experiments are carried out at varying speed of stirring, starting from 0 rpm (without stirring) to about 2500 rpm. It is observed from the figures that initially adsorption increases considerably with the increase in the stirrer speed due to an enhanced mass transfer effect. However, after increasing the stirrer speed to a certain range (about 2000 rpm), the rate of increase in the adsorption decreases. With further increase in the stirrer speed, adsorption increases slightly. Hence, the effect of external mass transfer becomes less at the higher stirring speeds (above 2000 rpm). Therefore, all the experiments are carried out at a higher stirrer speed (2500 rpm).

3.4.4 Modeling of Adsorption Kinetics

In adsorption mainly two resistances prevail – the external liquid film resistance and the resistance in the adsorbent particles. The intraparticle diffusion resistance may be neglected for solutes exhibiting strong solid to liquid-phase equilibrium solute distribution in the initial period of operation. However, even for such systems, the above assumption leads to substantial errors beyond the first few minutes if the agitation is high (Mathews and Weber 1976). Hence, both the resistances are comparable for kinetic study.

The external liquid film resistance is characterized by the external liquid-phase mass transfer coefficient (k_f). The mass transport within the adsorbent particles is

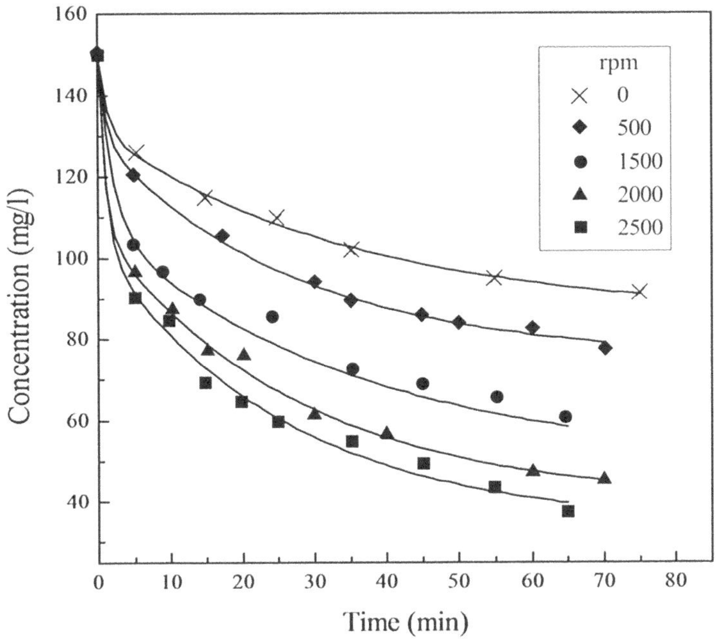

Fig. 3.16 Effect of stirrer speed on the adsorption of crystal violet ($C_o = 150$ mg/l, $d_p = 0.044$ mm, $T = 298$ K, $M_a = 0.5$ g)

assumed to be a pore diffusion (McKay 1984) or homogeneous solid diffusion process (Chatzopoulos et al. 1993).

The kinetic model used in the present work is based on the unreacted shrinking core mechanism (Jena et al. 2003). In case of single component system, the model details are presented in Chap. 2 (refer to Sect. 2.4)

3.4.5 Two-Component System

For the two-component system, the modified bisolute Langmuir isotherms given in Eq. (3.9) and (3.11) are used. Considering the spherical adsorbent particles, following rate equations are developed for components 1 and 2, respectively.

$$N_1(t) = 4\pi R_p{}^2 k_{f1}(C_{1t} - C_{e1t}) \quad (3.11a)$$

$$N_2(t) = 4\pi R_p{}^2 k_{f2}(C_{2t} - C_{e2t}) \quad (3.11b)$$

The diffusion in pore liquid for two components, according to Fick's law can be written as follows:

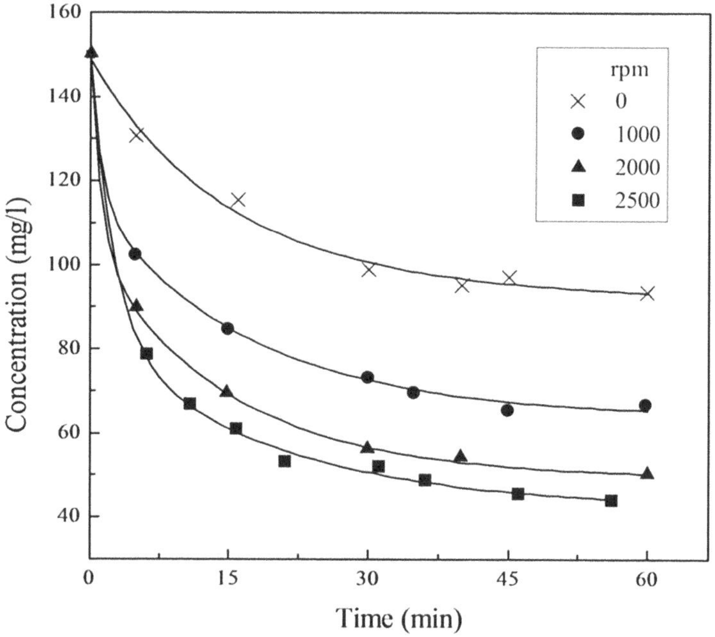

Fig. 3.17 Effect of stirrer speed on the adsorption of MB (C_o = 150 mg/l, d_p = 0.044 mm, T = 298 K, M_a = 0.5 g)

$$N_1(t) = \frac{4\pi D_{p1} C_{e1t}}{\left[\frac{1}{R_{f1}} - \frac{1}{R_p}\right]} \tag{3.12a}$$

$$N_2(t) = \frac{4\pi D_{p2} C_{e2t}}{\left[\frac{1}{R_{f2}} - \frac{1}{R_p}\right]} \tag{3.12b}$$

The mass balance on a spherical element of adsorbent particle can be written as

$$N_1(t) = -4\pi R_{f1}^2 Y_{e1t}\, \rho_p \left[\frac{dR_{f1}}{dt}\right] \tag{3.13a}$$

$$N_2(t) = -4\pi R_{f2}^2 Y_{e2t}\, \rho_p \left[\frac{dR_{f2}}{dt}\right] \tag{3.13b}$$

The average concentration on adsorbent particle can thus be written, similar to Eq. (2.12). The differential mass balance over the system by equating the decrease in the adsorbate concentrations in the solution with the accumulation of the adsorbates in the adsorbent can be written as

$$N_1(t) = -V_1\left(\frac{dC_{1t}}{dt}\right) = M_a\left(\frac{d\overline{Y}_{1t}}{dt}\right) \tag{3.14a}$$

$$N_2(t) = -V_1\left(\frac{dC_{2t}}{dt}\right) = M_a\left(\frac{d\overline{Y}_{2t}}{dt}\right) \tag{3.14b}$$

The dimensionless terms used are used are similar to the case of single component system, $C_{it}^* = \frac{C_{it}}{C_{0i}}, R_i^* = \frac{R_{fi}}{R_p}, Bi_i = \frac{k_{fi}R_p}{D_{pi}}, X_i = \frac{M_a}{V_1 C_{0i}}, C_{eit}^* = \frac{C_{eit}}{C_{0i}}$ and $\tau = \frac{D_{p1}t}{{R_p}^2}$ (i denotes the ith component). Combining Eqs. (3.11) and (3.12), the following expression is obtained.

$$C_{e1t}^* = \frac{Bi_1\left(1 - R_1^*\right)C_{1t}^*}{R_1^* + Bi_1\left(1 - R_1^*\right)} = g_{11}\left(C_{1t}^*, R_1^*\right) \tag{3.15a}$$

$$C_{e2t}^* = \frac{Bi_2\left(1 - R_2^*\right)C_{2t}^*}{R_2^* + Bi_2\left(1 - R_2^*\right)} = g_{12}\left(C_{2t}^*, R_2^*\right) \tag{3.15b}$$

Differentiating Eq. (3.15) with respect to τ gives the following expressions for component 1 and 2 respectively.

$$\frac{dC_{e1t}^*}{d\tau} = \underbrace{\frac{Bi_1\left(1 - R_1^*\right)}{\left[R_1^* + Bi_1\left(1 - R_1^*\right)\right]}}_{A_{11}}\frac{dC_{1t}^*}{d\tau} - \underbrace{\frac{Bi_1 C_{1t}^*}{\left[R_1^* + Bi_1\left(1 - R_1^*\right)\right]^2}}_{B_{11}}\frac{dR_1^*}{d\tau} = A_{11}\frac{dC_{1t}^*}{d\tau} - B_{11}\frac{dR_1^*}{d\tau} \tag{3.16a}$$

and,
$$\frac{dC_{e2t}^*}{d\tau} = \underbrace{\frac{Bi_2\left(1 - R_2^*\right)}{\left[R_2^* + Bi_2\left(1 - R_2^*\right)\right]}}_{A_{12}}\frac{dC_{2t}^*}{d\tau} - \underbrace{\frac{Bi_2 C_{2t}^*}{\left[R_2^* + Bi_2\left(1 - R_2^*\right)\right]^2}}_{B_{12}}\frac{dR_2^*}{d\tau} = A_{12}\frac{dC_{2t}^*}{d\tau} - B_{12}\frac{dR_2^*}{d\tau} \tag{3.16b}$$

In terms of nondimensional concentrations, the equilibrium relationship can be expressed as,

$$Y_{e,1t} = \frac{Y_{es,1}C_{e,1}^*/\eta_{1,1}}{1 + K_{o,1}^* C_{e,1}^*/\eta_{1,1} + K_{o,2}^* C_{e,2}^*/\eta_{2,1}} \tag{3.17a}$$

$$Y_{e,2t} = \frac{Y_{es,2}C_{e,2}^*/\eta_{2,2}}{1 + K_{o,1}^* C_{e,1}^*/\eta_{1,1} + K_{o,2}^* C_{e,2}^*/\eta_{2,1}} \tag{3.17b}$$

where, $Y_{es.1} = Y_{s,1}.C_{o,1}$; $Y_{es.2} = Y_{s,2}.C_{o,2}$; ${K_{o,1}}^* = K_{o,1}.C_{o,1}$; ${K_{o,2}}^* = K_{o,2}.C_{o,2}$.

Substituting the expressions of ${C_{e,1t}}^*$ and ${C_{e,2t}}^*$ from Eq. (3.15) in Eq. (3.17), $Y_{e1,t}$ and $Y_{e2,t}$ can be expressed in terms of ${C_{1t}}^*$, ${C_{2t}}^*$, ${R_1}^*$, and ${R_2}^*$.

$$Y_{e,1t} = \frac{Y_{es,1}g_{11}/\eta_{1,1}}{1 + K^*_{o,1}g_{11}/\eta_{1,1} + K^*_{o,2}g_{12}/\eta_{2,1}} = S_1\langle C^*_{1t}, C^*_{2t}, R^*_1, R^*_2\rangle \tag{3.18a}$$

$$Y_{e,2t} = \frac{Y_{es,2}g_{12}/\eta_{2,2}}{1 + K^*_{o,1}g_{11}/\eta_{1,2} + K^*_{o,2}g_{12}/\eta_{2,2}} = S_2\langle C^*_{1t}, C^*_{2t}, R^*_1, R^*_2\rangle \tag{3.18b}$$

Differentiating Eq. (3.18) and using Eqs. (3.15 and 3.16), the time derivatives of $Y_{e1,t}$ and $Y_{e2,t}$ are obtained as follows.

$$\frac{dY_{e,1t}}{d\tau} = h_{11}A_{11}\frac{dC^*_{1t}}{d\tau} - h_{12}A_{12}\frac{dC^*_{2t}}{d\tau} - h_{11}B_{11}\frac{dR^*_1}{d\tau} + h_{12}B_{12}\frac{dR^*_2}{d\tau} \tag{3.19a}$$

$$\frac{dY_{e,2t}}{d\tau} = -h_{21}A_{11}\frac{dC^*_{1t}}{d\tau} + h_{22}A_{12}\frac{dC^*_{2t}}{d\tau} + h_{21}B_{11}\frac{dR^*_1}{d\tau} - h_{22}B_{12}\frac{dR^*_2}{d\tau} \tag{3.19b}$$

where

$$h_{11} = \frac{Y_{es,1}}{\eta_{1,1}\eta_{2,1}} \frac{\left(\eta_{2,1} + K^*_{o,2}C^*_{e,2}\right)}{\left(1 + K^*_{o,1}C^*_{e,1}/_{\eta_{1,1}} + K^*_{o,2}C^*_{e,2}/_{\eta_{2,1}}\right)^2}$$

$$h_{12} = \frac{Y_{es,1}}{\eta_{,11}\eta_{2,1}} \frac{\left(K^*_{o,2}C^*_{e,1}\right)}{\left(1 + K^*_{o,1}C^*_{e,1}/_{\eta_{1,1}} + K^*_{o,2}C^*_{e,2}/_{\eta_{2,2}}\right)^2}$$

$$h_{21} = \frac{Y_{es,2}}{\eta_{1,2}\eta_{2,2}} \frac{\left(\eta_{1,2} + K^*_{o,1}C^*_{e,1}\right)}{\left(1 + K^*_{o,1}C^*_{e,1}/_{\eta_{1,2}} + K^*_{o,2}C^*_{e,2}/_{\eta_{2,2}}\right)^2}$$

$$h_{22} = \frac{Y_{es,2}}{\eta_{2,2}\eta_{1,2}} \frac{\left(K^*_{o,1}C^*_{e,2}\right)}{\left(1 + K^*_{o,1}C^*_{e,1}/_{\eta_{1,2}} + K^*_{o,2}C^*_{e,2}/_{\eta_{2,2}}\right)^2}$$

Combining Eqs. (3.11) and (3.13) and after nondimensionalization, an expression is obtained giving the rate of shrinkage of the adsorbate particle due to component 1 and 2,

$$\left(\frac{dR^*_1}{d\tau}\right) = \frac{-Bi_1\left(\frac{C_{0,1}}{\rho_p Y_{e,1t}}\right)\left(C^*_{1t} - C^*_{e,1t}\right)}{R^{*2}_1} \tag{3.20a}$$

$$\left(\frac{dR^*_2}{d\tau}\right) = \frac{-Bi_2 D_{p2}\left(\frac{C_{02}}{\rho_p Y_{e,2t}}\right)\left(C^*_{2t} - C^*_{e,2t}\right)}{D_{p1}R^{*2}_2} \tag{3.20b}$$

Using Eq. (3.15) for $C_{e1,t}{}^*$, $C_{e2,t}{}^*$ and Eq. (3.17) for $Y_{e1,t}$ and $Y_{e2,t}$, Eq. (3.20) can be expressed as,

$$\frac{dR_1^*}{d\tau} = d_{11}\langle C_{1t}^*, C_{2t}^*, R_1^*, R_2^* \rangle \tag{3.21a}$$

$$\frac{dR_2^*}{d\tau} = d_{22}\langle C_{1t}^*, C_{2t}^*, R_1^*, R_2^* \rangle \tag{3.21b}$$

Combining Eqs. (2.12, 3.14), the following expressions are obtained for component 1 and 2 respectively.

$$\left(\frac{dC_{1t}^*}{d\tau}\right) + X_1\left(1 - R_1^{*3}\right)\left(\frac{dY_{e,1t}}{d\tau}\right) = 3\,X_1\,Y_{e,1t}R_1^{*2}\left(\frac{dR_1^*}{d\tau}\right) \tag{3.22a}$$

$$\left(\frac{dC_{2t}^*}{d\tau}\right) + X_2\left(1 - R_2^{*3}\right)\left(\frac{dY_{e,2t}}{d\tau}\right) = 3\,X_2\,Y_{e,2t}R_2^{*2}\left(\frac{dR_2^*}{d\tau}\right) \tag{3.22b}$$

Using Eq. (3.22), following two governing equations of bulk concentration are obtained.

$$X_{11}\frac{dC_{1t}^*}{d\tau} - X_{12}\frac{dC_{2t}^*}{d\tau} = Z_{11} \tag{3.23a}$$

$$-X_{21}\frac{dC_{1t}^*}{d\tau} + X_{22}\frac{dC_{2t}^*}{d\tau} = Z_{22} \tag{3.23b}$$

where

$$X_{11} = 1 + X_1\left(1 - R_1^{*3}\right)h_{11}A_{11}$$

$$X_{12} = X_1\left(1 - R_1^{*3}\right)h_{12}A_{12}$$

$$Z_{11} = \left[X_1\left(1 - R_1^{*3}\right)h_{11}B_{11} + 3X_1Y_{e,1t}R_1^{*2}\right]d_{11} - X_1\left(1 - R_1^{*3}\right)h_{12}B_{12}d_{22}$$

$$X_{21} = X_2\left(1 - R_2^{*3}\right)h_{21}A_{11}$$

$$X_{22} = 1 + X_2\left(1 - R_2^{*3}\right)h_{22}A_{12}$$

$$Z_{22} = \left[X_2\left(1 - R_2^{*3}\right)h_{22}B_{12} + 3X_2Y_{e,2t}R_2^{*2}\right]d_{22} - X_2\left(1 - R_2^{*3}\right)h_{21}B_{11}d_{11}$$

3.4.6 *Model Predictions of the Kinetic Data*

The model equations are solved to find the dye concentration of crystal violet and methylene blue at any time. The model parameters (k_f and D_p) are estimated by optimizing the fitting of the experimental concentration profiles with those predicted from the model. For this, following three experimental conditions are selected:

Crystal violet,

$C_o = 200$ mg/l; $d_p = 0.044$ mm; $T = 298$ K; $M_a = 0.5$ g
$C_o = 100$ mg/l; $d_p = 0.044$ mm; $T = 298$ K; $M_a = 0.5$ g
$C_o = 150$ mg/l; $d_p = 0.044$ mm; $T = 323$ K; $M_a = 0.5$ g

and Methylene blue,

$C_o = 250$ mg/l; $d_p = 0.044$ mm; $T = 298$ K; $M_a = 0.5$ g
$C_o = 150$ mg/l; $d_p = 0.044$ mm; $T = 298$ K; $M_a = 0.5$ g
$C_o = 150$ mg/l; $d_p = 0.044$ mm; $T = 323$ K; $M_a = 0.5$ g

where, C_o is the initial concentration of dye, d_p is the particle diameter, and M_a is the adsorbent mass. The converged values of k_f and D_p are as follows:

For crystal violet, $k_f = 5.65 \times 10^{-4}$ m/s; $D_p = 3.52 \times 10^{-11}$ m^2/s and for methylene blue,

$k_f = 4.03 \times 10^{-4}$ m/s; and $D_p = 5.77 \times 10^{-11}$ m^2/s, respectively.

These parameter values are used to predict the concentration profiles at other conditions of initial concentration, temperature, particle size, and adsorbent loading. The decay of concentration with time is shown in Figs. 3.18, 3.19, 3.20, 3.21, 3.22, 3.23, 3.24, and 3.25, where the symbols represent experimental points and the continuous lines are the model predicted values. It is observed that the theoretical values are close with the experimental results in most of the cases. The average deviation of the predicted values from the experimental data is within ±10%, in all the cases.

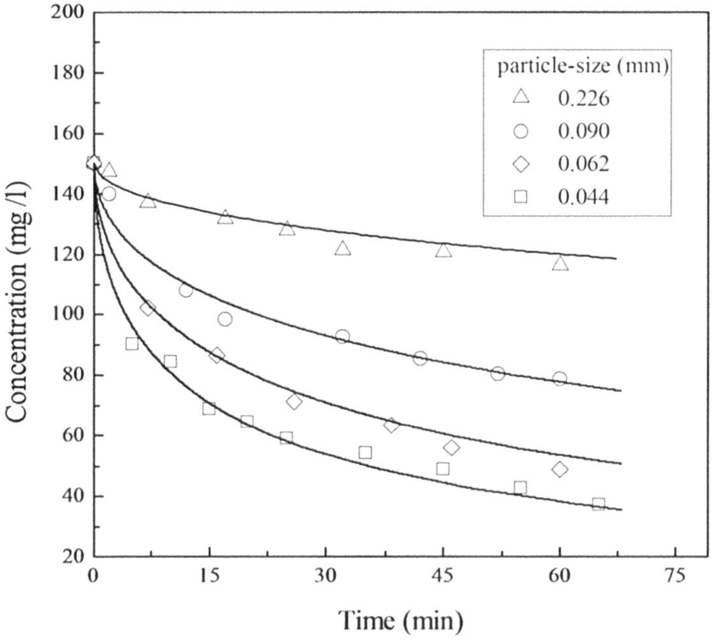

Fig. 3.18 Effect of particle sizes on the concentration decay of CV ($C_o = 150$ mg/l, $T = 298$ K, $M_a = 0.5$ g) (Reproduced from Chakraborty et al. (2005) with permission from Elsevier)

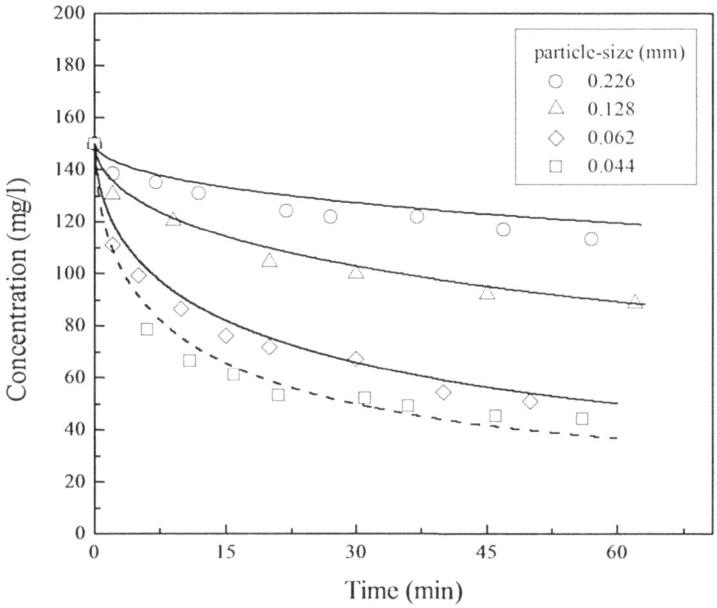

Fig. 3.19 Effect of particle sizes on the concentration decay of MB ($C_o = 150$ mg/l, $T = 298$ K, $M_a - 0.5$ g)

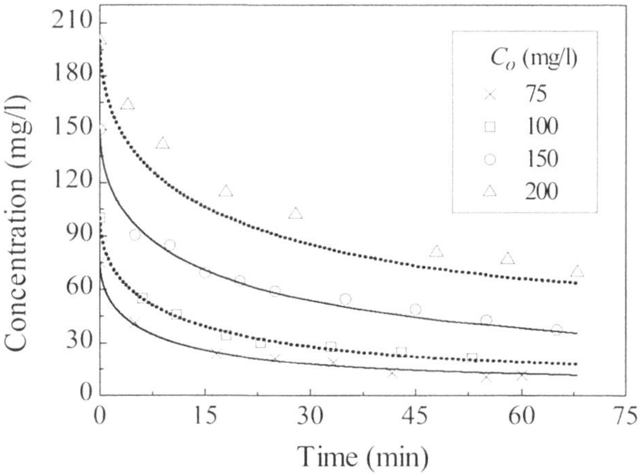

Fig. 3.20 Effect of initial concentrations on the concentration decay of CV ($d_p = 0.044$ mm, $T = 298$ K, $M_a = 0.5$ g) (Reproduced from Chakraborty et al. (2005) with permission from Elsevier)

3.4.7 *Effect of Particle Size*

The effect of adsorbent particle sizes on the concentration decay is shown in Figs. 3.18 and 3.19 for crystal violet and methylene blue, respectively. The results show that there is a gradual increase of adsorption with the decrease in particle size.

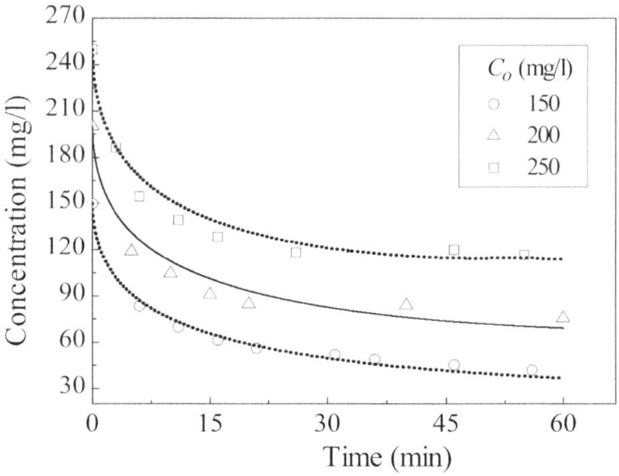

Fig. 3.21 Effect of initial concentrations on the concentration decay of MB ($d_p = 0.044$ mm, $T = 298$ K, $M_a = 0.5$ g)

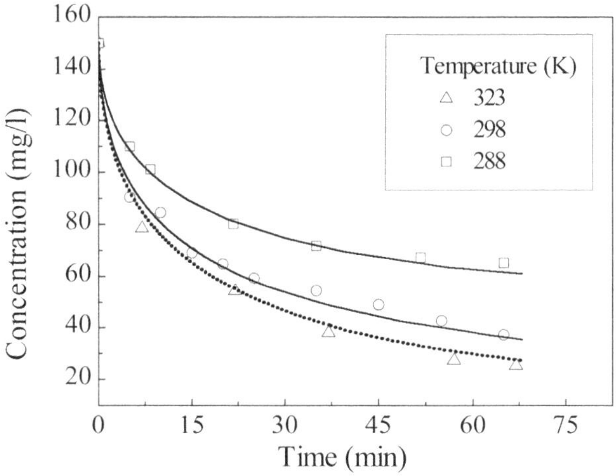

Fig. 3.22 Effect of temperature on the concentration decay of CV ($C_o = 150$ mg/l, $d_p = 0.044$ mm, $M_a = 0.5$ g) (Reproduced from Chakraborty et al. (2005) with permission from Elsevier)

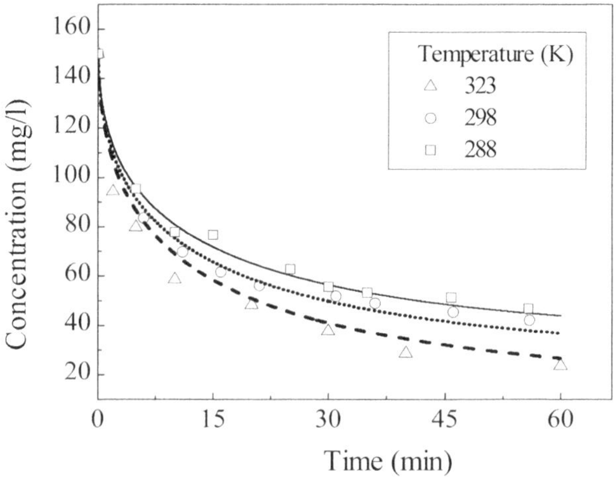

Fig. 3.23 Effect of temperature on the concentration decay of MB ($C_o = 150$ mg/l, $d_p = 0.044$ mm, $M_a = 0.5$ g)

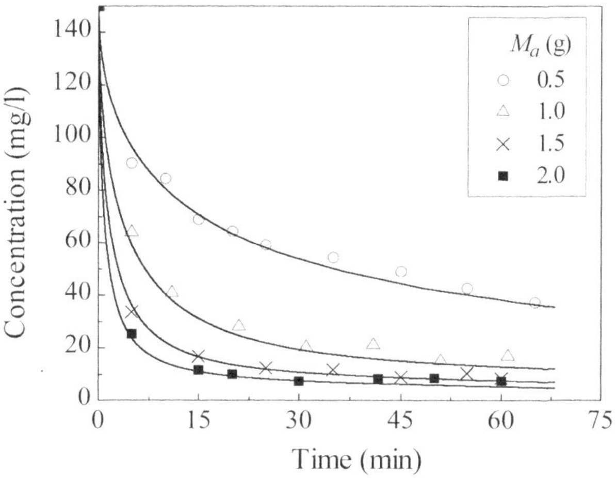

Fig. 3.24: Effect of adsorbent loading on the concentration decay of CV ($C_o = 150$ mg/l, $d_p = 0.044$ mm, $T = 298$ K) (Reproduced from Chakraborty et al. (2005) with permission from Elsevier)

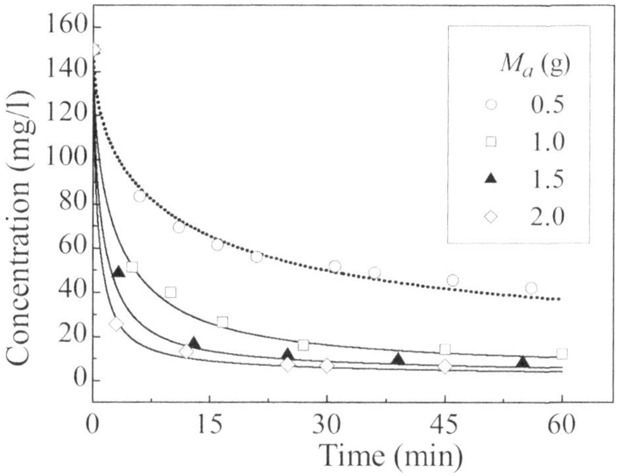

Fig. 3.25 Effect of adsorbent loading on the concentration decay of MB ($C_o = 150$ mg/l, $d_p = 0.044$ mm, $T = 298$ K)

It is obvious that the smaller particles, which have higher solid-liquid interfacial area, will have the greater adsorption rates. In these figures, the symbols represent experimental points, and the solid lines are the model predicted values obtained from the solution of Eqs. (2.22) and (2.23). In Fig. 3.19, the concentration profiles for all the four particle sizes are the predicted values (solid lines), and in Fig. 3.19, the concentration profile for the particle size 0.044 mm is the model fitted line (dashed line); the profiles for the other three particle sizes (solid lines) are the model predicted values. From the figures, it is observed that the predicted concentration profiles are close with the experimental results. The average deviation of the predicted values is within ±3% and ±5% in Figs. 3.18 and 3.19, respectively.

3.4.8 Effect of Initial Dye Concentration

Figures 3.20 and 3.21 show the effect of initial dye concentration on the concentration decay of crystal violet and methylene blue, respectively, with different initial concentrations. For crystal violet, four initial concentrations of 75, 100, 150, and 200 mg/l and, for methylene blue, three initial concentrations of 150, 200, and 250 mg/l are chosen. In Fig. 3.20, the experimental data for 100 and 200 mg/l of initial concentrations are used for estimation of the parameters (dashed lines). Using the estimated values of k_f and D_p, the concentration profile is predicted for the other two initial concentration values (150 and 75 mg/l). It is observed here that the predicted curves almost coincide with the experimental points for the other two concentrations, with the average deviation of $\pm 7\%$. The concentration decay with time for methylene blue is shown in the Fig. 3.21. The data for the initial concentrations of 150 and 250 mg/l are taken for the estimation of parameters k_f and D_p and are shown by dashed lines, while the solid lines are the predicted concentration profiles for the initial concentration 200 mg/l. The average deviation is about $\pm 7\%$.

From Fig. 3.20, it is also observed that after a process time of 1 h, the concentration of crystal violet decreases to about 12.45, 19.5, 42.75, and 67.25 mg/l, respectively, for the initial concentrations of 75, 100, 150, and 200 mg/l feed solution, respectively. Hence the percentage removal of dye gradually decreases with the increase in feed concentration (from 83.4% for 75 mg/l to 66.37% for 200 mg/l at the end of 1 h). In Fig. 3.21 also, the same trend is observed. Removal of methylene blue is about 71% after 55 min of adsorption for 150 mg/l initial concentration, which decreases to about 53% for the 250 mg/l initial concentration.

3.4.9 Effect of Temperature

The effect of temperature on the concentration decay of the dye solution is shown in Figs. 3.22 and 3.23 for three temperatures 323, 298, and 288 K for crystal violet and methylene blue, respectively. In these figures, the symbols represent the experimental data. The curve represented by dashed line in Fig. 3.22 (at 323 K) is the fitted one for the estimation of the parameters k_f and D_p. The concentration profiles for 298 and 288 K (continuous lines in Fig. 3.23) are predicted from the model. It is evident from the figure that the predicted curves are close with the experimental curves for all three temperatures (average deviation lies within $\pm 5\%$). In Fig. 3.23, the concentration profiles for 323 K and 298 K are the model fitted curves (dashed lines), while the continuous line (for 288 K) is the model predicted curve. Here the average deviation is $\pm 7\%$.

Figure 3.22 shows that adsorption of crystal violet increases with the increase in temperature. At 323 K, concentration reduces to about 27.8 mg/l after a process time of 1 h, whereas at 288 K this value is 65.2 mg/l. In Fig. 3.23 also, the same trend is observed. Concentration of methylene blue at 55 min drops from 150 to about 47 mg/l at a process temperature of 288 K, which for 323 K decreases to about 27 mg/l from the same initial concentration 150 mg/l. Hence, the increase in temperature favors the adsorption process which may be due to the chemisorption of the dyes in the present adsorbent (McKay 1996).

3.4.10 Effect of Adsorbent Loading

Figures 3.24 and 3.25 show the effect of mass of adsorbent (M_a) on the concentration decay, adding 0.5, 1.0, 1.5, and 2.0 g of adsorbent per liter solution. In Fig. 3.24, all the continuous lines are the model predicted values. Here the average deviation is within ±10%. In Fig. 3.25, the concentration profile for 0.5 g/l adsorbent dosing is the model fitted line, and the rest all are the model predicted lines. Here the average deviation observed is within ±10%.

From the figures, it is observed that initially with increasing the adsorbent amount (from 0.5 to 1.0 g), adsorption increases considerably. The initial rate of adsorption becomes almost independent of M_a at higher values of M_a. The dye content of the solution decreases drastically and reaches an almost constant value. As the initial rate becomes very high, it is practically independent of M_a.

3.4.11 Kinetic Study of the Reactive Dye System

Adsorption kinetic study is performed with the aqueous solutions of reactive red (RR), reactive black (RB), as well as the mixture of the two dyes, to observe the effects of various process parameters on the rate of adsorption. For the single component systems, the effect of pH of the solution, particle size of the adsorbent, and initial dye concentration is observed. In the two-component system, experiments are carried out with the synthetic solution of the dyes, at different ratio of the initial concentration of each dye. The concentration profiles of the dyes with time are predicted from the model for two-component system. The original industrial effluent is also treated with varying amount of the adsorbent, and the experimental data are compared with the data predicted from the model.

3.4.12 Effect of pH

Figures 3.26 and 3.27 show the effect of pH on the concentration decay of reactive red and reactive black, respectively. It is evident from the figures that acidic pH is favorable for adsorption of both the dyes. This is due to the electrostatic attraction between the dye particle and adsorbent surface charge during the sorption process.

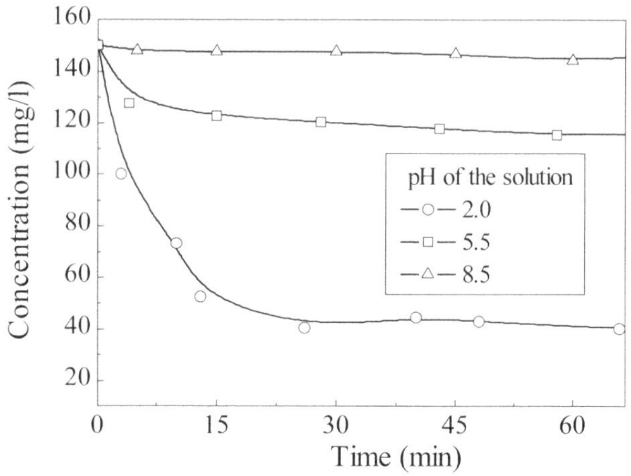

Fig. 3.26 Effect of pH on the concentration decay of reactive red ($C_o = 150$ mg/l; $d_p = 0.044$ mm; $T = 298$ K)

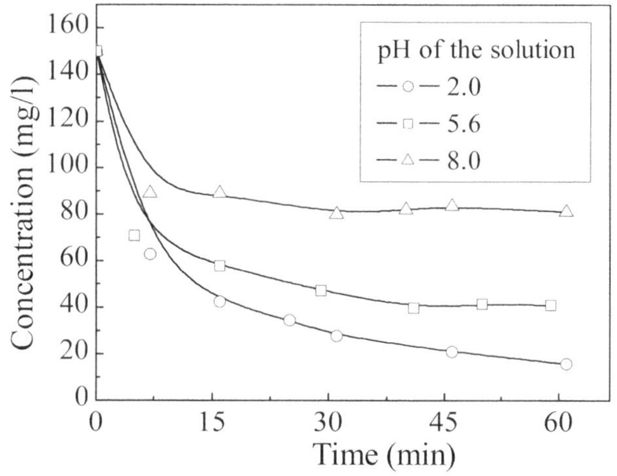

Fig. 3.27 Effect of pH on the concentration decay of reactive black ($C_o = 150$ mg/l; $d_p = 0.044$ mm; $T = 298$ K)

Reactive dyes are anionic in nature, and adsorption capacity is poor on the negatively charged adsorbent surface. This has been noted in the previous literature (Al-Degs et al. 2001), where it was observed that anionic dyes show less affinity toward the biological adsorbents like maize cob, bagasse pith, and wood-based materials in comparison to cationic dyes. With the decrease in the solution pH, the surface positive charge increases, leading to greater adsorption. In Fig. 3.26, concentration of reactive red decreases from 150 mg/l to 42 mg/l at acidic pH (2.0), whereas it reduces to 115 mg/l at a pH of 5.5 and almost no adsorption (145 mg/l) at basic pH (8.5) after an hour of processing time. From Fig. 3.27, adsorption of reactive black also shows same trend, although adsorption is slightly higher than that of reactive red. Here the concentration of reactive black decreases from 150 mg/l to 15 mg/l at an acidic pH, whereas it drops to about 81 mg/l at basic pH.

3.4.13 Effect of Particle Size of Adsorbent

Figures 3.28 and 3.29 show the particle size effect of the adsorbent on the adsorption of reactive red and reactive black, respectively. As expected, adsorption increases considerably with the decrease in particle size of the adsorbent, as the effective surface area of the adsorbent increases. Figure 3.28 shows that after 1 h of experiment, concentration of reactive red decreases from 150 mg/l to 87 mg/l for 0.114 mm particle sizes, which reduces to 63 and 42 mg/l when particle sizes are 0.062 mm and 0.044 mm, respectively. From Fig. 3.29, concentration of reactive black decreases from 150 mg/l to 105, 69, and 15 mg/l with the decrease in the particle size from 0.114 to 0.062 and 0.044 mm, respectively.

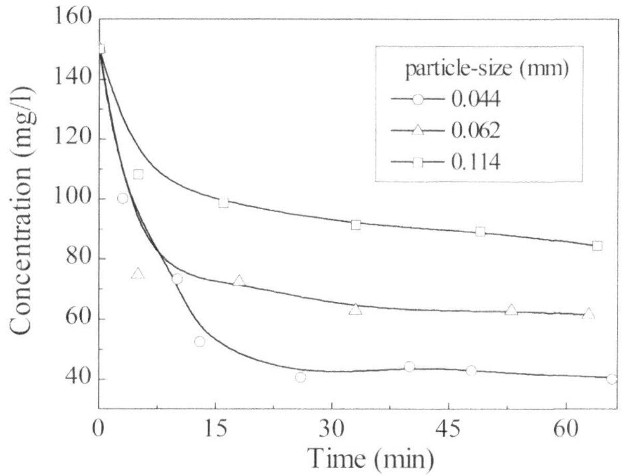

Fig. 3.28 Effect of particle size of the adsorbent on the concentration decay of reactive red ($C_o = 150$ mg/l; pH = 2.0; $T = 298$ K)

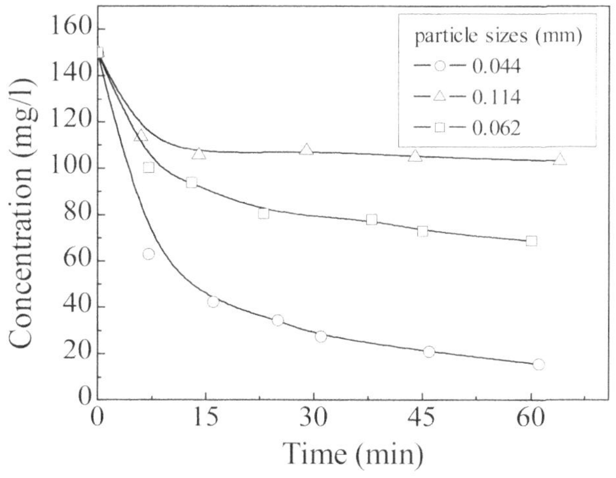

Fig. 3.29 Effect of particle size of the adsorbent on the concentration decay of reactive black ($C_o = 150$ mg/l; pH = 2.0; $T = 298$ K)

3.4.14 Effect of Initial Concentration of Dye

Figures 3.30 and 3.31 show the effect of initial concentration of dye on adsorption. It is observed from these figures that for both the dyes, the dye removal rate decreases with the increase in the initial dye concentration, although the net dye removal increases. In Fig. 3.30, after 1 h of adsorption, 89% dye removal is achieved for an initial concentration of 100 mg/l for reactive red. The removal rate is 70% for an initial concentration of 300 mg/l. Similarly for reactive black (Fig. 3.31), about 94% dye removal is achieved after 1 h, for an initial concentration of 100 mg/l, which reduces to 82% with an increase in the initial dye concentration at 300 mg/l (Fig. 3.32).

Fig. 3.30 Effect of initial dye concentration on the concentration decay of reactive red (pH = 2.0; $d_p = 0.044$ mm; $T = 298$ K)

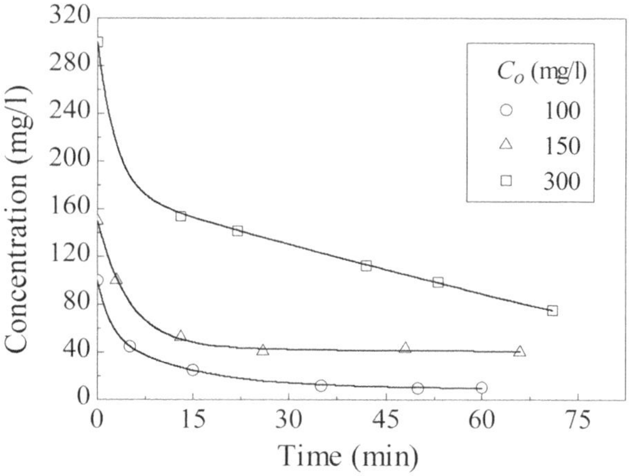

Fig. 3.31 Effect of initial dye concentration on the concentration decay of reactive black (pH = 2.0; $d_p = 0.044$ mm; $T = 298$ K)

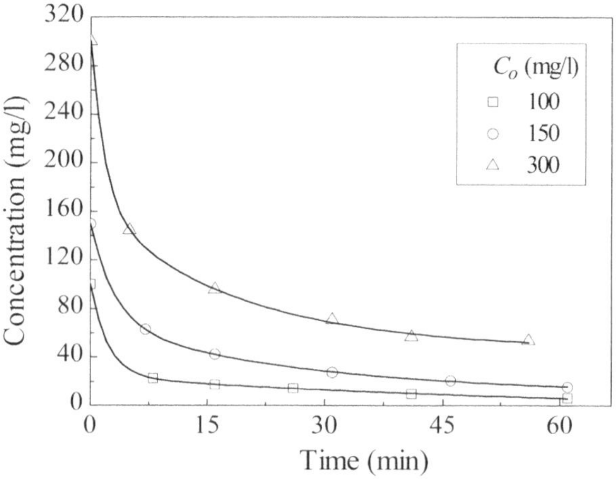

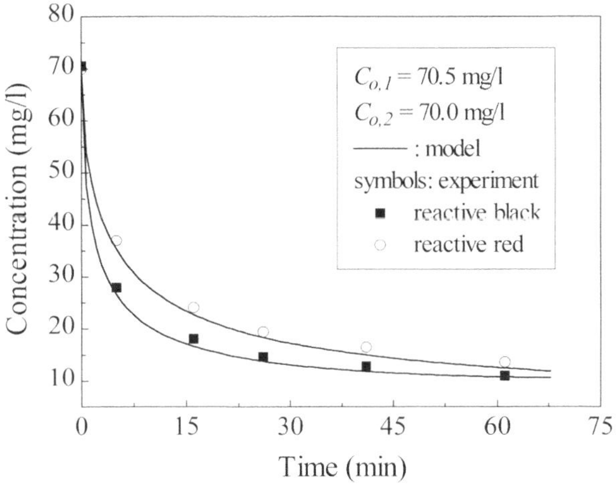

Fig. 3.32 Experimental and the model fitted concentration profiles in a synthetic mixture of reactive black and reactive red ($C_{o,1}$:$C_{o,2}$ = 70.5: 70 mg/l; M_a = 1.0 g) (Reproduced from Chakraborty et al. (2006) with permission from American Chemical Society)

3.4.15 Model Predictions of the Kinetic Data for the Two-Component System

The set of equations, Eqs. (3.21 and 3.23) described earlier, are numerically solved to predict the concentration profiles for both the reactive dyes. In order to solve the equations, the four unknown process parameters, i.e., the external mass transfer coefficients (k_{f1} and k_{f2}) and the internal effective diffusivities (D_{p1} and D_{p2}), are estimated using a nonlinear optimization technique. For this, following three experimental data at different initial concentration ratios are selected:

(i) $C_{o,1}$ = 70 mg/l and $C_{o,2}$ = 70.53 mg/l; (ii) $C_{o,1}$ = 54.18 mg/l and $C_{o,2}$ = 92.26 mg/l; $C_{0,1}$ = $C_{0,2}$ = 100 mg/l; and (iii) $C_{o,1}$ = 100.63 mg/l and $C_{o,2}$ = 46.84 mg/l for reactive black and reactive red, respectively. In Figs. 3.32, 3.33, 3.34, and 3.35, the concentration profiles of both the dyes are represented, where the symbols represent experimental data, and the solid lines are the model fitted lines. The values of the model parameters are given as: $k_{f1} = 9.46 \times 10^{-4}$ m/s; $k_{f2} = 1.67 \times 10^{-4}$ m/s; $D_{p1} = 1.12 \times 10^{-11}$ m^2/s; $D_{p2} = 1.02 \times 10^{-10}$ m^2/s.

The estimated values of the parameters are used for prediction of the concentration profiles with time at different initial concentrations (Figs. 3.35, 3.36, and 3.37). It is observed from the figures that the predicted values are close with the experimental data in most of the cases. So the present model is also applicable to predict the concentration profile for the two-component mixture of dyes.

3.4.16 Sensitivity Analysis of the Model Parameters

A sensitivity analysis of the model parameters is carried out and presented in Figs. 3.38 and 3.39. It is observed that the accuracy of the model predicted

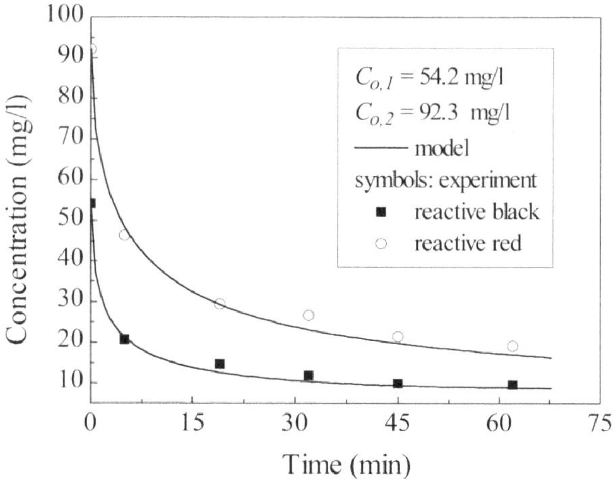

Fig. 3.33 Experimental and the model fitted concentration profiles in a synthetic mixture of reactive black and reactive red ($C_{o,1}$:$C_{o,2}$ = 54: 92 mg/l; M_a = 1.0 g) (Reproduced from Chakraborty et al. (2006) with permission from American Chemical Society)

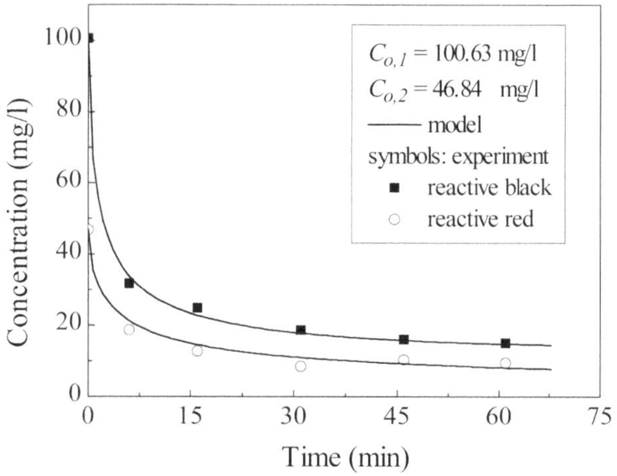

Fig. 3.34 Experimental and the model fitted concentration profiles in a synthetic mixture of reactive black and reactive red ($C_{o,1}$:$C_{o,2}$ = 101: 47 mg/l; M_a = 1.0 g) (Reproduced from Chakraborty et al. (2006) with permission from American Chemical Society)

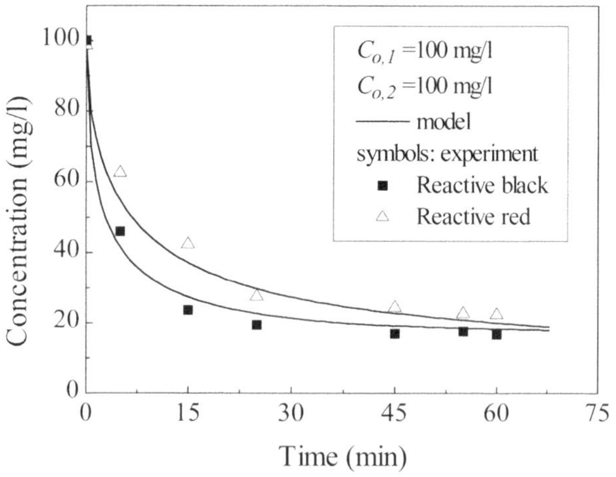

Fig. 3.35 Prediction of concentration profiles of reactive black and reactive red in a synthetic mixture ($C_{o,1}$:$C_{o,2}$ = 100: 100 mg/l; M_a = 1.0 g) (Reproduced from Chakraborty et al. (2006) with permission from American Chemical Society)

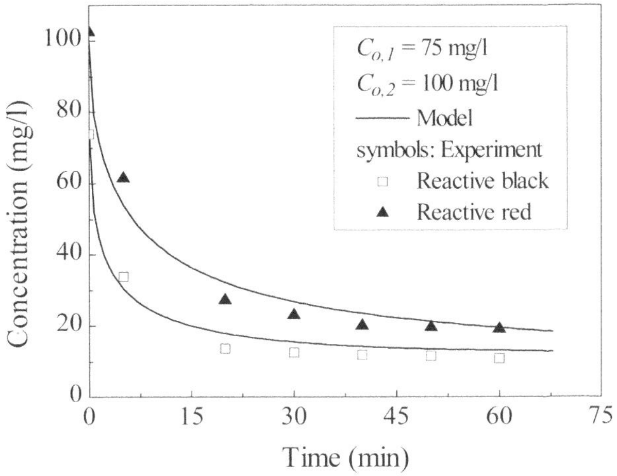

Fig. 3.36 Prediction of concentration profiles of reactive black and reactive red in a synthetic mixture ($C_{o,1}$:$C_{o,2}$ = 75: 100 mg/l; M_a = 1.0 g) (Reproduced from Chakraborty et al. (2006) with permission from American Chemical Society)

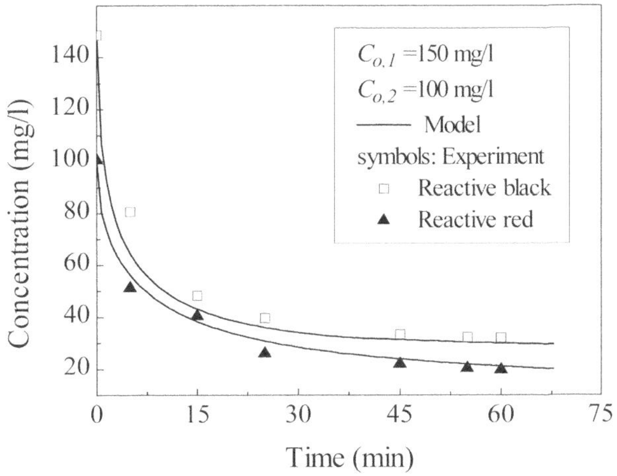

Fig. 3.37 Prediction of concentration profiles of reactive black and reactive red in a synthetic mixture ($C_{o,1}$: $C_{o,2}$ = 150: 100 mg/l; M_a = 1.0 g) (Reproduced from Chakraborty et al. (2006) with permission from American Chemical Society).

concentration of both the dyes is insensitive to the estimated values of k_{f1} and k_{f2}. Figures 3.38 and 3.39 show the effect of change in the parameters D_{p1} and D_{p2} on the predicted concentration profiles of reactive black (dye 1) and reactive red (dye 2), respectively, at an initial concentration of 100:100 mg/l and adsorbent loading of 1.0 g/l. Here variation of −50% to +50% are introduced into the value of D_{p1} and D_{p2}, and the concentration profiles are predicted for the corresponding changes in the parameters. The variations in the predicted concentration of dye 1 due to variations in D_{p1} are determined to be in the range of −32.75% to +15.73%, respectively. For dye 2, the variations in the predicted concentration due to variations in D_{p2} are in the range of −13.18% to +4.09%, respectively. Hence, the variations in the value of effective diffusivity have a profound effect on the prediction of concentration profile for dye 1, and the effect is relatively less for dye 2.

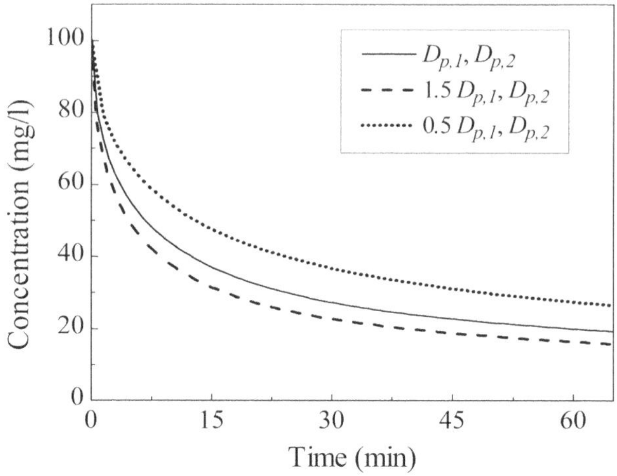

Fig. 3.38 Effect of variations in D_{p1} on the prediction of concentration profile of reactive black ($C_{o,1}$: $C_{o,2}$ = 100: 100 mg/l; M_a = 1.0 g) (Reproduced from Chakraborty et al. (2006) with permission from American Chemical Society)

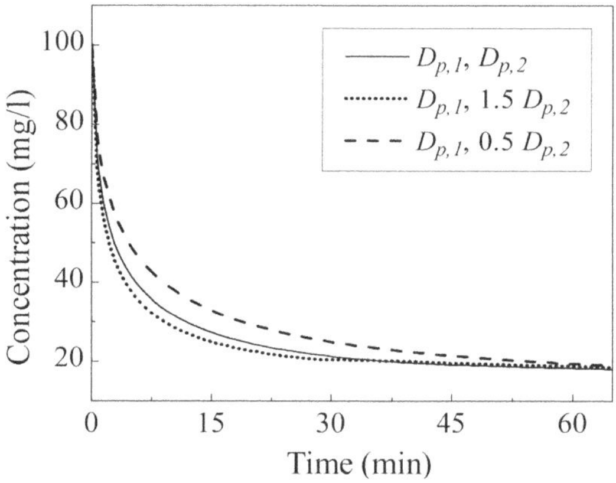

Fig. 3.39 Effect of variations in D_{p2} on the prediction of concentration profile of reactive red ($C_{o,1}$: $C_{o,2}$ = 100: 100 mg/l; M_a = 1.0 g) (Reproduced from Chakraborty et al. (2006) with permission from American Chemical Society)

3.5 Adsorption Studies Using Industrial Effluent

The present textile effluent, which contains a mixture of the two reactive dyes (reactive black and reactive red), is treated using varying amounts of the adsorbent (1.0–2.0 g/l). The initial concentrations of reactive black and reactive red in the effluent are 173 and 115 mg/l, respectively. The pH of the solution is maintained at 2.0, and smaller particles of adsorbent (0.044 mm) are used. The concentration decay of the dyes is shown in Figs. 3.40 and 3.41 for reactive black and reactive red, respectively. In these figures, the symbols are the experimental points, and the solid lines are the concentration profiles predicted from the model. It can be observed from Fig. 3.41 that, using an adsorbent dose of 1.0 g/l, about 80% removal of reactive black is possible in 1 h and the removal rate increases as the adsorbent dosing increases. Using an adsorbent dose of 2 g/l, about 89% dye removal is possible. The decline in the concentration of reactive red in the effluent is shown in

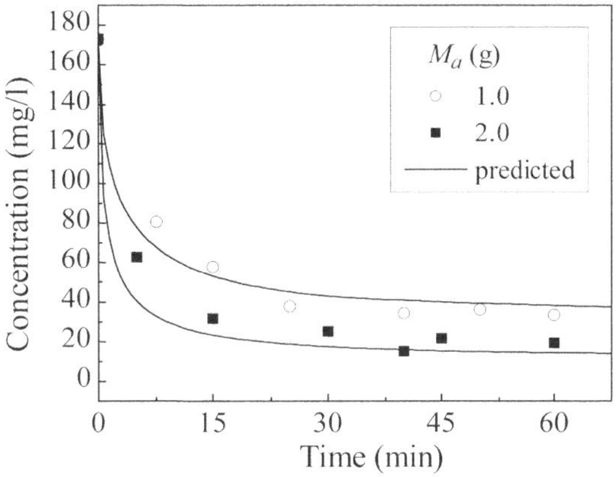

Fig. 3.40 Concentration decay of reactive black in the industrial effluent using varying amount of the adsorbent (pH = 2.0; $d_p = 0.044$ mm; $T = 298$ K) (Reproduced from Chakraborty et al. (2006) with permission from American Chemical Society)

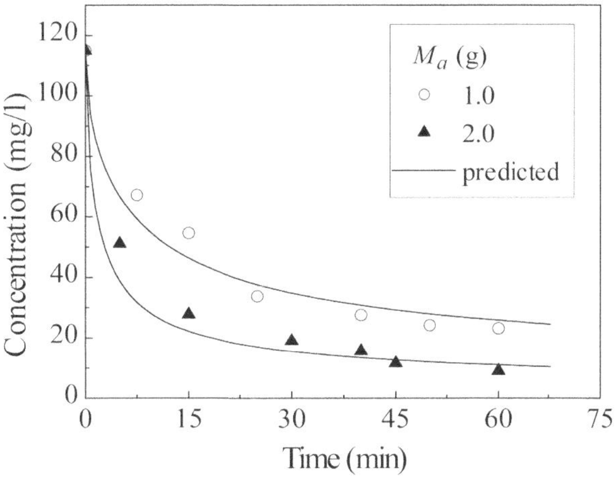

Fig. 3.41 Concentration decay of reactive red in the industrial effluent using varying amount of the adsorbent (pH = 2.0; $d_p = 0.044$ mm; $T = 298$ K) (Reproduced from Chakraborty et al. (2006) with permission from American Chemical Society)

Fig. 3.41. It is observed from the figure that using 1.0 g/l adsorbent, about 79% removal of reactive red is possible. Using an adsorbent dose of 2.0 g/l, about 92% removal is achieved. From the figures, it is observed that the experimental points are close with the predicted lines in most of the cases. Overall, it can be concluded that the present two-component model is able to predict the concentration profile of dyes with time in the adsorption of an industrial effluent.

The measure of COD in the industrial effluent is the important requirement as per pollution board. So the removal of dyes is seen in terms of COD removal. Figure 3.42 shows the COD removal (%) in adsorption of the effluent stream.

Removal of methylene blue and Congo red mixture in the textile dye effluent is presented in Fig. 3.43 (Rao and Rao 2006). It may be observed that using 4 g of adsorbent (both activated carbon and flyash), activated carbon shows higher removal efficiency. In case of activated carbon, the equilibrium time is around 70 minutes for both the dyes, whereas it is only half for flyash. However, using

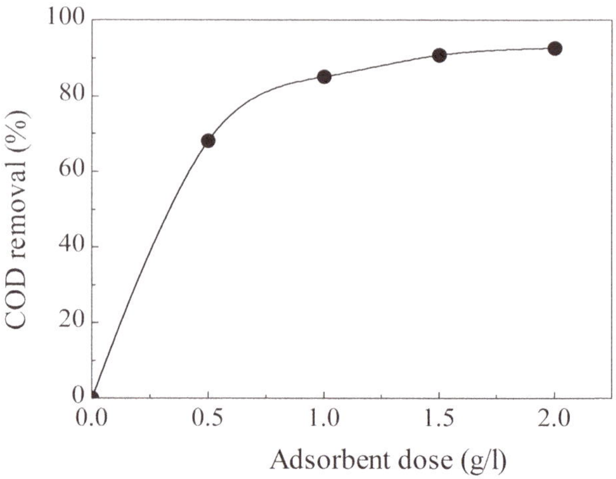

Fig. 3.42 COD removal (%) with variation in the adsorbent amount (pH = 2.0; d_p = 0.044 mm; T = 298 K)

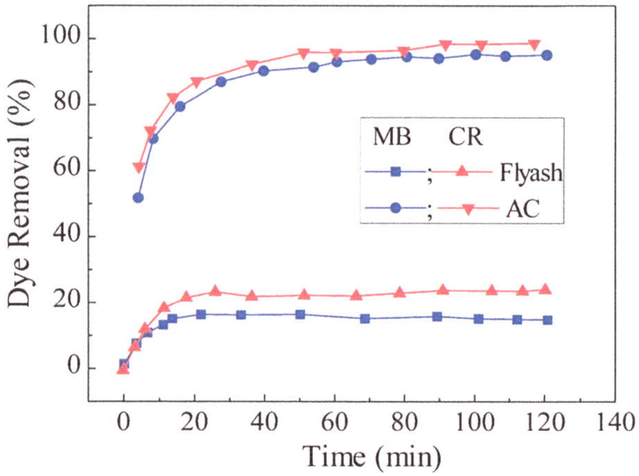

Fig. 3.43 Removal of Congo red (*CR*) and methylene blue (*MB*) using flyash and activated carbon. The feed concentration is 0.2 mg/l and solution pH is 7.5

higher amount of flyash adsorbent, the removal efficiency can be improved and may be higher than that of activated carbon. The dosing of appropriate amount of adsorbent is necessary achieving a particular extent of removal efficient. The cost of flyash with respect to activated carbon is quite inexpensive. So the dosing amounts with respect to cost per removal efficiency may be lower in case of flyash as compared to activated carbon. The removal efficiency using 12 g of flyash is similar to that of 4 g of activated carbon. Thus, the flyash may be a better substitute considering economic feasibility options over activated carbon.

The removal of an reactive dye (acid yellow 6) from a textile effluent is studied using activated carbon prepared from rice husk and saw husk (Malik 2003). The risk husk and saw dust materials were first carbonized in an inert atmosphere at 400 °C and 500 °C, respectively, for 1 h. The carbonized materials were activated by superheated steam at 600 °C and 800 °C, respectively, for 1 h under 1.5 atm

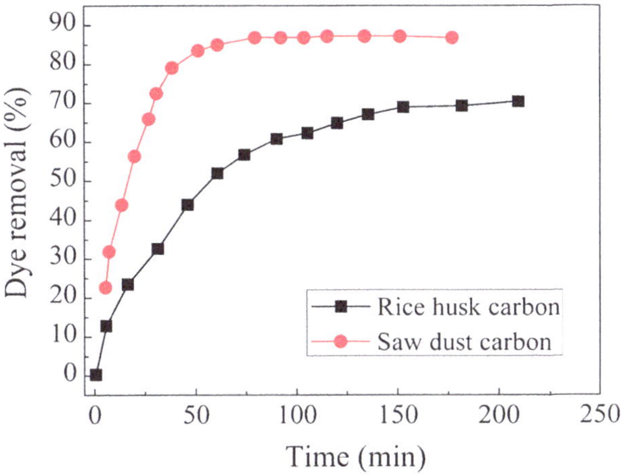

Fig. 3.44 Removal of acid yellow 36 (C.I. 13,065) using rice husk carbon and saw dust carbon

steam pressure. The resultant activated material was washed with deionized water adequately and dried overnight at 100 °C. The details of the physicochemical characteristics can be obtained from the literature (Malik 2003). The adsorption kinetic study for removal of acid yellow 36 present in a dye-house effluent is presented in Fig. 3.44. The adsorption capacity of the sawdust carbon is higher than carbonized rice husk. This is in conformation with the basicity levels of the surface together with the available pore surface area, which are greater for the sawdust compared to rice husk. The process of attaining equilibrium is gradual for both the different types of activated carbon. One of the probable reasons for such an observation is due to the presence of the micro- and macropores in the adsorbent. The mechanism of adsorption is coupled with diffusion, where the process of diffusion is the rate determining step. The dye molecule (adsorbate) has to diffuse through the boundary layer film over the adsorbent surface and then diffusion into the pore morphology of the adsorbent, responsible for large contact times. The adsorption equilibrium time takes around 1 h for the sawdust carbon, while it is almost three times (3 h) in the case of rice husk. The adsorption process is primarily physisorption, with some possibility of occurrence of chemisorption too.

The results show that saw dust carbon and rice husk carbon can be a better cost-effective alternative for removal of reactive dyes from effluent compared to GAC.

Removal of mixture of dyes from textile effluent from agricultural residues (or by-products) is shown in Fig. 3.45. Wood chips, corn cob, and wheat straw were selected because these are widely available residues in most agricultural countries and are mostly inexpensive. The adsorption studies were carried out at room temperature corresponding to an equivalent mixture of dye solution with 400 mg/l concentration. As observed from the figure, the adsorption capacity of wood chips is the lowest among the three adsorbents investigated. The equilibration time is quite high these adsorbents compared to carbon adsorbents. The exhausted dye-saturated adsorbents are degraded by the process of solid-state fermentation through microbial action for environmental disposal (Nigam et al. 2000).

Fig. 3.45 Removal of Remazol Red and Remazol Black B mixture (total concentration of 400 mg/l) using agricultural residues

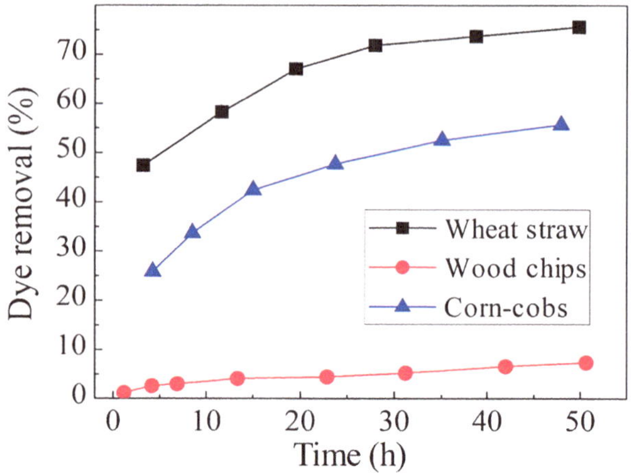

References

Ajmal M, Khan AH, Ahmad S, Ahmad A (1998) Role of saw dust in the removal of copper (II) from industrial wastes. Water Res 32:3085–3091

Al-Degs Y, Khraisheh MAM, Allen SJ, Ahmad MNA (2001) Sorption behavior of cationic and anionic dyes from aqueous solution on different types of activated carbons. Sep Sci Technol 36:91–102

Chakraborty S, De S, Dasgupta S, Basu JK (2005) Adsorption study for the removal of a basic dye: experimental and modeling. Chemosphere 58:1079–1086

Chakraborty S, Basu JK, De S, Dasgupta S (2006) Adsorption of reactive dyes from a textile effluent using sawdust as the adsorbent. Ind Eng Chem Res 45:4732–4741

Chatzopoulos D, Verma A, Irvine RL (1993) Activated carbon adsorption and desorption of toluene in the aqueous phase. AICHE J 39:2027–2041

Garg VK, Gupta R, Yadav A, Kumar R (2003) Dye removal from aqueous solution by adsorption on treated saw dust. Bioresour Technol 89:121–124

Hamdaouia O, Naffrechoux E (2007) Modeling of adsorption isotherms of phenol and chlorophenols onto granular activated carbon: Part II. Models with more than two parameters. J Hazard Mater 147:401–411

Ho YS, Mckay G (1999) Competitive sorption of copper and nickel ions from aqueous solution using peat. Adsorption 5:409–417

IS:877 (1977) Methods and sampling and test for activated carbons, powdered and granular. ISI, New Delhi

Jena PR, De S, Basu JK (2003) A generalized shrinking core model applied to batch adsorption. Chem Eng J 95:143–154

Levenspiel O (1972) Chemical reaction engineering. Wiley, New York

Leyva-Ramos R, Bernal-Jacome LA, Guerrero-Coronado RM, Luentes-Rubio L (2001) Competitive adsorption of Cd(II) and Zn(II) from aqueous solution onto activated carbon. Sep Sci Technol 36:3673–3687

Malik PK (2003) Use of activated carbons prepared from sawdust and rice-husk for adsorption of acid dyes: a case study of acid yellow 36. Dyes Pigments 56:239–249

Mathews P, Weber WJ Jr (1976) Effects of external mass transfer and intraparticle diffusion on adsorption rates in slurry reactors, in: physical, chemical wastewater treatment. AIChE Symp Ser water 91

Mckay G (1982) Adsorption of dyestuffs from aqueous solutions with activated carbon, Part I Equilibrium and batch contact-time studies. J Chem Technol Biotechnol 32:759–772
Mckay G (1984) Analytical solution using a pore diffusion model for the adsorption of basic dye on silica. AICHE J 30:692–697
Mckay G (1996) Use of adsorbents for the removal of pollutants from wastewater. CRC Press, New York
Namasivayam C, Yamuna RT (1995) Adsorption of chromium(VI) by a low-cost adsorbent: biogas residual slurry. Chemosphere 30:561–578
Nigam P, Armour G, Banat IM, Singh D, Marchant R (2000) Physical removal of textile dyes from effluents and solid-state fermentation of dye-adsorbed agricultural residues. Bioresour Technol 72:219–226
Rao VVB, Rao SRM (2006) Adsorption studies on treatment of textile dyeing industrial effluent by flyash. Chem Eng J 116:77–84
Vogel AI (1970) Textbook of practical organic chemistry. Longmans, London, p 818

Chapter 4
Surfactant-Enhanced Carbon Regeneration

Abstract To make a process sustainable, it is often important to regenerate the adsorbent for reuse. One of the commonly used technique in this case is surfactant-enhanced regeneration which is inexpensive and low-energy intensive. This chapter presents a comprehensive study of this case, the surfactant regeneration process using different types of surfactants, including a simple theoretical model for the adsorbent regeneration.

Keywords Regeneration • Surfactant • Surfactant-enhanced carbon regeneration

Adsorption beds containing activated carbon are widely used in industry to remove organics from water and air. One of the major disadvantages of this method is the difficulty of regeneration of the saturated carbon. When the organic is highly volatile, in situ thermal regeneration (e.g., hot steam) can be used. However, often the carbon must be removed from the bed and the organic is burned off in regeneration furnaces. A universal in situ regeneration process would be valuable.

Surfactant-enhanced carbon regeneration (SECR) is an original, in situ, energy inexpensive process of regenerating an exhausted adsorbent. In the process of SECR, a strong surfactant solution, much above the critical micellar concentration, is passed through the saturated adsorption column. The organic dye (or contaminant) desorbs through the bed into the regenerant solution, and gets solubilized in the micelles. A concentrated solution containing organic solutes is produced, for disposal or recovery of the surfactant and the solute. The residual surfactant in the carbon bed is washed by flushing with water. Since the surfactant is nontoxic and biodegradable, the flush stream can be safely disposed without any risk of environmental hazard.

There are several well-established methods for the regeneration of spent commercial activated carbon (CAC) that can be classified in three broad groups: thermal, chemical, and biological regeneration (Perrich 1981). In spite of being the most widely used regeneration technology, thermal regeneration has an important energy demand since temperature must be kept above 1100 K. Moreover, the carbon needs to be removed from the bed and transported to the hearth furnace, where the organics are volatilized and carbonized. The process is labor intensive, energy inefficient, and time consuming. Also, considerable amount of carbon (5–15%) is usually lost by attrition, burn-off, and washout (Miguel et al. 2001).

S. Mondal et al., *Advances in Dye Removal Technologies*, Green Chemistry and Sustainable Technology, DOI 10.1007/978-981-10-6293-3_4

Chemical regeneration can be carried out by desorption of adsorbates using specific solvents or by decomposition of adsorbed species using oxidizing chemical agents. Extraction with solvents requires further purification of the solvent, and it is only recommended when a valuable product is recovered (Wedeking et al. 1987). Oxidative chemical regeneration is usually accomplished by wet oxidation (WO) under either subcritical or supercritical conditions. Although these techniques have been successfully applied to a number of adsorbates, regeneration efficiency depends on the solubility of the sorbed substances and the effect of pressure on the chemical structure of carbon. Also, such a large investment in high-pressure equipment makes the method rather expensive (Mishra et al. 1995). Biological regeneration is usually slow and requires biodegradable adsorbed species which is not the case in many water pollutants (Scholz and Martin 1998). Because of these drawbacks, a number of other alternative CAC regeneration methods are subject of current investigation. Among them, steam regeneration (Kim et al. 2001), catalytic oxidation (Matatov-Meytal and Sheintuch 1997), microwave regeneration (Fang and Lai 1996), electrochemical methods (Zhang 2002), or extraction with supercritical fluids (Ryu et al. 2000) can be highlighted.

4.1 Basics of Surfactant-Enhanced Carbon Regeneration

In SECR, a concentrated surfactant solution is passed through the saturated carbon bed. The adsorbed organic solute desorbs and is solubilized in the micelles formed by the surfactant. Micelles are self-assembled aggregates that typically consist of 50–200 monomer units. The micelle has a hydrophobic interior that favors the solubilization of the organic molecules. The concentrated surfactant solution may have a large solubilizing capacity for the organic. Therefore, a much smaller volume of surfactant solution can be used than the volume of the water originally treated for aqueous-phase applications. After the organic is removed from the carbon, water is used to rinse residual surfactant from the carbon. An advantage of SECR over solvent regeneration is the relatively innocuous nature of surfactants from an environmental point of view. Thus, the product of the rinse step can be directed to the normal sewage treatment system, since it contains only surfactant. The regenerated carbon would then be ready for reuse for liquid-phase operations. If the carbon is being used for gas-phase applications, the carbon can be dried after the rinse step before reuse. A schematic of the SECR process is shown in Fig. 4.1.

4.2 Experimental Study of Desorption

The effects of pH and concentration of different surfactants (both cationic and anionic) on the percentage desorption of chrysoidine and eosin are studied. Before each desorption study, the amount of each dye adsorbed on a known amount (1.0 g)

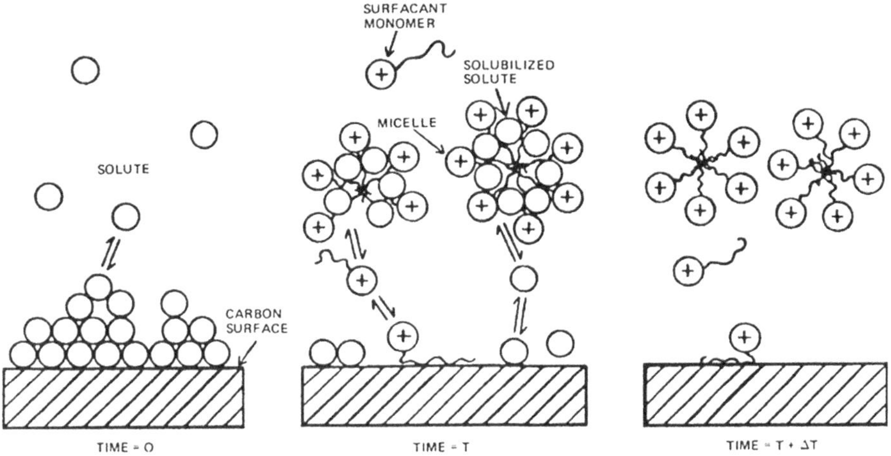

Fig. 4.1 Schematic of the different stages of the SECR process

of activated carbon is measured. This is done by keeping 1.0 g of activated carbon and 250 ml of a solution of dye with a concentration of 5000 mg/L in a glass bottle and agitated by a shaker for 12 h at room temperature. The amount of dye adsorbed per gram of activated carbon is calculated by measuring the concentration of dye solution before and after adsorption. After filtration, the carbon is dried in an oven. For the desorption study, 1.0 g of carbon with adsorbed dye is mixed with 250 ml of the desorbing medium. Distilled water at different pH and both the anionic and cationic surfactant at various concentrations (CPC, TTAB, AOT) are used as the desorbing medium. The amount of desorbed dye is estimated from the mass balance as in the case of adsorption study.

In this section the efficacy of SECR technique is discussed in terms of percentage of chrysoidine and eosin desorption from activated carbon. The use of different surfactant has been explored and compared with the conventional regeneration techniques (e.g., changing the pH of the solution).

4.2.1 Desorption Kinetic Model

It has been proposed that the desorption rate at any instant would be proportional to the driving force, i.e., difference between the initial (at $t = 0$) amount of the adsorbed dye and the dye concentration in the solution, at any time t. The dye concentration in the solution, in turn, would be related to the amount of the dye still remaining adsorbed through a mass balance. This is mathematically represented by the following equation.

$$\frac{dq}{dt} = \alpha(q_0 - kq) \tag{4.1}$$

where, α and k are the constants ($k \neq 1$); q_0 and q are the amount of adsorbed dye present per gm of CAC at time $t = 0$ and at any time $t = t$, respectively. The expression for the adsorbed dye remaining in the adsorbent at any time t can be obtained by integrating Eq. (4.1) as follows,

$$q = \frac{q_0}{k}\left[1 - (1 - k)e^{-k\alpha t}\right] \tag{4.2}$$

Using Eq. (4.2), the expression of the percentage desorption at any time is expressed as,

$$D = 100 \times \left(\frac{k - 1}{k}\right)\left[1 - e^{-k\alpha t}\right] \tag{4.3}$$

The values of k and α can be determined from the experimental data under various operating conditions.

4.3 Effect of pH

The effects of pH without adding any surfactants on the percent desorption of dyes are studied and shown in Figs. 4.2 and 4.3 for chrysoidine and eosin, respectively. At lower pH, active sites are protonated and adsorption of dyes is restricted to the active site of charcoal as discussed earlier. Thus with the decrease of pH, desorption increases. From Fig. 4.2, it has been found that after 50 min of operation, desorption of chrysoidine increases from about 0.18 to 1.6% when pH decreases from 11.2 to 2.7. Chrysoidine changes its original red color when pH of the solution changes from acidic to basic. Therefore, it is appropriate to desorbs chrysoidine by setting the pH as acidic (in order to maintain its color). From Fig. 4.3, it may be observed that after 40 min of operation, desorption of eosin increases from about 0 to 46.5% when pH increases from 2 to 12. The desorption kinetic is described by Eq. (4.3). The value of k varies within 1.01–1.2 for chrysoidine and that of 1.01 to 1.7 for eosin in a pH range of 2–12. On the other hand, α varies from 4.89×10^{-2} to 8.67×10^{-2} (min^{-1}) for chrysoidine and 7.4×10^{-2} to 7.8×10^{-2} (min^{-1}) for eosin in the same pH range. It may be noted here that the continuous curves in both the figures are the fitting obtained from Eq. (4.3) and the symbols are the experimental data. The r^2 values of all the fits are more than 0.98.

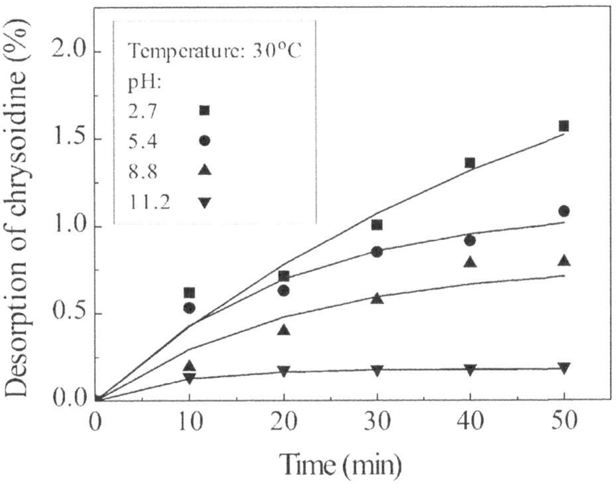

Fig. 4.2 Effect of pH on desorption of chrysoidine from spent charcoal. *Solid lines* are model fitted results and symbols are experimental data (Reproduced from Purkait et al. (2004) with permission from Taylor & Francis Ltd)

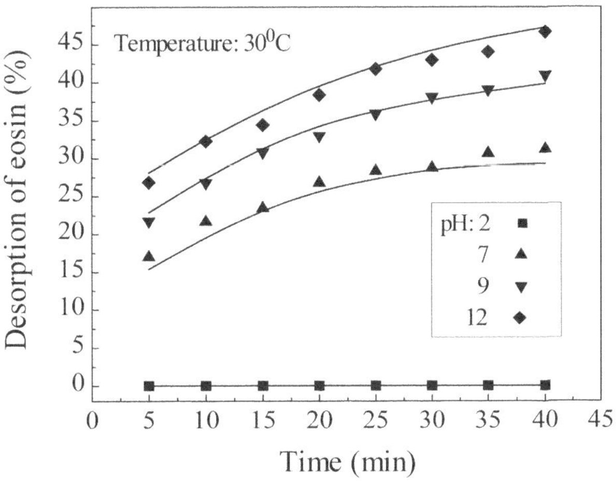

Fig. 4.3 Effect of pH on desorption of eosin from spent charcoal. *Solid lines* are model fitted results and the symbols are the experimental data

4.4 Effect of Different Surfactants

4.4.1 Desorption of Chrysoidine

Figures 4.4, 4.5, and 4.6 show the variations of percent desorption of chrysoidine using SDS, TTAB, and CPC, respectively, at various concentrations higher than the corresponding critical micellar concentrations (CMC). CMC of SDS, TTAB, and CPC are 2.33, 1.345, and 0.322 g/L (Rosen 1978), respectively. Desorption increases with increase in surfactant concentration and also with operating time for all the cases. In case of SDS (anionic surfactant), desorption is only about 1.7%, whereas for TTAB and CPC (cationic surfactant), desorption increases up to about 35% and 24%, respectively, for the feed surfactant concentration of 2000 mg/L

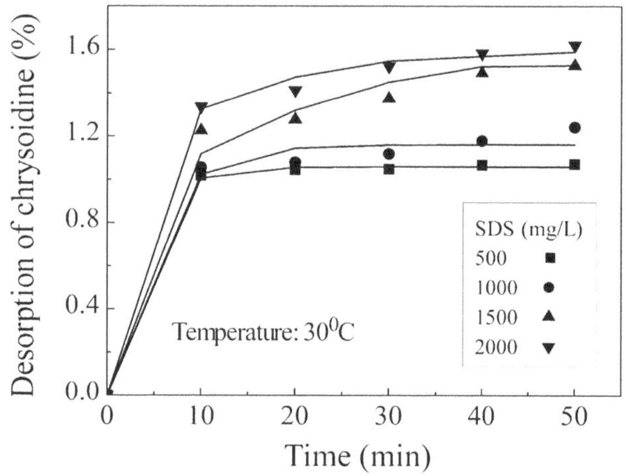

Fig. 4.4 Effect of SDS on desorption of chrysoidine from spent charcoal. *Solid lines* are model fitted results and symbols are the experimental data (Reproduced from Purkait et al. (2004) with permission from Taylor & Francis Ltd)

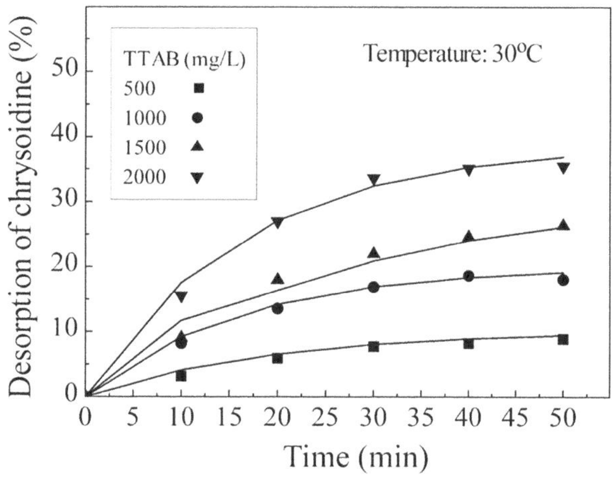

Fig. 4.5 Effect of TTAB on desorption of chrysoidine from spent charcoal. *Solid lines* are model fitted results and symbols are the experimental data (Reproduced from Purkait et al. (2004) with permission from Taylor & Francis Ltd)

within 50 min of operation. It may also be observed from Figs. 4.4, 4.5 and 4.6 that desorption is more for TTAB (cationic surfactant) and least for SDS (anionic surfactant). Therefore, a suitable cationic surfactant may be a better agent for desorption of chrysoidine dye from activated charcoal.

Better performance of cationic surfactants (TTAB and CPC) compared to anionic one (SDS) may be explained as follows. During adsorption, chrysoidine dye gets adsorbed on the surface or inside the pores of activated charcoal. Activated charcoal itself is negatively charged. Micelles of the cationic surfactants, TTAB and CPC, are positively charged, and there is high possibility of cationic surfactant getting adsorbed on the surface of negatively charged activated charcoal by replacing the adsorbed dye. Some dyes may still get solubilized inside the hydrophobic core of the aggregated surfactant molecule (micelles). On the other hand, micelles of anionic surfactant (SDS) are negatively charged, and therefore it is

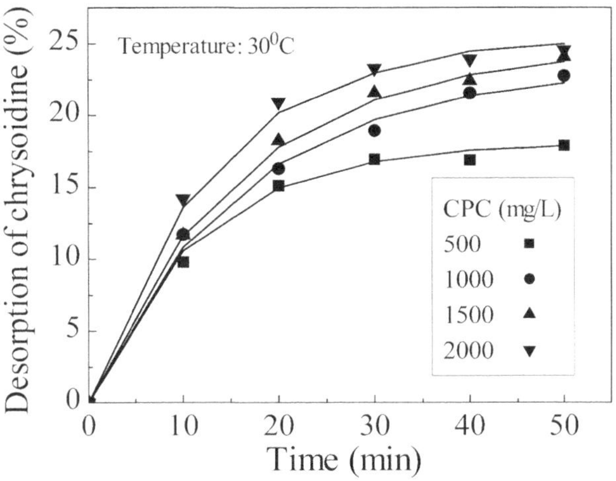

Fig. 4.6 Effect of CPC on desorption of chrysoidine from spent charcoal. *Solid lines* are model fitted results and symbols are the experimental data (Reproduced from Purkait et al. (2004) with permission from Taylor & Francis Ltd)

unlikely that micelles are adsorbed on charcoal as such due to electrostatic repulsion. Therefore, when activated charcoal adsorbed with chrysoidine dye is exposed to SDS solution, dye molecules remain on the surface of the adsorbent. These results in very less dye desorption for anionic surfactants. The desorption kinetic using the surfactants is described by Eq. (4.3). For each feed surfactant concentration, the constants k and α are estimated using curve fitting the corresponding experimental data. The value of k is found to be 1.02 for SDS, 1.35 for TTAB, and 1.3 for CPC. The values for α varies in a narrow range from 0.13 to 0.0.3 for SDS, 0.03 to 0.04 for TTAB, and 0.04 to 0.07 for CPC corresponding to various feed surfactant concentrations. The continuous curves in Figs. 4.4, 4.5, and 4.6 are the fitting obtained from Eq. (4.3) and the symbols are the experimental data. The r^2 values of all the curves in Figs. 4.4, 4.5, and 4.6 are more than 0.99.

4.4.2 Desorption of Eosin

Effects of different concentrations of CPC, SDS, and AOT on the desorption of eosin are presented in Figs. 4.7, 4.8, 4.9, and 4.10. In this case, the concentrations of surfactants are greater than their CMC. CMC of AOT is 0.246 g/L (Rosen 1978). Like chrysoidine, desorption of eosin increases with surfactant concentration and operating time. For CPC (cationic surfactant), desorption is only 23%, whereas for SDS and AOT (anionic surfactant), desorption increases up to 70% and 63%, respectively, for feed surfactant concentration of 2000 mg/L. It is found from Fig. 4.8 that desorption may be up to 78% for SDS concentration of 5000 mg/L after 40 min of operation. Whereas, for 5000 mg/L of AOT, desorption is about 80% (Fig. 4.9). The comparison of the different techniques of desorption of dye is presented in Fig. 4.10. It is clear from the figure that at higher pH, desorption is only about 46% after 40 min. A comparison of the effectiveness of the three surfactants

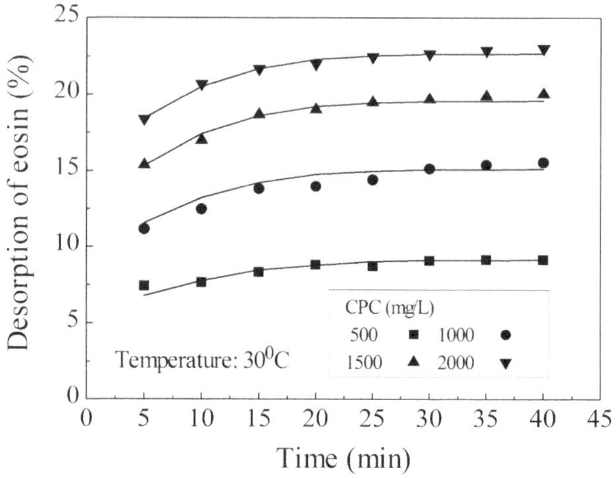

Fig. 4.7 Effect of CPC on desorption of eosin from spent charcoal. *Solid lines* are model fitted results and symbols are the experimental data

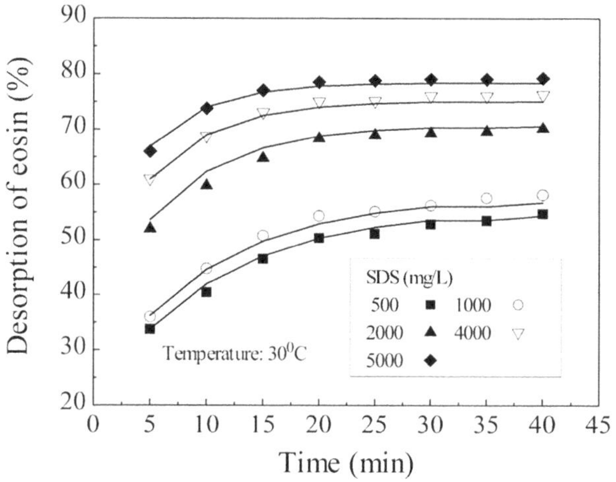

Fig. 4.8 Effect of SDS on desorption of eosin from spent charcoal. *Solid lines* are model fitted results and symbols are the experimental data

at the same concentration level is also shown in Fig. 4.10. After 40 min of operation, dye desorption is about 22% for CPC (2000) mg/L, whereas, for both SDS and AOT (2000 mg/L), it is about 70% and 63%, respectively. It is evident from figure that apart from the last point (40 min), desorption is more for SDS compared to AOT. Therefore, a suitable anionic surfactant may be a better agent for desorption of eosin dye from activated charcoal.

Better performance of anionic surfactant (SDS and AOT) compared to cationic one (CPC) needs further discussion. During adsorption, negatively charged eosin dye gets adsorbed on the surface or may be inside the pores of activated charcoal. Activated charcoal is negatively charged. The cationic surfactant, CPC micelles are positively charged, and they themselves get adsorbed on the surface of negatively charged activated carbon, although some dyes still get solubilized inside the

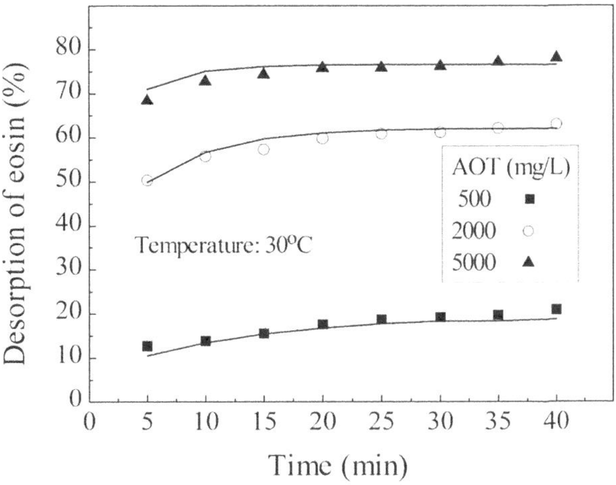

Fig. 4.9 Effect of AOT on desorption of eosin from spent charcoal. *Solid lines* are model fitted results and symbols are the experimental data

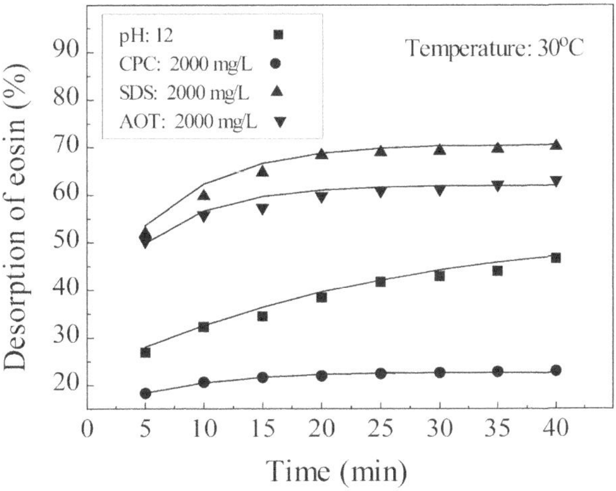

Fig. 4.10 Comparison of the performance of various surfactants for the desorption of eosin. Concentration of each surfactant is 2000 mg/L

hydrophobic core of CPC micelles. On the other hand, micelles of anionic surfactant (SDS) are negatively charged and unlikely to get adsorbed on charcoal. Therefore, with SDS or AOT micelles, the dye molecules desorbed from the surface of the adsorbent and get solubilized in the micelles of the anionic surfactants. This result in superior dye desorption characteristics substantially. For each surfactant feed concentration, the constants k and α are estimated from the corresponding experimental data. The value of α is found to be 12.23×10^{-2} (min^{-1}) for CPC, 4.14×10^{-2} (min^{-1}) for SDS, and 6.17×10^{-2} (min^{-1}) for AOT. The values for k vary in a range from 1.10 to 1.26 for CPC, 2.26 to 4.64 for SDS, and 1.25 to 4.27 for AOT corresponding to various feed surfactant concentrations. The continuous curves in Figs. 4.7, 4.8, and 4.9 are the fitting obtained from Eq. (4.3), and the

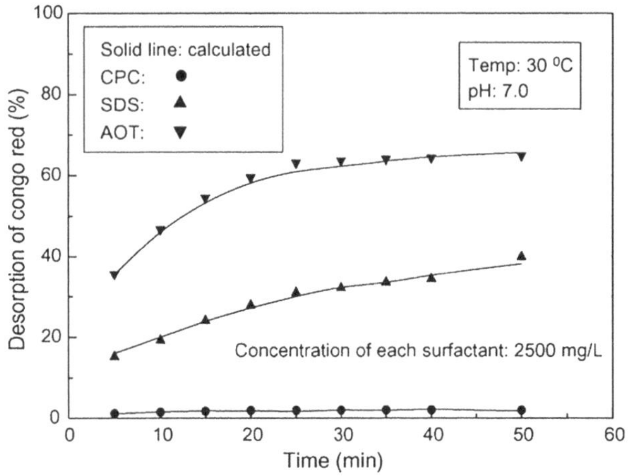

Fig. 4.11 Comparison of the performance of various surfactants for the desorption of Congo red. Concentration of each surfactant is 2500 mg/L (with surface loading 376 mg dye/g CAC) (Reproduced from Purkait et al. (2007) with permission from Elsevier)

symbols are the experimental data. The r^2 values of all the cases in Figs. 4.7, 4.8, and 4.9 are more than 0.99.

4.4.3 Desorption of Congo Red

Figure 4.11 shows the variations of the percentage desorption of Congo red using CPC, SDS, and AOT at 2500 mg/L (higher than the corresponding critical micellar concentrations (CMC)). CMC of CPC, SDS, and AOT are 350, 2307, and 1182 mg/L at 25 °C temperature in aqueous medium, respectively. In case of CPC (cationic surfactant), desorption is only about 2%, whereas for SDS and AOT (anionic surfactant), the desorption increases up to 40 and 65%, respectively, for a feed surfactant concentration of 2500 mg/L. Better performance of anionic surfactants (SDS and AOT) compared to cationic one (CPC) needs discussion. During adsorption, anionic Congo red (at pH 7.0) gets adsorbed on the surface or may be trapped inside the pores of activated carbon. During desorption, CPC micelles are positively charged and get adsorbed on the negatively charged surface of activated carbon ($pH_{ZPC} \approx 6.6$), although some dye molecules may still get solubilized inside the hydrophobic core of CPC micelles. Therefore, solubilization of dye in the CPC micelle is less. On the other hand, micelles of the anionic surfactants (SDS and AOT) are negatively charged and are unlikely to get adsorbed on the activated carbon. Due to strong hydrophobic attraction between micellar core and dye molecules, the adsorbed dye gets solubilized in the micelle. This result in better dye desorption characteristics using anionic surfactants. From Fig. 4.11, it is observed that performance of AOT is better than SDS.

The extent of dye solubilization in the micelles depends on the concentration of the micelles which depends on the CMC of surfactant. As stated earlier, AOT has a much lower CMC than SDS. Therefore, it is expected that the micellar concentration is more in case of AOT at a feed surfactant concentration of 2500 mg/L. Therefore, dye solubilization capacity of AOT micelles is more than those of SDS at same surfactant feed concentration. The desorption study is carried out using a number of surfactant concentrations in the range of 500–1000 mg/L. The kinetic data are then fitted to Eq. (4.3), and the parameters k and α, of these three surfactants, are estimated separately. The average value of α is found to be 4.34×10^{-2} (min^{-1}) for CPC, 5.23×10^{-2} (min^{-1}) for SDS, and 8.23×10^{-2} (min^{-1}) for AOT. The values for k vary from 1.35 to 1.66 for CPC, 2.36 to 3.44 for SDS, and 2.66 to 4.12 for AOT corresponding to various feed surfactant concentrations. The continuous curves in Fig. 4.11 represent the fit obtained from Eq. (4.3) and the symbols are the experimental data. The r^2 values in all the cases are more than 0.99.

References

Fang CS, Lai PMC (1996) Microwave regeneration of spent powder activated carbon. Chem Eng Comm 147:17–27

Kim JH, Ryu RK, Haam S, Lee CH, Kim WS (2001) Adsorption and steam regeneration of n-hexane, MEK and toluene on activated carbon. Sep Sci Technol 36:263–281

Matatov-Meytal YI, Sheintuch M (1997) Abatement of pollutants by adsorption and oxidative catalytic regeneration. Ind Eng Chem Res 36:4374–4380

Miguel GS, Lambert SD, Graham NJD (2001) The regeneration of field spent activated carbons. Water Res 35:740–748

Mishra VS, Mahajani VV, Joshi JB (1995) Wet air oxidation. Ind Eng Chem Res 34:2–48

Perrich JR (1981) Handbook of activated carbon adsorption for wastewater treatment. CRC Press, Boca Raton

Purkait MK, Gusain DS, Dasgupta S, De S (2004) Adsorption behavior of chrysoidine dye on activated charcoal and its regeneration characteristics by using different surfactants. Sep Sci Technol 39:2419–2440

Purkait MK, Maiti A, Dasgupta S, De S (2007) Removal of Congo red using activated carbon and its regeneration. J Hazard Mater 145:287–295

Rosen MJ (1978) Surfactants and interfacial phenomena. Wiley, New York

Ryu YK, Kim KL, Lee CH (2000) Adsorption and desorption of n-hexane, methyl ethyl ketone, and toluene on activated carbon fiber from supercritical carbon dioxide. Ind Eng Chem Res 39:2510–2518

Scholz M, Martin RJ (1998) Control of bio-regenerated granular activated carbon by spreadsheet modeling. J Chem Technol Biotechnol 71:253–261

Wedeking CA, Snoeyink VL, Larson RA, Ding J (1987) Wet air regeneration of PAC: comparison of carbons with different surface oxygen characteristics. Water Res 21:929–937

Zhang HP (2002) Regeneration of exhausted activated carbon by electrochemical method. Chem Eng J 85:81–85

Chapter 5
Nanofiltration of Dyes

Abstract Membrane separation process is perhaps one of the promising alternatives in water treatment over adsorption, being a greener mode of separation. Nanofiltration of dye-contaminated wastewater is a potential and viable membrane separation technology to effectively remove dyes from the water stream. Various nanofiltration systems (different flow configurations and types of membrane) are discussed, and the filtration efficiencies are compared. Besides, a simple transport phenomenon-based model is presented to explain the experimental observations and the underlying mechanisms.

Keywords Nanofiltration • Membrane separation • Diffusion • Textile effluent • Cross flow system

Membrane-based separation processes are gradually emerging as technically significant and commercially viable process in the treatment of wastewater. The major applications of the pressure-driven membrane processes – reverse osmosis (RO), nanofiltration (NF), ultrafiltration (UF), and microfiltration (MF) – include separation, purification, and fractionation of various process streams (Cheryan 1986). Nanofiltration is somewhere between reverse osmosis and ultrafiltration (Table 5.1), considering the driving force and the rejection of organic components in aqueous solutions (Rautenbach et al. 1996).

The less energy consumption in NF (compared to RO) makes its use more frequent for the treatment of various industrial effluents. The advantages of NF over other conventional separation processes are (i) generally less energy intensive, (ii) operable under ambient temperature, (iii) no phase change, and (iv) usually causes no damage to the species under processing. Considering these advantages, the process can play a major role in replacing many of the conventional separation processes. Both NF and RO are widely used in the textile industries. These processes have the potential to remove the dyestuff and allow the reuse of auxiliary chemicals for dyeing.

The textile dyeing industry demands large quantities of water, and hence, the different steps of dyeing processes result in large amounts of wastewater streams. Membrane filtration is used for recycling processes in textile industry for a long time. Reverse osmosis was widely used to treat the textile dyehouse effluent

S. Mondal et al., *Advances in Dye Removal Technologies*, Green Chemistry and Sustainable Technology, DOI 10.1007/978-981-10-6293-3_5

Table 5.1 Pressure-driven membrane processes

	Transmembrane pressure (kPa)	Molecular weight cutoff (g/mol)
Ultrafiltration	200–1000	> 1000
Nanofiltration	1000–2000	> 200–1000
Reverse osmosis	2000–4000	< 200

(El-Nasher 1980; Brandon et al. 1981; Treffry-Goatley et al. 1983; Porter and Goodman 1984; Buckley et al. 1985).

Brandon et al. (1981) and Porter and Goodman (1984) used dynamically formed zirconium oxide-polyacrylate membrane in tubular form to treat textile dyehouse effluent and observed more than 95% of dye recovery. El Nasher (1980) studied economic feasibility of reverse osmosis to treat dye-containing effluent. Goatley et al. (1983) designed a pilot plant with two stages consisting of RO membranes. These RO membranes were applied in desalination of brackish water (first stage) and seawater (second stage), respectively. The retentate was recycled in the second stage in order to obtain higher water recoveries. Nowak et al. (1986) studied cross flow ultrafiltration of an actual textile dye bath in a tubular membrane and reported up to 97% dye removal (of molecular weight greater than 780). Schrig and Widmer (1992) performed the nanofiltration of a mixture of dye salt and sodium chloride in a spiral wound module. Desizing wastewaters from the bleaching and dyeing industry were treated by NF in the work of Chen et al. (1997). Permeate flux and COD retention were investigated in relation to transmembrane pressure drop, temperature, and feed solution concentration. Yazhen et al. (1999) used nanofiltration for the treatment of textile dye plant effluent. The process was proved to be potential in treating the synergistic as well as industrial dye solutions. Yu et al. (2001) applied nanofiltration for the desalting and concentrating the aqueous dye at a dye-producing plant. The process was continuous and not labor intensive and produced a high-purity product of consistent quality. Dhale and Mahajani (2000) reported a combined process of nanofiltration and wet oxidation to treat dye bath waste stream. About 99% of color rejection was reported in nanofiltration, and the retentate was treated by wet oxidation process. In the work of Van der Bruggen et al. (2001), different samples from the wastewater treatment sequence of a textile factory were filtrated with different nanofiltration membranes. The study showed nanofiltration to be a realistic method in the textile industry for direct treatment of dye baths as well as treatment of the activated sludge effluent. In this process, the flux decline due to adsorption of organic material on the membrane decreased the membrane capacity by up to 73%. However, the process water flux reached a stable value in all experiments. In a separate study by Marcucci et al. (2001), the possibility of reusing the textile wastewater using pilotscale membrane-based separation processes has been analyzed. The pretreatment step for the nanofiltration or reverse osmosis process involves a sandbed or microfiltration.

Wide applications of NF in the treatment of textile wastewater were reported in recent works (Voigt et al. 2001; Frank et al. 2002; Akbari et al. 2002; Koyuncu 2002). Voigt et al. (2001) used new TiO_2-NF membranes in order to treat only the colored parts of the wastewater and to recycle the hot permeate as process water in the textile finishing process. The pilot plant was tested with 30 types of different colored wastewater over a period of 6 weeks to detect optimal parameters for the filtration with a focus on high flux and low operating costs. In another work, a two-step NF was applied to remove color for the recycling of an industrial wastewater (Frank et al. 2002). This process was able to recover 90% of the waste stream and 99.8% reduction of color content. In the work of Akbari et al. (2002), a nanofiltration membrane was developed by UV photografting. The grafted membranes were evaluated for the removal of five different dyes with an aim to reuse water in the process house. Dye retention was higher than 97%. In another study, nanofiltration was used to recycle the reactive dye bath effluents (Koyuncu 2002). Three reactive dyes and NaCl were used in the experiments to prepare the synthetic dye and salt mixtures. Effects of feed concentration, pressure and cross flow velocity on the permeate flux, and color removal were investigated.

5.1 Theoretical Description of Membrane Filtration of Dyes

An unsteady state mass transfer model is developed for single component (Chakraborty et al. 2004a) as well as the two-component (Pastagia et al. 2003a) system.

5.1.1 Single Component System

In the batch cell, the dye solution is pressurized over the membrane surface to facilitate the water to pass through the membrane. Considering an overall component balance in the cell results,

$$V\frac{dC_{\mathrm{b}}}{dt} = v_{\mathrm{w}}(t)A\left[C_0 - C_{\mathrm{p}}\right] \tag{5.1}$$

where V is the solution volume, C_{b} is the concentration in the batch cell, A is the surface area of the membrane, C_0 is the initial concentration, C_{p} is the permeate concentration, and v_{w} is the permeate flux. The initial condition for Eq. (5.1) at $t = 0$ and $C_{\mathrm{b}} = C_0$. On nondimensionalization, Eq. (5.1) becomes,

$$\frac{dC_{\mathrm{b}}^*}{d\tau} = Pe_{\mathrm{w}}(\tau) \quad \gamma \quad \left[1 - C_{\mathrm{p}}^*\right] \tag{5.2}$$

where τ is the nondimensional time and Pe_w the nondimensional permeate flux. The dimensionless quantities are defined as,

$C_b^* = C_b/C_0$; $\tau = tD/R^2$; $Pe_w(\tau) = v_w(\tau)\,R/D$; $C_p^* = C_p/C_0$; and $\gamma = AR/V$. The nondimensionalized initial condition of Eq. (5.2) is

$$\text{at}\,\tau = 0, C_b{}^* = 1 \tag{5.3}$$

The mass transport through the porous membrane is represented by the Darcy's law as,

$$v_w = L_p(\Delta P - \Delta\pi) \tag{5.4}$$

$$\text{where}\ \Delta\pi = (\pi_m - \pi_p) \tag{5.5}$$

For a dilute dye solution, osmotic pressure can be expressed using Van't Hoff's relationship,

$$\pi = (R_g T/M_w)C \tag{5.6}$$

where R_g is the universal gas constant, M_w is the molecular weight, and T is the temperature in Kelvin. For crystal violet of molecular weight (M_w) 408 Da, the osmotic pressure can be expressed as,

$$\pi = 6072\ C = a.C \tag{5.7}$$

The above value of a is obtained by taking the temperature to be equal to 298 K.

The membrane-solute system is characterized by a parameter real retention (R_r), which is constant for a particular membrane-solute combination (Opong and Zydney 1991) that is defined as,

$$R_r = 1 - C_p/C_m \tag{5.8}$$

where C_p and C_m are the permeate and membrane surface concentration of the solute, respectively. Using Eq. (5.8), Eq. (5.2) can be expressed in terms of $C_m{}^*$ as,

$$\frac{dC_b^*}{d\tau} = \gamma Pe_w(\tau)\left[1 - C_m^*(1 - R_r)\right] \tag{5.9}$$

where C_m^* is the dimensionless membrane surface concentration (C_m/C_0).

Using Eq. (5.4) and (5.7), the permeate flux can be obtained from Eq. (5.4) in its nondimensional form as,

$$Pe_w(\tau) = \beta\left[1 - \alpha C_m^*\right] \tag{5.10}$$

where $\beta = L_p\,(\Delta P)R/D$ and $\alpha = aC_0R_r/\Delta P$.

In order to solve Eq. (5.9), the membrane surface concentration (C_m) has to be estimated. This is done by taking a component balance in the thin concentration boundary layer above the membrane surface as,

$$\frac{\partial C}{\partial t} - v_{\rm w}\frac{\partial C}{\partial y} = D\frac{\partial^2 C}{\partial y^2} \tag{5.11}$$

In the nondimensional form, the above equation can be expressed as,

$$\frac{\partial C^*}{\partial \tau} - Pe_{\rm w}\frac{\partial C^*}{\partial y^*} = \frac{\partial^2 C^*}{\partial y^{*2}} \tag{5.12}$$

where C^* is the dimensionless concentration (C/C_0), y^* is the dimensionless distance (y/R) from the membrane surface, and R is the radius of the batch cell. The initial and boundary conditions for Eq. (5.11) in nondimensional terms are expressed as,

$$C^* = 1 \text{ at } \tau = 0 \tag{5.13}$$

$$C^* = C_{\rm b}{}^* \text{at } y^* = \delta^* \tag{5.14}$$

$$\frac{\partial C^*}{\partial y^*} + Pe_{\rm w}R_{\rm r}C^* = 0 \text{ at y}^* = 0 \tag{5.15}$$

where δ^* is the dimensionless concentration boundary layer thickness (δ/R). It has to be noted that Eq. (5.12) is coupled with C_b^* and Pe_w and therefore has to be solved in conjugation with Eqs. (5.2) and (5.10). Considering an integral method of solution, the following parabolic concentration profile is assumed within the concentration boundary layer,

$$C^* = a_1 + a_2\left(\frac{y^*}{\delta^*}\right) + a_3\left(\frac{y^*}{\delta^*}\right)^2 \tag{5.16}$$

where a_1, a_2, and a_3 are constants. The above concentration profile must satisfy the following conditions,

$$\text{at } y^* = 0,\ \ C^* = C^*{}_{\rm m}(\tau) \tag{5.17}$$

$$\text{at } y^* = \delta^*,\ \ \frac{\partial C^*}{\partial y^*} = 0 \tag{5.18}$$

$$\text{at } y^* = \delta^*,\ \ C^* = C^*{}_{\rm b}(\tau) \tag{5.19}$$

Using conditions in Eq. (5.17), (5.18), and (5.19), the concentration profile in Eq. (5.16) can be expressed as,

$$C^* = C_{\rm m}^* - \left(C_{\rm m}^* - C_{\rm b}^*\right)\left[2\left(\frac{y^*}{\delta^*}\right) - \left(\frac{y^*}{\delta^*}\right)^2\right] \tag{5.20}$$

Substituting $\left.\frac{\partial C^*}{\partial y^*}\right|_{y*=0}$ in Eq. (5.15), a relationship between δ^*, C^*_{b}, and C^*_{m} can be obtained,

$$\delta^* = \frac{2}{\beta R_{\mathrm{r}}}\left(1 - \frac{C^*_{\mathrm{b}}}{C^*_{\mathrm{m}}}\right)\frac{1}{1 - \alpha C^*_{\mathrm{m}}} \tag{5.21}$$

Considering a derivative with respect to τ results into the following equation,

$$\frac{d\delta^*}{d\tau} = f_1\left(C^*_{\mathrm{m}}\right)\frac{dC^*_{\mathrm{b}}}{d\tau} + f_2\left(C^*_{\mathrm{m}} - C^*_{\mathrm{b}}\right)\frac{dC^*_{\mathrm{m}}}{d\tau} \tag{5.22}$$

where $f_1 = -\frac{2}{\beta R_r}\left\{\frac{1}{C^*_m\left(1 - \alpha C^*_m\right)}\right\}$ and $f_2 = -\frac{2}{\beta R_r}\left\{\frac{C^*_b + \alpha C^{*2}_m - 2\alpha C^*_m C^*_b}{C^{*2}_m\left(1 - \alpha C^*_m\right)^2}\right\}$.

The expressions of $\frac{\partial C^*}{\partial y^*}$, $\frac{\partial^2 C^*}{\partial y^{2*}}$, and $\frac{\partial C^*}{\partial \tau}$ for substitution in Eq. (5.12) are calculated from Eq. (5.20) and are presented,

$$\frac{\partial C^*}{\partial \tau} = \frac{dC^*_{\mathrm{m}}}{d\tau} + \frac{2y^*}{\delta^{*2}}\left(1 - \frac{y^*}{\delta^*}\right)\left(C^*_{\mathrm{m}} - C^*_{\mathrm{b}}\right)\frac{d\delta^*}{d\tau} - \frac{y^*}{\delta^*}\left(2 - \frac{y^*}{\delta^*}\right)\left(\frac{dC^*_{\mathrm{m}}}{d\tau} - \frac{dC^*_{\mathrm{b}}}{d\tau}\right) \tag{5.23}$$

$$\frac{\partial C^*}{\partial y^*} = -\left(C^*_{\mathrm{m}} - C^*_{\mathrm{b}}\right)\left(\frac{2}{\delta^*} - \frac{2y^*}{\delta^{*2}}\right) \tag{5.24}$$

$$\text{and} \quad \frac{\partial^2 C^*}{\partial y^{*2}} = \left(C^*_{\mathrm{m}} - C^*_{\mathrm{b}}\right)\frac{2}{\delta^{*2}} \tag{5.25}$$

Substituting these derivatives in Eq. (5.12) and considering a zeroth moment integral across the boundary layer thickness (from 0 to δ^*) and replacing $\frac{d\delta^*}{d\tau}$ by Eq. (5.22), the following simplified expression is obtained,

$$\begin{aligned}&\frac{dC^*_{\mathrm{m}}}{d\tau}\left\{\frac{f_2\left(C^*_{\mathrm{m}} - C^*_{\mathrm{b}}\right)}{3} + \frac{\delta^*}{3}\right\} + \frac{dC^*_{\mathrm{b}}}{d\tau}\left\{\frac{f_1\left(C^*_{\mathrm{m}} - C^*_{\mathrm{b}}\right)}{3} + \frac{2\delta^*}{3}\right\}\\&+ Pe_{\mathrm{w}}\left(C^*_{\mathrm{m}} - C^*_{\mathrm{b}}\right) = \frac{2\left(C^*_{\mathrm{m}} - C^*_{\mathrm{b}}\right)}{\delta^*}\end{aligned} \tag{5.26}$$

Equation (5.26) can be simplified by substituting the expression of $\frac{dC^*_{\mathrm{b}}}{d\tau}$ from Eq. (5.10) and Pe_{w} from Eq. (5.10), and an implicit expression describing the transient nature of $C_{\mathrm{m}}*$ $\left(\text{ODE of } \frac{dC^*_{\mathrm{m}}}{d\tau}\right)$ is obtained.

5.1.2 Solution Strategy

Equation (5.26) along with Eqs. (5.2), (5.10), and (5.21) is solved using fourth-order Runge-Kutta method. The initial condition for Eq. (5.26) is $C_m^* = 1$ at $\tau = 0$, but this condition makes Eq. (5.32) indeterminate. Therefore, an asymptotic solution of Eq. (5.26) is sought at the limit, $\tau \to 0$.

As $\tau \to 0$, $C_b^* \to 1$, $C_m^* \to 1$, $\delta^* \to 0$, $f_1 \to f_{10}$, $f_2 \to f_{20}$ and $Pe_w \to Pe_{w0}$ where $f_{10} = -\frac{2}{\beta R_r(1-\alpha)}$; $f_{20} = \frac{2}{\beta R_r(1-\alpha)}$ and $Pe_{w0} = \beta(1-\alpha)$.

Assuming $(C_m^* - C_b^*) = Z$, Eq. (5.26) becomes,

$$\frac{dC_m^*}{d\tau} = \frac{2(Z+C_b^*)^2 R_r^2 - 2Z(Z+C_b^*)R_r - \frac{8}{3}\left(\frac{dC_b^*}{d\tau}\right)\frac{Z}{Pe_w^2} - 2\frac{f_1 Z}{3}\frac{(Z+C_b^*)}{Pe_w}R_r\left(\frac{dC_b^*}{d\tau}\right)}{\frac{4}{3}\frac{Z}{Pe_w^2} + \frac{2f_2}{3}\frac{Z(C_b^*+Z)}{Pe_w}R_r} \tag{5.27}$$

As $\tau \to 0$, $Z \to 0$, applying L'Hospital's rule in Eq. (5.27), the following expression is derived,

$$\left|\frac{dC_m^*}{d\tau}\right|_{\tau\to 0} = \frac{4R_r^2 - 2R_r - \frac{8}{3}\left(\frac{dC_b^*}{d\tau}\right)_{\tau\to 0}\frac{1}{Pe_{w0}^2} - \frac{2f_{10}}{3Pe_{w0}}R_r\left(\frac{dC_b^*}{d\tau}\right)_{\tau\to 0}}{\frac{4}{3Pe_{w0}^2} + \frac{2f_{20}}{3Pe_{w0}}R_r} \tag{5.28}$$

which can be simplified as,

$$\left|\frac{dC^*{}_m}{d\tau}\right|_{\tau\to 0} = \frac{3}{4}R_r\beta^2(1-\alpha)^2(2R_r - 1) - \frac{1}{2}\beta(1-\alpha)\gamma \tag{5.29}$$

The solution of the above equation is,

$$C_m^* = 1 + \left[\frac{3}{4}R_r\beta^2(1-\alpha)^2(2R_r - 1) - \frac{1}{2}\beta(1-\alpha)\gamma\right]\tau \tag{5.30}$$

Since Eq. (5.30) is an asymptotic solution of C_m^* in the limit $\tau \to 0$, a small value of τ is selected to evaluate C_m^*. This selection of τ in the present case is 10^{-8}. It is observed that using this value of τ, C_m^* value is close to 1.0. A still smaller value of τ brings C_m^* more close to 1.0, but the calculation time in the optimization program (discussed later) increases considerably. Therefore, the initial value of τ is chosen as 10^{-8}, and the corresponding value of C_m^* is evaluated from Eq. (5.30). This served as the initial condition to Eq. (5.27). A large value of τ and the corresponding value of C_m^* as initial condition result in numerical instability. The numerical solutions provide the variations of C_m^* and Pe_w with time and, consequently, the profiles of the permeate concentration, C_p (from Eq. 5.8), and the permeate flux, v_w, as a function of time. However, the parameters, diffusivity (D) and real retention

(R_r), should be known, a priori. In this work, D and R_r values are optimized with respect to the experimental profiles of v_w and C_p for all experiments (single component crystal violet), using "direct complex search algorithm" (the optimization subroutine BCPOL from IMSL Math library of FORTRAN 90). The BCPOL library routine uses a function comparison to determine the minimum point of the n variable function. A set of $2n$ points is introduced in the function, and at every iteration, a new point is calculated which replaces a bad point x_j having the maximum function value among the $2n$ points. The new point is computed from the recursive relation:

$$x_k = c + r_c(c - x_j)$$

where $c = \frac{1}{2n-1}\sum_{i \neq j} x_i$ and r_c ($r_c > 0$) is the reflection coefficient. When x_k is a best point, which means for $f(x_k)$ [$f(x_i)$ for $i = 1, \ldots, 2n$], an expansion point is computed $x_e = c + b(x_k - c)$, where $b(b > 1)$. If the new point is a worst point, then the complex would be contracted to obtain a better new point. If the contraction step is unsuccessful, the complex is shrunk by moving the vertices halfway toward the existing best point. Whenever a new point generated is beyond the bounds, it will be set on the bound. This procedure is carried out until $f(x_{best})$- $f(x_{worst})$ is less than some given tolerance.

In the present system of aqueous crystal violet (single component), the optimized value of D obtained is 1.82×10^{-10} m^2/s and that for R_r is 0.99. The value of D obtained from the Wilke-Chang equation (Treybal 1981) is found to be in the same order, i.e., 3.72×10^{-10} m^2/s.

5.1.3 Two-Component System

The theoretical model developed for the two-component system of the reactive dyes is the extension of the single component model described in the earlier section. Henceforth, Cibacron Black B (molecular weight 924.5 Da) is referred as component 1, and Cibacron Red RB (molecular weight 855.5 Da) is referred as component 2. The material balance for component 1 in the unstirred cell is written as,

$$V\frac{dC_{b1}}{dt} = v_w(t)\ A\left[C_{01} - C_{p1}\right] \tag{5.31}$$

where A is the surface area of the membrane. The initial condition of Eq. (5.31) is $C_{b1} = C_{01}$ at $t = 0$. Similarly, the material balance for component 2 in the cell leads to,

$$V\frac{dC_{b2}}{dt} = v_w(t)\ A\left[C_{02} - C_{p2}\right] \tag{5.32}$$

Subject to the initial condition, $C_{b2} = C_{02}$ at $t = 0$, Eq. (5.31) is nondimensionalized as,

$$\frac{dC^*_{b1}}{d\tau} = Pe_w(\tau)\gamma\left[1 - C^*_{p1}\right] \tag{5.33}$$

and the dimensionless form of the initial condition is at $\tau = 0$ and $C^*_{b1} = 1.0$, where $C^*_{b1} = C_{b1}/C_{01}$ and $\tau = tD_0/R^2$. D_0 is a reference diffusivity, which is used to define dimensionless time irrespective of D_1 and D_2 (diffusivity of component 1 and 2, respectively). The value of D_0 selected for the present system is 2.5×10^{-10} m^2/s, $Pe_w = v_w R/D_0$, $C^*_{p1} = C_{p1}/C_{01}$, and $\gamma = AR/V$. Similarly, the nondimensional form of Eq. (5.32) is,

$$\frac{dC^*_{b2}}{d\tau} = Pe_w(\tau)\ \gamma\ \left[1 - C^*_{p2}\right] \tag{5.34}$$

where $C^*_{b2} = C_{b2}/C_{02}$ and $C^*_{p2} = C_{p2}/C_{02}$. The dimensionless initial condition is at $\tau = 0$ and $C^*_{b2} = 1$. For a multicomponent dilute solution, Darcy's equation (Eq. 5.4) can be rewritten as,

$$v_w = L_p\left(\Delta P - \sum_{i=1}^{i=n}\Delta\pi_i\right) \tag{5.35}$$

where n is number of components and $\sum_{i=1}^{i=n}\Delta\pi_i = \Delta\pi_1 + \Delta\pi_2 + \Delta\pi_3 + \ldots\ldots\ldots\ldots + \Delta\pi_n$. For a dilute solution, the osmotic pressure can be estimated using Van't Hoff's equation (Eqs. 5.6 and 5.7). Using Eqs. (5.7) and (5.8), the following expression for the mixture osmotic pressure can be obtained:

$$\begin{aligned}\sum_{i=1}^{i=n}\Delta\pi_i &= \Delta\pi_1 + \Delta\pi_2 = a_1\left(C_{m1} - C_{p1}\right) + a_2\left(C_{m2} - C_{p2}\right)\\ &= a_1 C_{01} C^*_{m1} R_{r1} + a_2 C_{02} C^*_{m2} R_{r2}\end{aligned} \tag{5.36}$$

Using Eq. (5.36) for component 1, Eq. (5.33) can be written as,

$$\frac{dC^*_{\mathrm{b1}}}{d\tau} = \gamma Pe_{\mathrm{w}}(\tau)\left[1 - C^*_{\mathrm{m1}}(1 - R_{\mathrm{r1}})\right] \tag{5.37}$$

Similarly for component 2, $$\frac{dC^*_{\mathrm{b2}}}{d\tau} = \gamma Pe_{\mathrm{w}}(\tau)\left[1 - C^*_{\mathrm{m2}}(1 - R_{\mathrm{r2}})\right] \tag{5.38}$$

Using Eqs. (5.8) and (5.36), Eq. (5.35) can be written in terms of nondimensional membrane surface concentrations,

$$Pe_{\mathrm{w}}(\tau) = \beta\left\lfloor 1 - \alpha_1 C^*_{\mathrm{m1}} - \alpha_2 C^*_{\mathrm{m2}}\right\rfloor \tag{5.39}$$

where $\beta = L_{\mathrm{p}}(\Delta P)R/D_0$, $\alpha_1 = a_1 C_{01} R_{\mathrm{r1}}/\Delta P$, and $\alpha_2 = a_2 C_{02} R_{\mathrm{r2}}/\Delta P$. The material balance for component 1 in the thin concentration boundary layer near the membrane surface is given

$$\text{by}\frac{\partial C_1}{\partial t} - v_{\mathrm{w}}\frac{\partial C_1}{\partial y} = D_1\frac{\partial^2 C_1}{\partial y^2} \tag{5.40}$$

In the nondimensional form, the above equation can be expressed as,

$$\frac{\partial C^*_1}{\partial \tau} - Pe_{\mathrm{w}}\frac{\partial C^*_1}{\partial y^*} = \frac{D_1}{D_0}\frac{\partial^2 C^*_1}{\partial y^{*2}} \tag{5.41}$$

Equation (5.40) has the following initial and boundary conditions:

$$C_1 = C_{01} \text{ at } t = 0 \tag{5.42}$$

$$\text{At the edge of the boundary layer}, C_1 = C_{\mathrm{b1}} \text{ at } y = \delta_1 \tag{5.43}$$

At the membrane surface, the net convective flux is equal to the net diffusive flux, which leads to the interface boundary condition as,

$$v_{\mathrm{w}}\left(C_1 - C_{\mathrm{p1}}\right) + D_1\frac{\partial C_1}{\partial y} = 0 \qquad \text{at} \quad y = 0 \tag{5.44}$$

Equations (5.42–5.44) can be nondimensionalized as,

$$C_1{}^* = 1 \text{ at } \tau = 0 \tag{5.45}$$

$$C_1{}^* = C_{\mathrm{b1}}{}^* \text{at } y^* = \delta_1{}^* \tag{5.46}$$

$$\frac{D_1}{D_0}\frac{\partial C^*_1}{\partial y^*} + Pe_{\mathrm{w}} R_{\mathrm{r2}} C^*_1 = 0 \quad at \quad y^* = 0 \tag{5.47}$$

Similarly the Eqs. (5.40), (5.41), (5.42), (5.43), (5.44), (5.45), (5.46), and (5.47) are developed for component 2. Considering an integral method of approach, the concentration profile for component 1 and component 2 is evaluated as described by

Eqs. (5.16), (5.17), (5.18), (5.19), and (5.20). Now, substituting $\frac{\partial C_1^*}{\partial y^*}$ from the parabolic concentration profile in Eq. (5.47), a relationship between δ_1^*, C^*_{b1}, and C^*_{m1} is obtained,

$$\delta_1{}^* = \left(\frac{2}{\beta Rr_1}\right)\left(\frac{D_1}{D_0}\right)\left(1-\frac{C^*_{\mathrm{b1}}}{C^*_{\mathrm{m1}}}\right)\left(\frac{1}{1-\alpha_1 C^*_{\mathrm{m1}}-\alpha_2 C^*_{\mathrm{m2}}}\right) \tag{5.48}$$

Similarly, for component 2, $$\delta_2{}^* = \left(\frac{2}{\beta R_{\mathrm{r2}}}\right)\left(\frac{D_2}{D_0}\right)\left(1-\frac{C^*_{\mathrm{b2}}}{C^*_{\mathrm{m2}}}\right)\left(\frac{1}{1-\alpha_1 C^*_{\mathrm{m1}}-\alpha_2 C^*_{\mathrm{m2}}}\right) \tag{5.49}$$

Derivative of Eq. (5.48) with respect to τ can be expressed as,

$$\frac{d\delta_1^*}{d\tau} = f_{1a}\left(C^*_{\mathrm{m1}}, C^*_{\mathrm{m2}}\right)\frac{dC^*_{\mathrm{b1}}}{d\tau} + f_{2\mathrm{a}}\left(C^*_{\mathrm{m1}}, C^*_{\mathrm{m2}}, C^*_{\mathrm{b1}}\right)\frac{dC^*_{\mathrm{m1}}}{d\tau} + f_{3\mathrm{a}}\left(C^*_{\mathrm{m1}}, C^*_{\mathrm{m2}}, C^*_{\mathrm{b1}}\right)\frac{dC^*_{\mathrm{m2}}}{d\tau} \tag{5.50}$$

where

$$f_{1\mathrm{a}} = -\left(\frac{2}{\beta R_{\mathrm{r1}}}\right)\left(\frac{D_1}{D_0}\right)\left(\frac{1}{C^*_{\mathrm{m1}}\left(1-\alpha_1 C^*_{\mathrm{m1}}-\alpha_2 C^*_{\mathrm{m2}}\right)}\right)$$

$$f_{2\mathrm{a}} = \left(\frac{2}{\beta R_{\mathrm{r1}}}\right)\left(\frac{D_1}{D_0}\right)\left(\frac{C^*_{\mathrm{m1}}\alpha_1\left(C^*_{\mathrm{m1}}-C^*_{\mathrm{m2}}\right)+C^*_{\mathrm{b1}}\left(1-\alpha_1 C^*_{\mathrm{m1}}-\alpha_2 C^*_{\mathrm{m2}}\right)}{C^{*2}_{\mathrm{m1}}\left(1-\alpha_1 C^*_{\mathrm{m1}}-\alpha_2 C^*_{\mathrm{m2}}\right)^2}\right)$$

$$f_{3\mathrm{a}} = \left(\frac{2}{\beta R_{\mathrm{r1}}}\right)\left(\frac{D_1}{D_0}\right)\left(\frac{\alpha_2\left(C^*_{\mathrm{m1}}-C^*_{\mathrm{b1}}\right)}{C^{*2}_{\mathrm{m1}}\left(1-\alpha_1 C^*_{\mathrm{m1}}-\alpha_2 C^*_{\mathrm{m2}}\right)^2}\right)$$

Similarly, the derivative of the Eq. (5.49) with respect to τ can be expressed, after algebraic simplification,

$$\frac{d\delta_2{}^*}{d\tau} = f_{4\mathrm{a}}\left(C^*_{\mathrm{m1}}, C^*_{\mathrm{m2}}\right)\frac{dC_{\mathrm{b2}}{}^*}{d\tau} + f_5\left(C^*_{\mathrm{m1}}, C^*_{\mathrm{m2}}, C^*_{\mathrm{b2}}\right)\frac{dC^*_{\mathrm{m1}}}{d\tau} + f_6\left(C^*_{\mathrm{m1}}, C^*_{\mathrm{m2}}, C^*_{\mathrm{b2}}\right)\frac{dC^*_{\mathrm{m2}}}{d\tau} \tag{5.51}$$

where, $$f_{4\mathrm{a}} = -\left(\frac{2}{\beta R_{\mathrm{r2}}}\right)\left(\frac{D_2}{D_0}\right)\left(\frac{1}{C^*_{\mathrm{m2}}\left(1-\alpha_1 C^*_{\mathrm{m1}}-\alpha_2 C^*_{\mathrm{m2}}\right)}\right);$$

$$f_5 = \left(\frac{2}{\beta R_{\mathrm{r1}}}\right)\left(\frac{D_2}{D_0}\right)\left(\frac{\alpha_1\left(C^*_{\mathrm{m2}}-C^*_{\mathrm{b2}}\right)}{C^*_{\mathrm{m2}}\left(1-\alpha_1 C^*_{\mathrm{m1}}-\alpha_2 C^*_{\mathrm{m2}}\right)^2}\right);$$

$$f_6 = \left(\frac{2}{\beta R_{\mathrm{r2}}}\right)\left(\frac{D_2}{D_0}\right)\left(\frac{C^*_{\mathrm{m21}}\alpha_2\left(C^*_{\mathrm{m2}}-C^*_{\mathrm{b2}}\right)+C^*_{\mathrm{b2}}\left(1-\alpha_1 C^*_{\mathrm{m1}}-\alpha_2 C^*_{\mathrm{m2}}\right)}{C^{2*}_{\mathrm{m2}}\left(1-\alpha_1 C^*_{\mathrm{m1}}-\alpha_2 C^*_{\mathrm{m2}}\right)^2}\right)$$

From Eq. (5.20), the derivative of C^*_1 with respect to τ and y^* is inserted in Eq. (5.41) (and also taking help of Eqs. 5.48 and 5.50) and integrated across the boundary layer thickness $\delta_1{}^*$. The simplified equation is,

$$\frac{dC^*_{\mathrm{m1}}}{d\tau} = \frac{f_8 f_9 - f_{3a} f_{10}}{f_7 f_9 - f_{3a} f_5} \tag{5.52}$$

Subject to the initial condition, at $\tau = 0$, $C^*_{\mathrm{m1}} = 1.0$.

Similarly the governing equation for C^*_{m2} is obtained,

$$\frac{dC^*_{\mathrm{m2}}}{d\tau} = \frac{f_7 f_{10} - f_5 f_8}{f_7 f_9 - f_{3a} f_5} \tag{5.53}$$

Subject to the initial condition, at $\tau = 0$, $C^*_{m2} = 1.0$.

The functions in the right-hand side of Eqs. (5.52) and (5.53) are,

$$f_7 = f_{2\mathrm{a}} + \left(\frac{2}{Pe_{\mathrm{w}} C^*_{\mathrm{m1}} R_{\mathrm{r1}}}\right)\left(\frac{D_1}{D_0}\right) \tag{5.54}$$

$$f_8 = \frac{3Pe_{\mathrm{w}} C^*_{\mathrm{m1}} R_{\mathrm{r1}}}{C^*_{\mathrm{m1}} - C^*_{\mathrm{b1}}} - 3Pe_{\mathrm{w}} - f_{1\mathrm{a}} \frac{dC^*_{\mathrm{b1}}}{d\tau} - \left(\frac{1}{Pe_{\mathrm{w}} C^*_{\mathrm{m1}} R_{\mathrm{r1}}}\right)\left(\frac{D_1}{D_0}\right)\left(\frac{dC^*_{\mathrm{b1}}}{d\tau}\right) \tag{5.55}$$

$$f_9 = f_6 + \left(\frac{2}{Pe_{\mathrm{w}} C^*_{\mathrm{m2}} R_{\mathrm{r2}}}\right)\left(\frac{D_2}{D_0}\right) \tag{5.56}$$

$$f_{10} = \frac{3Pe_{\mathrm{w}} C^*_{\mathrm{m2}} R_{\mathrm{r2}}}{C^*_{\mathrm{m2}} - C^*_{\mathrm{b2}}} - 3Pe_w - f_{4\mathrm{a}} \frac{dC^*_{\mathrm{b2}}}{d\tau} - \left(\frac{4}{Pe_{\mathrm{w}} C^*_{\mathrm{m2}} R_{\mathrm{r2}}}\right)\left(\frac{D_2}{D_0}\right)\left(\frac{dC^*_{\mathrm{b2}}}{d\tau}\right) \tag{5.57}$$

5.1.4 Solution Strategy

Equations (5.52), (5.53), (5.37), and (5.38) along with Eqs. (5.48) and (5.49) are solved simultaneously using IVPRK (FORTRAN 90, IMSL Math library) routine. The result gives C^*_{p1}, C^*_{p2}, and Pe_{w} (consequently C_{p1}, C_{p2}, and v_{w}) as a function of time simultaneously, provided the parameters like diffusivities (D_1 and D_2) and real retentions (R_{r1} and R_{r2}) are known, a priori. It can be seen from Eqs. (5.55) and (5.57) that f_8 and f_{10} are indeterminate because C^*_{m1}, C^*_{b1}, C^*_{m2}, and C^*_{b2} tend to be 1.0 as τ approaches zero. So an asymptotic solution is found to initiate the integration. The asymptotic solutions of Eqs (5.52) and (5.53) can be computed as follows: assuming $C^*_{\mathrm{m1}} - C^*_{\mathrm{b1}} = Z_1$ and $C^*_{\mathrm{m2}} - C^*_{\mathrm{b2}} = Z_2$, as $\tau \to 0$, each of $C_{m1}{}^*$, $C_{m2}{}^*$, $C_{b1}{}^*$, $C_{b2}{}^* \to 1.0$, and Z_1, $Z_2 \to 0$. Now applying L'Hospital's rule to Eq. (5.52),

$$\left|\frac{dC^*_{\mathrm{m1}}}{d\tau}\right|_{\tau\to 0} = \frac{d|f_8 f_9 - f_{3a} f_{10}|_{\tau\to 0}}{d|f_7 f_9 - f_{3a} f_5|_{\tau\to 0}} \tag{5.58}$$

Differentiating the numerator of the right-hand side of Eq. (5.58) with respect to Z_I and Z_2 and taking the limiting value (at $\tau \to 0$), the following expression is obtained:

$$P_{11} = \left[3Pe^0_{\mathrm{w}} f^0_6 R_{\mathrm{r1}} - 3Pe^0_{\mathrm{w}} f^0_6 - f^0_{1a} f^0_6 (Pe^0_{\mathrm{w}}\gamma) - \left(\frac{4f^0_6}{Pe^0_{\mathrm{w}} R_{r1}}\right)\left(\frac{D_1}{D_0}\right)(Pe^0_{\mathrm{w}}\gamma)\right] + \left[\frac{6R_{\mathrm{r1}}}{R_{\mathrm{r2}}}\left(\frac{D_2}{D_0}\right) - \frac{6}{R_{\mathrm{r2}}}\left(\frac{D_2}{D_0}\right)\right] - \frac{2f^0_{1a}}{Pe^0_{\mathrm{w}} R_{\mathrm{r2}}}\left(\frac{D_2}{D_0}\right)(Pe^0_{\mathrm{w}}\gamma) - \left(\frac{8}{Pe^{02}_{w} R_{\mathrm{r1}} R_{\mathrm{r2}}}\right)\left(\frac{D_1 D_2}{D_0}\right)(Pe^0_{\mathrm{w}}\gamma) \tag{5.59}$$

Similarly, differentiating the denominator of the right-hand side of Eq. (5.58) with respect to Z_1 and Z_2 and taking the limiting value (at $\tau \to 0$), the following expression is obtained:

$$P_{22} = f^0_{2a} f^0_6 + \frac{2f^0_6}{Pe^0_{\mathrm{w}} R_{\mathrm{r1}}}\left(\frac{D_1}{D_0}\right) + \frac{2f^0_{2a}}{Pe^0_{\mathrm{w}} R_{\mathrm{r2}}}\left(\frac{D_2}{D_0}\right) + \frac{4}{Pe^{02}_{\mathrm{w}} R_{\mathrm{r1}} R_{\mathrm{r2}}}\left(\frac{D_1 D_2}{D^2_0}\right) \tag{5.60}$$

where $Pe_{\mathrm{w}}{}^0, f_{1a}{}^0, f_{2a}{}^0, f_4{}^0$, and $f_6{}^0$ are the value of $Pe_{\mathrm{w}}, f_{1a}, f_{2a}, f_4$, and f_6, respectively, at $\tau \to 0$, i.e., Z_1 and $Z_2 \to 0$.

$$Pe^0_{\mathrm{w}} = \beta[1 - \alpha_1 - \alpha_2] \tag{5.61}$$

$$f^0_{1a} = -\left(\frac{2}{\beta R_{\mathrm{r1}}}\right)\left(\frac{D_1}{D_0}\right)\left(\frac{1}{1 - \alpha_1 - \alpha_2}\right) \tag{5.62}$$

$$f^0_{2a} = -\left(\frac{2}{\beta R_{\mathrm{r1}}}\right)\left(\frac{D_1}{D_0}\right)\left(\frac{1}{1 - \alpha_1 - \alpha_2}\right) \tag{5.63}$$

$$f^0_4 = -\left(\frac{2}{\beta R_{\mathrm{r2}}}\right)\left(\frac{D_2}{D_0}\right)\left(\frac{1}{1 - \alpha_1 - \alpha_2}\right) \tag{5.64}$$

$$f^0_6 = -\left(\frac{2}{\beta R_{\mathrm{r2}}}\right)\left(\frac{D_2}{D_0}\right)\left(\frac{1}{1 - \alpha_1 - \alpha_2}\right) \tag{5.65}$$

Hence, Eq. (5.58) can be expressed as,

$$\left|\frac{dC^*_{\mathrm{m1}}}{d\tau}\right|_{\tau\to 0} = \frac{P_{11}}{P_{22}} \tag{5.66}$$

The solution of the Eq. (5.66) after rearrangement is expressed as follows:

$$C^*_{\mathrm{m1}} = 1 + \left(\frac{f_{11}}{f_{12}}\right)\tau \tag{5.67}$$

where $f_{11} = 3Pe_{\mathrm{w}}^{0}R_{\mathrm{r1}}f_{6}^{0} - 3Pe_{\mathrm{w}}^{0}f_{6}^{0} - Pe_{\mathrm{w}}^{0}\gamma f_{6}^{0} \times \left[f_{1\mathrm{a}}^{0} + \frac{4D_1}{D_0Pe_{\mathrm{w}}^{0}R_{\mathrm{r1}}}\right] + \frac{6D_2}{D_0R_{\mathrm{r2}}}(R_{r1} - 1) - \frac{2D_2\gamma}{D_0R_{\mathrm{r2}}}\left[f_{1\mathrm{a}}^{0} + \frac{4D_1}{D_0Pe^{0}{}_{\mathrm{w}}R_{\mathrm{r1}}}\right]$ and $f_{12} = f_{2\mathrm{a}}^{0}f_{6}^{0} + \frac{2D_2f_{2\mathrm{a}}^{0}}{D_0Pe_{\mathrm{w}}^{0}R_{\mathrm{r2}}} + \frac{2D_1f_{6}^{0}}{D_0Pe_{\mathrm{w}}^{0}R_{\mathrm{r1}}} + \frac{4D_1D_2}{D_0^2Pe_{\mathrm{w}}^{0^2}R_{\mathrm{r1}}R_{\mathrm{r2}}}$. The asymptotic solution of Eq. (5.53) can be found in the similar way like that of Eq. (5.58). Applying L'Hospital's rule to Eq. (5.53),

$$\left|\frac{dC_{\mathrm{m2}}^{*}}{d\tau}\right|_{\tau\to 0} = \frac{d|f_7f_{10} - f_5f_8|_{\tau\to 0}}{d|f_7f_9 - f_{3\mathrm{a}}f_5|_{\tau\to 0}} \tag{5.68}$$

The solution of Eq. (5.68) after simplification is represented as,

$$C_{\mathrm{m2}}^{*} = 1 + \left(\frac{f_{13}}{f_{14}}\right)\tau \tag{5.69}$$

where $f_{13} = 3Pe_{\mathrm{w}}^{0}R_{\mathrm{r2}}f_{2}^{0} - 3Pe_{\mathrm{w}}^{0}f_{2}^{0} - Pe_{\mathrm{w}}^{0}\gamma f_{2}^{0} \times \left[f_{4}^{0} + \frac{4D_2}{D_0Pe_{\mathrm{w}}^{0}R_{\mathrm{r2}}}\right] + \frac{6D_1}{D_0R_{\mathrm{r1}}}(R_{r2} - 1) - \frac{2D_1\gamma}{D_0R_{\mathrm{r1}}}\left[f_{4}^{0} + \frac{4D_2}{D_0Pe^{0}{}_{\mathrm{w}}R_{\mathrm{r2}}}\right]$ and $f_{14} = f_{12}$. The parameters D_1, D_2, R_{r1}, and R_{r2} values are optimized using the same optimization technique discussed in the numerical solution of the single component system (using an optimization subroutine BCPOL/DBCPOL from IMSL Math library of FORTRAN 90) overall experimental conditions. Here the experimental flux (v_{w}) and the permeate concentration profiles of component 1 and 2 (C_{p1} and C_{p2}) are matched with the profiles estimated from the model, in order to estimate the parameter values.

The values of the diffusivities of reactive black dye and reactive red dye (presented in Table 5.2) are in the same order of magnitude as found from the Wilke-Chang equation (Treybal 1981). 95% confidence interval for each regressed parameter is also evaluated which gives a range of values that has a specified probability of containing the parameters being estimated (Constantinides and Mostoufi 1999). The optimized values of the parameters and their confidence intervals are presented in Table 5.2.

5.1.5 *Cross Flow System*

In the cross flow system, where the steady state is attained quickly (within a few minutes from start-up), stagnant film theory is used to explain the data. At steady state, the expression for estimation of the permeate flux based on film theory equation is given by,

Table 5.2 Model parameter values for the two-component system

Parameter	Parameter value with 95% confidence interval
R_{r1}	(0.9982 ± 0.00003)
R_{r2}	(0.9988 ± 0.00002)
D_1 (m^2/s)	$(4.8788 \pm 0.4784) \times 10^{-10}$
D_2 (m^2/s)	$(6.2835 \pm 0.5692) \times 10^{-10}$

$$v_{\mathrm{w}}^{s} = k \ln \left[\frac{C_{\mathrm{m}}^{s} - C_{\mathrm{p}}^{s}}{C_0 - C_{\mathrm{p}}^{s}} \right] \tag{5.70}$$

where the superscript "*s*" indicates the steady-state values of various quantities and k is the mass transfer coefficient. The permeate (C_p) and membrane surface concentration (C_m) are related by the definition of the real retention (Eq. 5.8). For the case of laminar flow regime, the mass transfer coefficient is calculated from the Leveque relationship defined by,

$$Sh = 1.86 \left(\mathrm{Re} Sc \frac{d_e}{L} \right)^{1/3} \tag{5.71}$$

where $\mathrm{Re} = \frac{\rho u_0 d_e}{\mu}$ and $Sc = \frac{\mu}{\rho D}$; u_0 is the cross-sectional average velocity in the channel; and ρ, μ, D are the solution density, viscosity, and the solute diffusivity, respectively.

5.2 Experiments in Unstirred Batch Cell

5.2.1 Single Component System

Figure 5.1 shows the variation of permeate flux with time for different feed concentrations of crystal violet at different pressures. It is evident from the figure that the permeate flux decreases with time of operation. This is due to the concentration polarization phenomenon. During the experimental run, the membrane surface concentration increases. This leads to an increase in the osmotic pressure near the membrane-solution interface and, thereby, a decrease in the available driving force (i.e., transmembrane pressure) resulting in a reduction in permeate flux. It may be observed from the figure that the decline in flux at the end of the operation is about 15% from its initial value for the feed concentration of 19.87 mg/l and 550 kPa pressure. At higher feed concentration, the solute buildup on the membrane surface is more, resulting in larger reduction in the permeate flux. On the other hand, for a fixed feed concentration of 19.87 mg/l, the increase in flux with operating pressures 415, 550 and 690 kPa is evident in the figure. It can also be

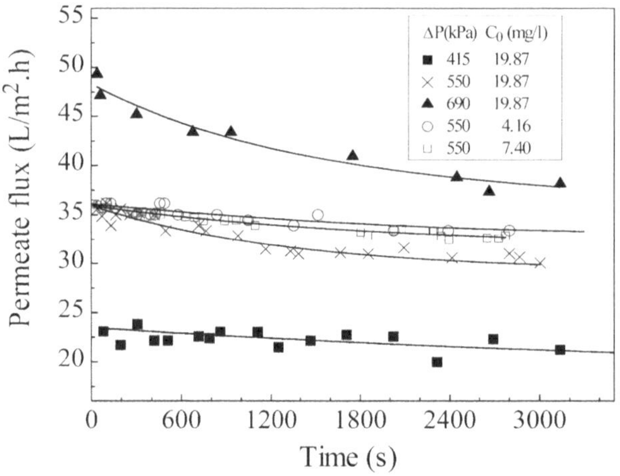

Fig. 5.1 Variation of permeate flux of crystal violet with time at different feed concentrations and operating pressure differences in the batch cell (the solid lines are guides for the reader's eyes) (Reproduced from Chakraborty et al. (2003b) with permission from Taylor & Francis Ltd)

observed that at a fixed pressure (550 kPa), permeate flux decreases with the feed concentration (hollow circle, square, and the cross symbols).

Figure 5.2 shows the variation in the observed retention ($R_0 = 1\text{-}C_p/C_0$) of crystal violet with pressure difference at a fixed feed concentration. It may be observed from the figure that at higher pressure, the permeate concentration increases leading to a decline in observed retention. At higher pressures, membrane surface concentration increases, thereby increasing the convection of the solute through the membrane. Thus, permeate concentration increases resulting in a reduction in observed retention. It may be noted that for a feed concentration of 17.6 mg/l, retention is about 67% for crystal violet at 415 kPa pressure at the end of operation. As pressure increases to 690 kPa, retention decreases to about 35% (at the end of the run). Therefore, a lower operating pressure results in a higher retention of crystal violet but at the cost of the flux.

Figure 5.3 shows the variation of observed retention of methylene blue with time at different feed concentrations at an operating pressure of 415 kPa. It is clear that the retention of methylene blue decreases with time for a fixed pressure and concentration. The retention is more for lower feed concentration. This trend is similar to the experimental results with crystal violet. Interestingly, it may be observed that at the end of operation, methylene blue retention $[(1-C_p / C_0)]$ is 20% (for 14.31 mg/l feed concentration) and 15% (for 23.5 mg/l), whereas for crystal violet, the maximum observed retention (for $C_0 = 17.6$ mg/l) is about 67% at the same pressure of 415 kPa (Fig. 5.2). Therefore, the present membrane retains more crystal violet compared to methylene blue.

As discussed in the theory section, the model equations (Eqs. 5.2, 5.10, 5.21, and 5.26) are solved to determine the model parameters, namely, D and R_r, by optimizing the experimental permeate flux and concentration profiles. These values of the parameters are then used to estimate the values of the permeate flux and concentration at different operating conditions.

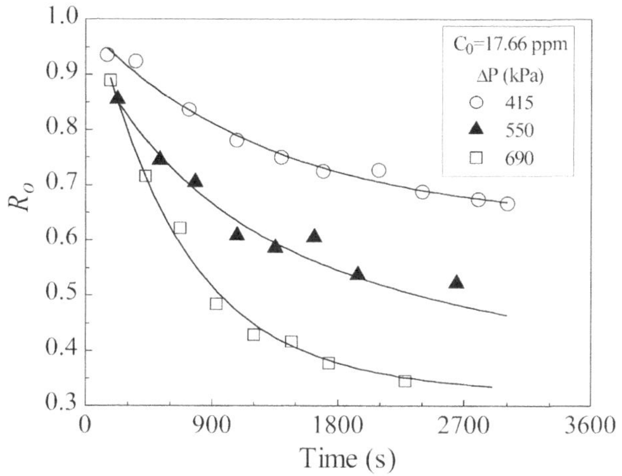

Fig. 5.2 Variation of observed retention of crystal violet with time at a fixed feed concentration (17.6 mg/l) in the batch cell (the solid lines are guides for the reader's eyes)

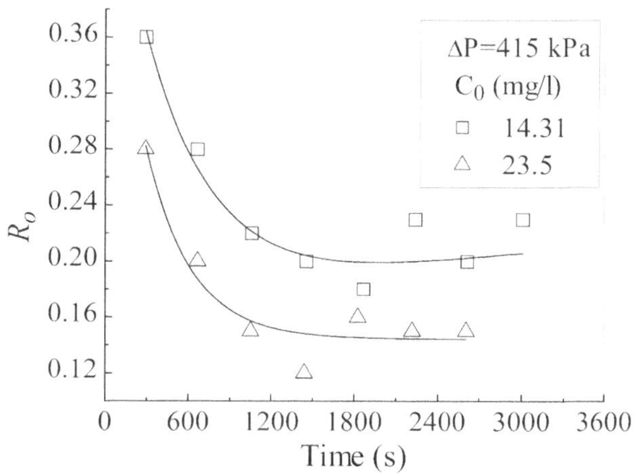

Fig. 5.3 Variation of observed retention of methylene blue with time at a fixed operating pressure of 415 kPa in the batch cell (the solid lines are guides for the reader's eyes) (Reproduced from Chakraborty et al. (2003b) with permission from Taylor & Francis Ltd)

Figure 5.4 shows the variations of permeate concentration of crystal violet with time at different feed concentrations for a transmembrane pressure of 690 kPa. The symbols are the experimental data, and the continuous lines are the model estimations. It is observed from the figure that initially the permeate concentration increases rapidly and gradually thereafter, for a fixed feed concentration. Concentration polarization leads to a buildup of the solute particles over the membrane surface, enhancing the solute permeation by convection through the membrane. Therefore, initially the permeate concentration increases rapidly. During extended operations, the flux due to backward diffusion from the membrane surface toward the bulk solution becomes competitive with the convective flux through the membrane, and an increase in permeate concentration becomes gradual. It may also be observed from the figure that the permeate concentration is higher for higher feed concentrations. At higher feed concentrations, the concentration polarization is

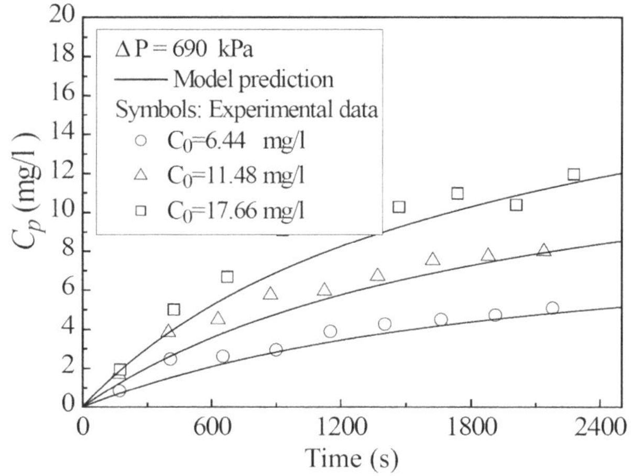

Fig. 5.4 Comparison between the experimental and estimated permeate concentration profiles in the batch cell at a fixed pressure

more severe leading to an enhanced convective flux through the membrane, and hence, permeate concentration increases.

Figure 5.5 shows the variation of permeate concentration of crystal violet with pressure difference at a fixed feed concentration. The symbols are the experimental data, and the continuous lines are the model estimations. It may be observed from the figure that with increase in operating pressure difference for a fixed feed concentration, the permeate concentration increases at any point of time of operation. Increase in the operating pressure leads to more severe polarization, and membrane surface concentration increases. Increased pressure also facilitates the convective flux through the membrane due to availability of enhanced driving force. As a result, the permeate concentration increases with pressure.

Figure 5.6 presents the profile of membrane surface concentration (C_m) in the batch cell for different feed concentrations at a fixed pressure. As observed from the figure, the membrane surface concentration increases with feed concentration at a particular time. The concentration polarization increases with feed concentration, resulting in higher osmotic pressure and, thereby, increasing the value of C_m. It is also evident from the figure that for a fixed feed concentration, C_m increases steeply with time and gradually thereafter. This is because of the fact that as C_m increases, the back diffusion from the membrane surface also increases (due to the increase in concentration gradient) and the difference between convective flux away from the membrane (to permeate) and diffusive flux from the membrane surface to the bulk of the solution decreases. Therefore, increase in C_m with time becomes gradual.

The concentration boundary layer profile for different feed concentrations is shown in Fig. 5.7. For a fixed feed concentration, boundary layer thickness increases steeply with time initially and gradually thereafter. Also, for higher feed concentrations, concentration boundary layer thickness is more at a fixed time. These trends are explained by the concentration polarization phenomenon as discussed in the preceding paragraph.

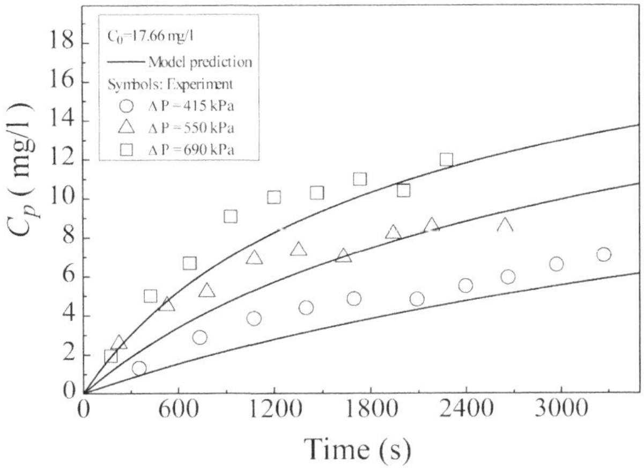

Fig. 5.5 Comparison between the experimental and estimated permeate concentration profiles in the batch cell at a fixed feed concentration (Reproduced from Chakraborty et al. (2004) with permission from Elsevier)

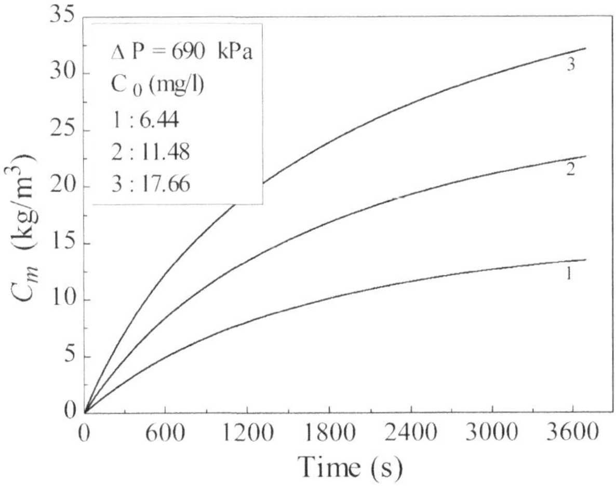

Fig. 5.6 Variation of membrane surface concentration profiles in the batch cell at a fixed pressure (Reproduced from Chakraborty et al. (2004) with permission from Elsevier)

The variation of permeate flux with time for different feed concentrations of crystal violet at a transmembrane pressure of 690 kPa is presented in Fig. 5.8. The symbols are the experimental data, and the continuous lines are the model estimations. It is evident from the figure that the permeate flux decreases with time of operation. As explained earlier, during the experimental run, the membrane surface concentration goes on increasing due to the concentration polarization phenomenon. This leads to an increase in osmotic pressure near the membrane-solution interface and, thereby, decreases in the available driving force (i.e., the transmembrane pressure). This leads to a decline in flux. It may be observed from the figure that decline in flux at the end of the operation is about 15% from its initial value for the feed concentration 17.66 mg/l.

Experiments were also carried out at 276, 415, 550, and 690 kPa, and the curves obtained in each case show the same trend. From Fig. 5.8, it may also be observed

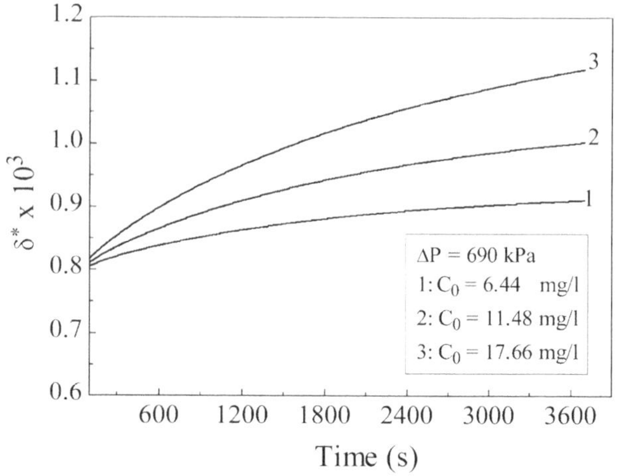

Fig. 5.7 Development of nondimensional concentration boundary layer with time at a fixed pressure in the batch cell (Reproduced from Chakraborty et al. (2004) with permission from Elsevier)

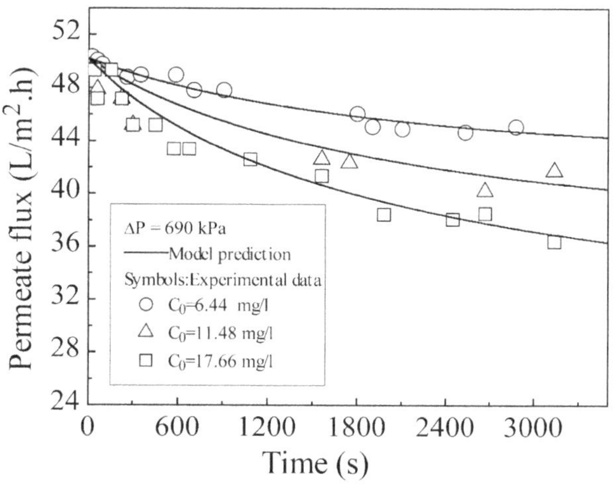

Fig. 5.8 Comparison between the experimental and estimated permeate flux profiles in the batch cell at a fixed pressure (Reproduced from Chakraborty et al. (2004) with permission from Elsevier)

that flux decline is severe for higher feed concentrations due to the increased effect of concentration polarization.

The variation of permeate flux values of crystal violet with transmembrane pressure at a fixed feed concentration (17.66 mg/l) is presented in Fig. 5.9. The symbols are for the experimental data, and the continuous lines are the model predictions. It is clear from the figure that the flux is higher for higher operating pressure. With increase in operating pressure (at a fixed feed concentration), the driving force across the membrane increases leading to an enhancement of permeate flux.

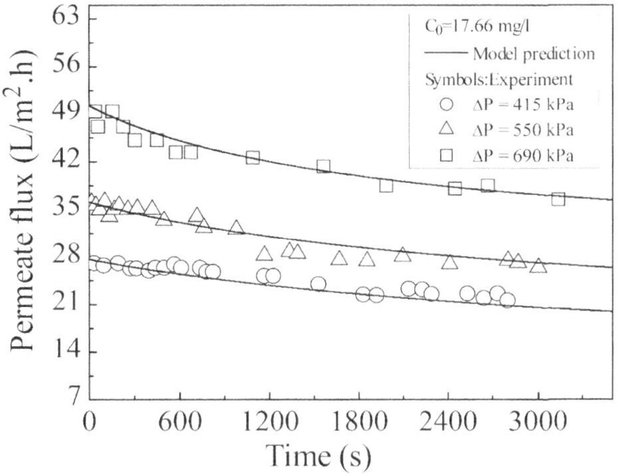

Fig. 5.9 Comparison between the experimental and estimated permeate flux profiles in the batch cell at different operating pressures (Reproduced from Chakraborty et al. (2004) with permission from Elsevier)

5.2.2 *Two-Component System*

The selective permeation rate of crystal violet and methylene blue through the membrane, as discussed earlier, necessitates the study of the fractionation of the dye mixture (Chakraborty et al. 2003c). The concentration level achieved after the fractionation in nanofiltration process is indeed very low, and an economic recovery for reuse may not be possible for the dyes selected herein. But the study is general in nature, and the fractionation concept probed herein ensures its applicability for a variety of operational setups involving costly dyes for which the economics of the reuse may be favorable.

The effect of pressure on dye fractionation in terms of observed retention is shown in Fig. 5.10, with feed concentration 10 mg/l each. At 415 kPa, average retention for methylene blue is 30% and that of crystal violet is 53%. At 276 kPa pressure, R_0 for crystal violet is 60% and that of methylene blue is 40%. These observations corroborate the fact that the observed retention of a species decreases with increase in pressure. It is apparent that the permeate stream is richer (compared to feed) in methylene blue.

The comparison between the experimental flux profiles at 415 kPa of single component and the mixture at the same overall feed concentration level is presented in Fig. 5.11. It is clear from the figure that in between two single component systems, the permeate flux for methylene blue is higher than that for crystal violet. The mixture of crystal violet and methylene blue (10 mg/l each) is also presented in the figure. The flux profile for the mixture lies in between the flux profiles of methylene blue and crystal violet. Since methylene blue has a molecular weight (373 Da) lower than the MWCO of the present membrane (400 Da), the flux value for methylene blue is expectedly higher than that for crystal violet (molecular weight: 408 Da).

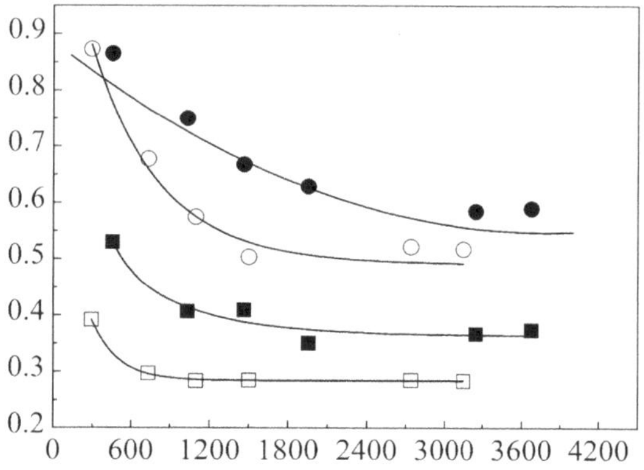

Fig. 5.10 Variation of observed retention of crystal violet and methylene blue with time from a mixture of 10 mg/l each in the batch cell (solid lines are guides for the reader's eyes) (Reproduced from Chakraborty et al. (2003) with permission from Taylor & Francis Ltd)

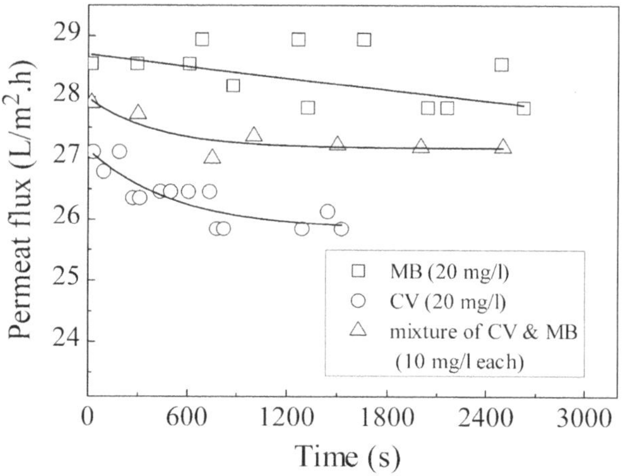

Fig. 5.11 Comparison between the experimental values of permeate flux at a pressure of 415 kPa in the batch cell (solid lines are guides for the reader's eyes) (Reproduced from Chakraborty et al. (2003b) with permission from Taylor & Francis Ltd)

It is observed during the experiments that the steady state in the cross flow (CF) system is attained within a few minutes from the start of the operation. Therefore, the steady-state values of the permeate flux and concentration are reported for the cross flow nanofiltration experiments. The effect of various operating conditions, viz., the applied pressure differences, feed concentrations, and the cross flow velocities, was observed on the permeate flux and the observed retention of crystal violet and methylene blue. The variations of observed retention of crystal violet with operating pressure at different cross flow velocities are presented in Fig. 5.12. It may be observed that the observed retention increases with cross flow velocities due to minimization of concentration polarization. Interestingly, it may be noted that even for $C_0 = 20$ mg/l, the observed retention for crystal violet was more than 90% for all operating pressures and cross flow velocities which is a marked improvement over the batch cell experiments (Fig. 5.2).

The variations of permeate flux of crystal violet with operating pressure for different combinations of feed concentration and cross flow velocities are presented in Fig. 5.13. As expected, the permeate flux increases with pressure for any

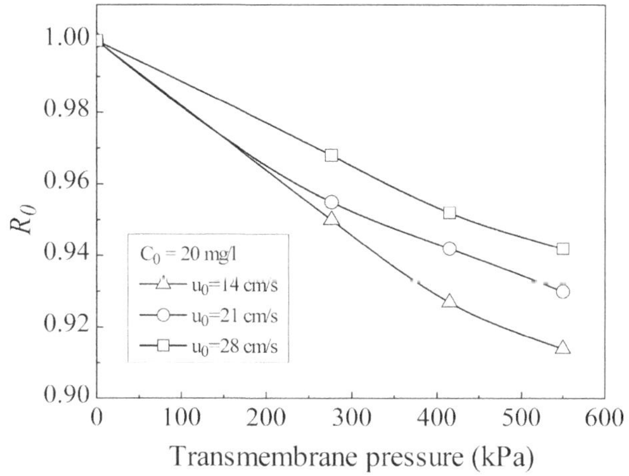

Fig. 5.12 Variation of observed retention of crystal violet with pressure difference at different bulk velocity in the cross flow cell (solid lines are guides for the reader's eyes) (Reproduced from Chakraborty et al. (2003b) with permission from Taylor & Francis Ltd)

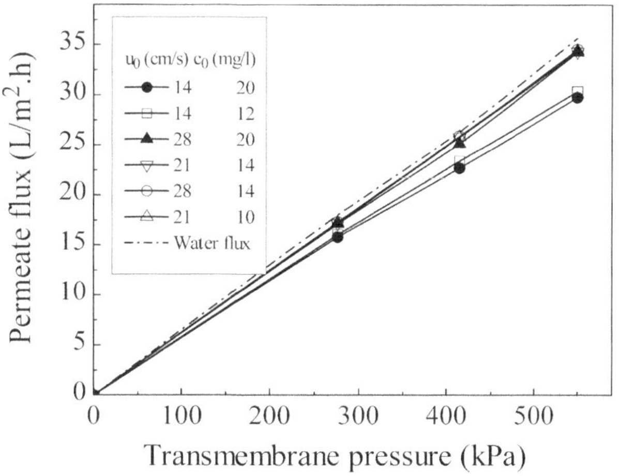

Fig. 5.13 Variation of permeate flux with pressure at different feed concentrations and bulk velocities (solid lines are guides for the reader's eyes) (Reproduced from Chakraborty et al. (2003b) with permission from Taylor & Francis Ltd)

combination of feed concentration and cross flow velocity. Also, the permeate flux follows the expected trends with feed concentration at a fixed operating pressure, i.e., flux decreases with feed concentration. At a fixed feed concentration, flux increases with cross flow velocity due to reduction in concentration polarization. However, the difference is marginal as observed from the figure because the surface concentration on the membrane cannot rise to a high value, due to the effect of the cross flow on the membrane surface.

The film theory is used to explain the results of cross flow nanofiltration experiments. The parameters D and R_r obtained from the profile optimization of the batch cell data ($D = 1.82 \times 10^{-10}$ m^2/s; $R_r = 0.99$) are used initially for the prediction of the permeate flux and permeate concentration in the cross flow nanofiltration. Figures 5.14 and 5.15 show the comparison between the experimental and the predicted values of the permeate flux and permeate concentration,

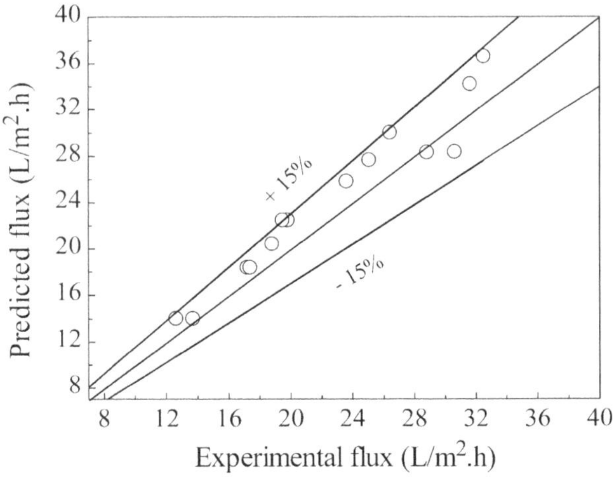

Fig. 5.14 Prediction of the steady-state permeate flux in CF NF from film theory using D and R_r obtained from the batch cell results (Reproduced from Chakraborty et al. (2004) with permission from Elsevier)

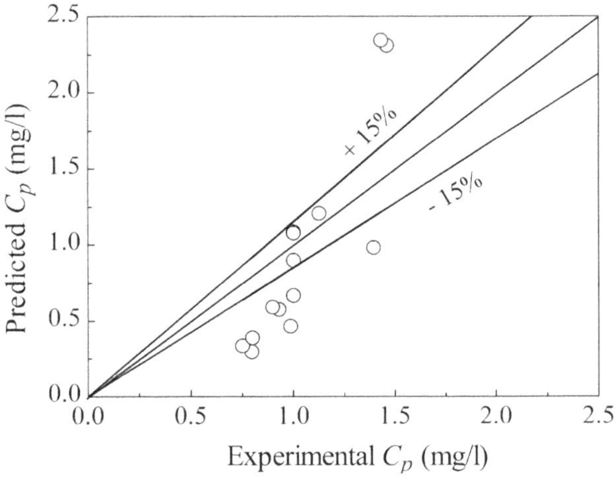

Fig. 5.15 Prediction of the steady-state permeate concentration in CF NF from film theory using D and R_r obtained from the batch cell results (Reproduced from Chakraborty et al. (2004) with permission from Elsevier)

respectively, at different conditions of feed concentrations and cross flow velocities. From the figures, it is observed that the predicted flux values are within $\pm 15\%$ to the experimental results, whereas most of the permeate concentrations lie beyond $\pm 15\%$.

In an alternative approach, the current set of cross flow data is used to obtain values of D and R_r, in situ, specifically for the cross flow runs. In this method, Eqs. (3.102) and (3.11) are solved, and the parameters (D and R_r) are estimated by optimizing the experimental values of the flux and permeate concentration of the cross flow system only (using the same optimizer BCPOL of IMSL library). The optimized value of D obtained is 8.51×10^{-10} m^2/s, and R_r is 0.96, and interestingly the value of diffusivity is in the same order of magnitude with the result obtained from the batch cell data and that estimated from the Wilke-Chang equation. Using these D and R_r, the permeate flux and concentration values are measured

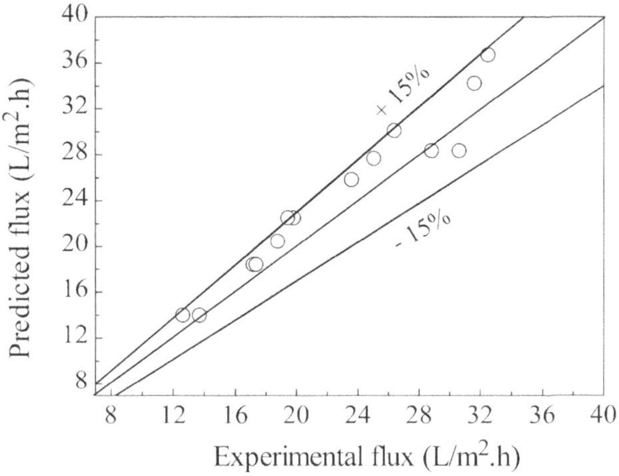

Fig. 5.16 Prediction of the steady-state permeate flux in CF NF from film theory using D and R_r obtained from the optimization of CF experimental data (Reproduced from Chakraborty et al. (2004) with permission from Elsevier)

using the film theory. The results are shown in Figs. 5.16 and 5.17. From Fig. 5.16, it can be observed that the predicted permeate fluxes are still well within the ±15% limit. From Fig. 5.17, it can be observed that most of the permeate concentrations (apart from three experimental data) lie within ±15%. The changes in the permeate flux values are marginal using the new D and R_r (see Figs. 5.16 and 5.17). But it is observed from Fig. 5.17 that there is a marked improvement in the prediction of permeate concentration, compared to that in Fig. 3.18. Since the same membrane and solute-solvent system are used in both the batch and cross flow experiments, D and R_r obtained from both the systems should be close. The prediction of the permeate flux is equally good in both the set of these parameters (Figs. 5.14 and 5.16). This indicates that the membrane surface concentration is close enough in both cases. Therefore, the deviation for the prediction of permeate concentration is due to the value of R_r (refer Eq.3.9). The R_r value obtained is 0.96 using cross flow data, and it is 0.99 using batch cell data. Because of this difference, the predictions for the permeate concentration presented in Fig. 5.15 have more deviation from the experimental data. Since, the optimized values of the parameters, namely, D and R_r are obtained using the experimental flux and permeate concentration data, measurement inaccuracies will certainly affect their estimations. Therefore, the deviations in Fig. 5.15 may be due to the inaccuracies in the measurement of the permeate concentration values at around 1 mg/l level. An instrumental error of about 7% (average) is observed during the measurement of concentrations of 1 mg/l level, shown in Table 3.2.

Figure 5.18 represents the variation of selectivity of methylene blue (between the two dyes) with cross flow velocity at two different operating pressure levels, namely, 276 and 415 kPa. The selectivity (S) is defined as $S = \frac{\left[{}^{c_p}/_{c_o}\right]_{methylene\ blue}}{\left[{}^{c_p}/_{c_o}\right]_{crystal\ violet}}$.

The figure reveals two trends. First, the selectivity of methylene blue increases with cross flow velocity at any pressure, and second, for a fixed cross flow velocity, selectivity increases with the operating pressure difference. The first trend can be

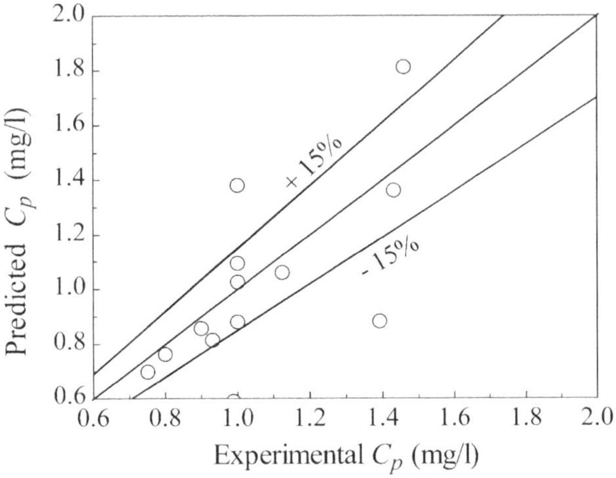

Fig. 5.17 Prediction of the steady-state permeate concentration in CF NF from film theory using D and R_r obtained from the optimization of CF experimental data (Reproduced from Chakraborty et al. (2004) with permission from Elsevier)

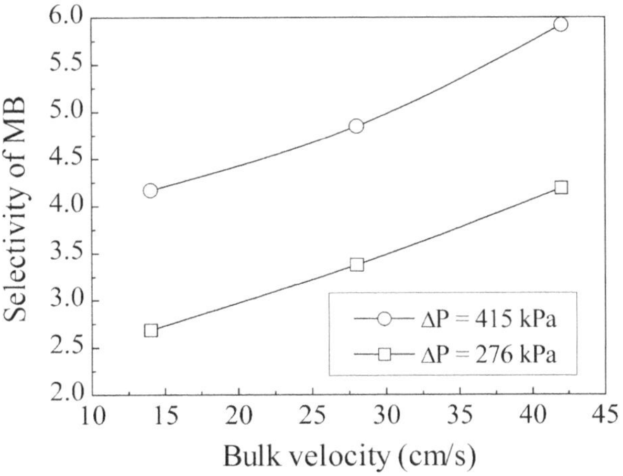

Fig. 5.18 Variation of selectivity of MB with bulk velocity in a mixture of CV and MB (16 mg/l each) in the cross flow cell (solid lines are guides for the reader's eyes) (Reproduced from Chakraborty et al. (2003b) with permission from Taylor & Francis Ltd)

explained by the fact that the membrane surface concentration decreases with cross flow velocity, and therefore, more of methylene blue permeates through the membrane, whereas the crystal violet is retained by the membrane due to size exclusion (crystal violet has molecular weight 408, whereas the membrane is a sharp 400 molecular weight cutoff). Thus, it can be said that the effect of increase in the cross flow velocity will be more on the permeate concentration of methylene blue compared to that of crystal violet. This would result in a relative increase of the permeation of methylene blue resulting in an increase in its selectivity. Similarly increase in pressure at a fixed cross flow velocity results in relative increase in the permeation of methylene blue compared to crystal violet and a subsequent increase in selectivity results. The selectivity can be as high as 6.0 for $\Delta P = 415$ kPa, for a cross flow velocity of 42 cm/s. Therefore, with a suitable selection of membrane,

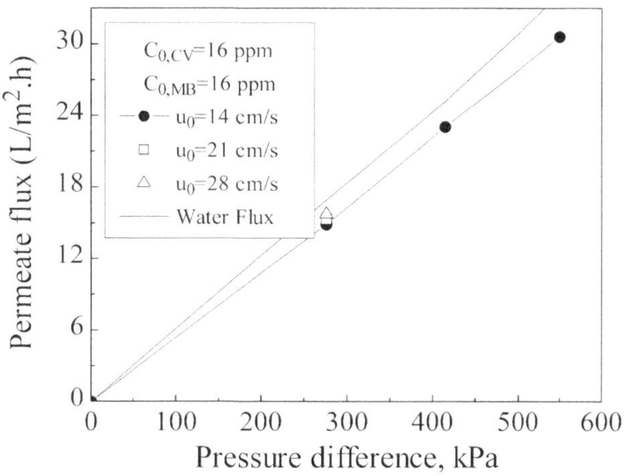

Fig. 5.19 Variation of permeate flux with pressure and cross flow velocity in the cross flow cell in a 50:50 mixture of CV and MB (Reproduced from Chakraborty et al. (2003b) with permission from Taylor & Francis Ltd)

operating pressure and cross flow velocity, fractionation of a specific dye from a dye mixture would be possible and may be used in practice.

The variation of permeate flux with pressure for a 50:50 mixture (16 mg/l each) at cross flow velocity of 14 cm/s is presented in Fig. 5.19. The effect of the cross flow velocity at a particular pressure of 276 kPa is presented in the figure. The figure indicates an increase in permeate flux with pressure as expected. It may be noted that the permeate flux increases with cross flow velocity due to reduction in concentration polarization. But the increase in flux values with cross flow velocity is not much, as the osmotic pressure of the present dye solution is not significant.

5.3 Nanofiltration of the Textile Effluent

Nanofiltration study is performed with the textile effluent containing the mixture of the reactive dyes (Chakraborty et al. 2003a). The concentrations of the dyes in the effluent are brought to a lower range prior to the NF study. In this study, the reactive black dye (Cibacron black B) is denoted by 1, and the reactive red dye (Cibacron red RB) is denoted by 2.

Figures 5.20 and 5.21 represent the variation of observed retention ($R_0 = 1 - C_p/C_0$) with time of dye 1 and 2, respectively, at different conditions of feed concentration and operating pressures. For dye 1 the expression for the observed retention is $R_{0,1} = 1 - C_{p,1}/C_{0,1}$ and that for dye 2 is $R_{0,2} = 1 - C_{p,2}/C_{0,2}$. It is observed from the figures that for a fixed feed concentration, initially retention decreases sharply which becomes gradual during extended operation. This can be explained by the phenomena of concentration polarization which is discussed earlier in the prediction of permeate concentration profile of single component crystal violet.

It must be pointed out here that in the case of a batch cell, the membrane surface concentration will keep on increasing. Thus, a steady state, as far as the permeate

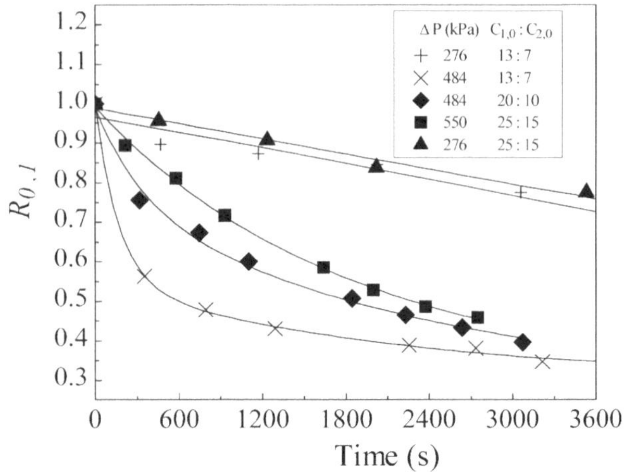

Fig. 5.20 Variation of observed retention of dye 1 in a mixture of dyes with time in the batch cell for different conditions of ΔP and C_0 (solid lines are guides for the reader's eyes) (Reproduced from Chakraborty et al. (2003) with permission from Elsevier)

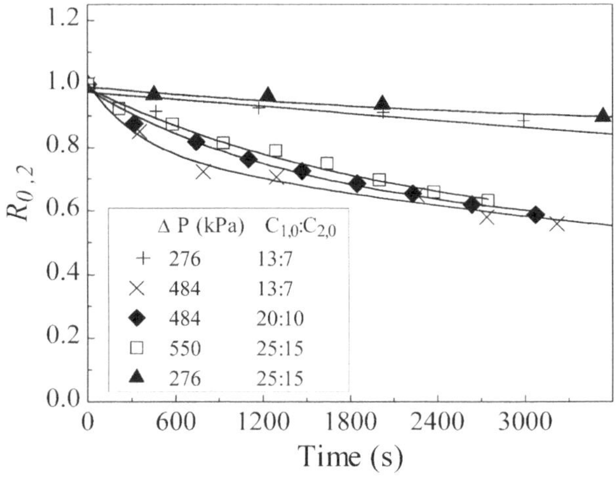

Fig. 5.21 Variation of observed retention of dye 2 in a mixture of dyes with time in the batch cell for different conditions of ΔP and C_0 (solid lines are guides for the reader's eyes) (Reproduced from Chakraborty et al. (2003) with permission from Elsevier)

concentrations are concerned, will never be attained. It may also be observed from the figure that at same feed concentration (25:15 mg/l), R_0 is lower for higher operating pressure (square and triangle symbols in the figure). At higher operating pressure, permeate concentration is more due to an enhanced convective flux through the membrane (because of higher driving force), and permeate concentration increases with a decrease in R_0.

At a fixed applied pressure difference, permeate concentration increases with increasing feed concentrations (C_0). For example, at a pressure of 276 kPa, the permeate concentrations of dye 1 after a time of 1200 sec. are 2.30 mg/l and 1.50 mg/l, respectively, for feed concentrations of 25:15 mg/l and 13:7 mg/l, respectively. But the change in retention values with feed concentrations is not apparent here ($R_{0,1}$ is 0.907 and 0.873, respectively). As ($R_0 = 1 - C_p/C_0$), hence, the effect of feed concentration (C_0) is normalized, therefore, in spite of a higher

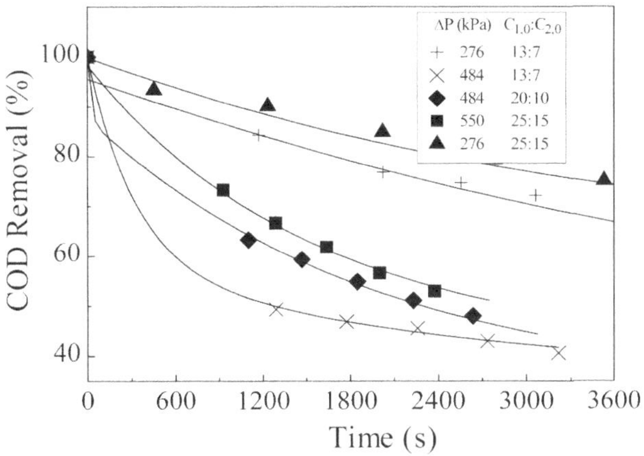

Fig. 5.22 Variation of percentage COD removal of effluent mixture with time in the batch cell for different conditions of ΔP and C_0 (solid lines are guides for the reader's eyes) (Reproduced from Chakraborty et al. (2003) with permission from Elsevier)

permeate concentration, retention value is sometimes higher for the higher feed concentration compared to the lower feed concentration.

Figure 5.22 shows the COD removal (%) with time for the same conditions of feed concentration and operating pressures as in Figs. 3.23 and 3.24. As COD is directly proportional to the concentration, the profile of the removal of COD shows the same trend as in Figs. 5.20 and 5.21. It is observed from the figure that at a feed concentration of 25:15 mg/l and operating pressure of 276 kPa, the maximum removal of COD can be up to 90% after a time of 1230 s from the starting of the experiment.

Figure 5.23 shows the variation of flux with time for the same conditions of feed concentrations and operating pressures as in Figs. 5.20 and 5.21. It is observed from the figure that the permeate flux decreases with time of operation. At 276 kPa, the permeate flux for a 25:15 mg/l feed concentrations is about 3.97×10^{-6} m^3/m^2.s after 5 min of operation. The flux value decreases to about 3.86×10^{-6} m^3/m^2.s after 55 min of operation. As the operating feed concentrations are of low range, the decrease in the permeate flux values is not substantial, and the curves appear almost flat. From the figure, it is observed that permeate flux is higher for higher operating pressure (due to the enhancement in the driving force) if feed concentration remains same. With increase in feed concentration, permeate flux decreases, which is the expected trend.

5.3.1 Estimations from Model for Two-Component System

Using the values of the four parameters presented in Table 5.2, the governing equations (Eqs. (3.45), (3.46), (3.74), (3.75), (3.84), and (3.85)) are solved as described in the theory section, and the system performance is simulated. Figures 5.24 and 5.25 show plots of permeate concentration of dye 1 and dye 2 with respect to time at operating pressure of 276 kPa and 550 kPa, respectively. The

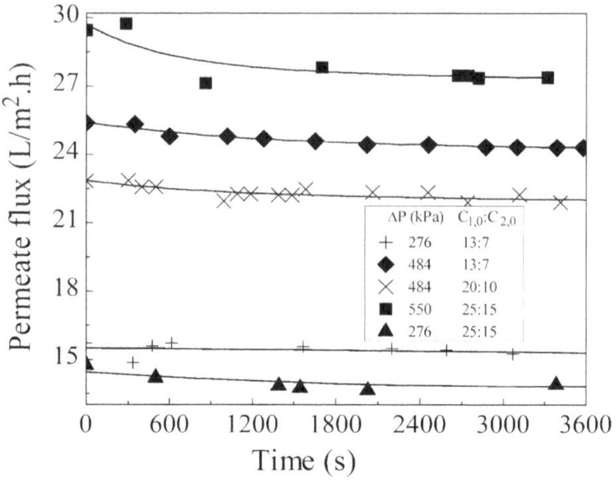

Fig. 5.23 Variation of permeate flux of effluent mixture with time in the batch cell for different conditions of ΔP and C_0 (solid lines are guides for the reader's eyes) (Reproduced from Chakraborty et al. (2003) with permission from Elsevier)

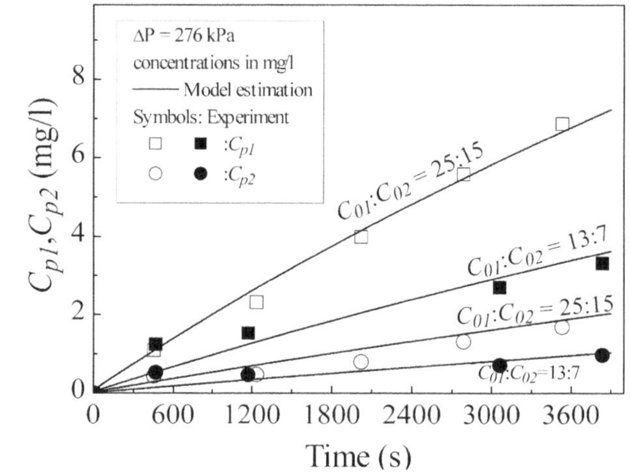

Fig. 5.24 Comparison between experimental and estimated value of concentration of dye 1 and 2, respectively, at a fixed pressure of 276 kPa for two different initial concentration in batch cell (Reproduced from Pastagia KMet al. (2003) with permission from Elsevier)

symbols represent the experimental data, and the continuous lines are the model predictions. It is clear that as feed concentration increases, permeate concentration also increases. At higher feed concentrations, the concentration polarization is more severe leading to an enhanced convective flux through the membrane and the permeate concentration increases.

Figure 5.26 shows the effect of concentration ratio of dye 1 and 2 in feed on the permeate concentration at constant pressure of 415 kPa. From the figure, it can be observed that the permeate concentration increases with time. This is due to the concentration polarization as explained earlier. With increase in feed concentration of one of the components, the permeate concentration of that component also increases. For example, in curves 2 and 4, the feed concentration of component 2 increases from 10 to 15 mg/l. This leads to an increase in osmotic pressure at the

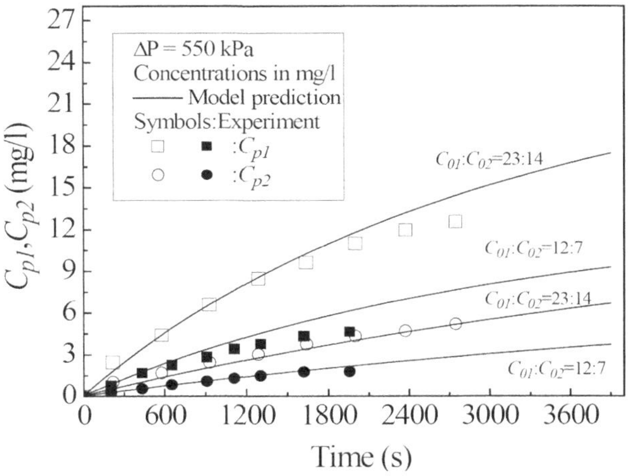

Fig. 5.25 Comparison between experimental and predicted values of permeate concentration of dye 1 and 2, respectively, at a fixed pressure of 550 kPa for two different initial concentrations in batch cell (Reproduced from Pastagia et al. (2003) with permission from Elsevier)

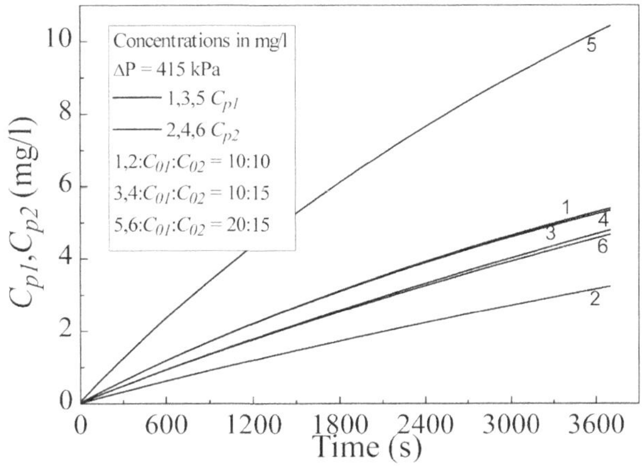

Fig. 5.26 Effect of ratio of initial feed concentration on permeate concentration for constant pressure difference (Reproduced from Pastagia et al. (2003) with permission from Elsevier)

membrane-solution interface leading to an enhancement of concentration polarization. Therefore, membrane surface concentration of this component increases leading to an increase in convective flux of this component through the membrane. Therefore, permeate concentration of component 2 increases as shown by curves 2 and 4 as the feed concentration of component 2 increases from 10 to 15 mg/l. Since feed concentration of component 1 is kept constant in these two cases, permeate concentration profiles of component 1 are almost overlapping as shown in curve 1 and 3. In curves 3 and 5, feed concentration of component 1 is increased from 10 to 20 mg/l, keeping the concentration of component 2 at the same level of 15 mg/l. This results in corresponding increase in permeate concentration of component 1 as evident from the curves 3 and 5.

Figure 5.27 shows the permeate concentration profiles of dye 1 and 2 estimated from the model, at various operating pressures keeping the initial feed concentration ratio constant (10:10 (mg/l)). It is clear from the figure that for a fixed operating

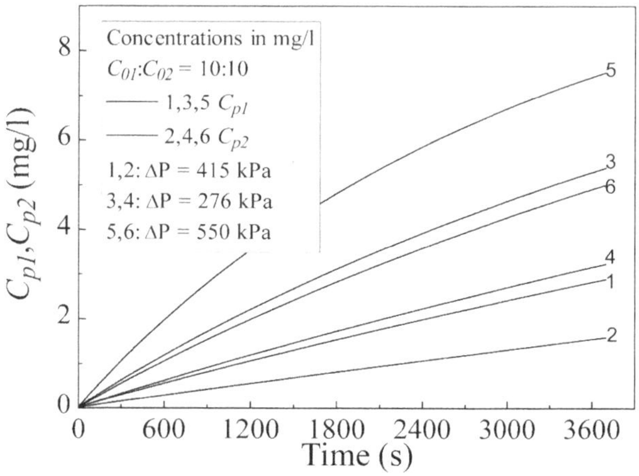

Fig. 5.27 Effect of operating pressure difference on permeate concentration for constant initial feed concentration (model estimated) (Reproduced from Pastagia et al. (2003) with permission from Elsevier)

pressure, permeate concentration of component 1 is more than that of component 2. This is due to the fact that component 2 has a higher real retention compared to that of component 1.

For a fixed pressure, permeate concentration increases with time because as the time of filtration increases, the concentration polarization becomes severe (as the membrane surface concentration increases with time as shown in Fig. 5.28). This leads to an enhancement of convective flux through the membrane, and consequently permeate concentration increases. Figure 5.28 represents the effect of the operating pressure on the membrane surface concentration profiles of dye 1 and 2, estimated from the model. It can be observed from the figure that membrane surface concentration increases with time for a fixed feed concentration and operating pressure. This is due to the more accumulation of solute particles over the membrane surface with time, which in turn increases the concentration polarization.

Figure 5.29 shows the effect of concentration ratio of dye 1 and dye 2 in the feed on the membrane surface concentrations at constant pressure of 415 kPa. This figure shows that the membrane surface concentration increases with time due to concentration polarization as described earlier. With increase in feed concentration of one of the components, the membrane surface concentration of that component increases. For example, in curves 2 and 4, the feed concentration of component 2 increases from 10 to 15 mg/l. This leads to an increase in osmotic pressure at the membrane-solution interface leading to enhanced concentration polarization. Therefore, membrane surface concentration of component 2 increases as shown by curves 2 and 4. Since feed concentration of component 1 is kept constant in these two cases, membrane surface concentration of component 1 is almost invariant as shown in curves 1 and 3. In curves 3 and 5, feed concentration of component 1 is increased from 10 to 20 mg/l, keeping the concentration of component 2 at the same level of 15 mg/l. This results in corresponding increase in membrane surface concentration of component 1 as evident from curves 3 and 5.

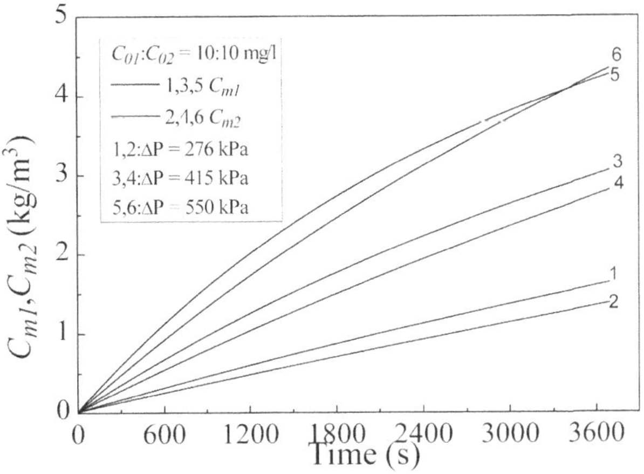

Fig. 5.28 Effect of operating pressure difference on membrane surface concentration for constant initial feed concentration ratio (model estimated) (Reproduced from Pastagia et al. (2003) with permission from Elsevier)

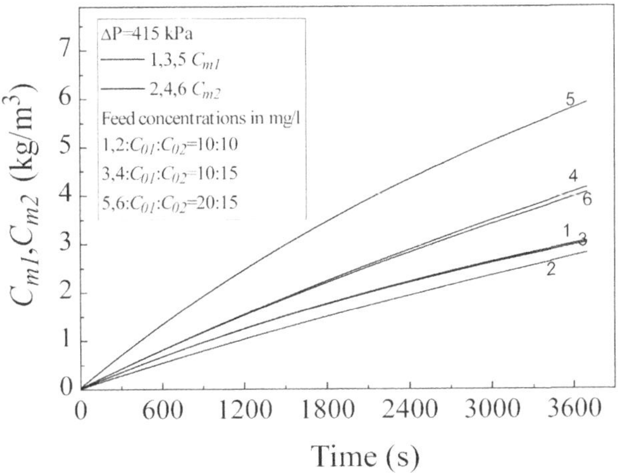

Fig. 5.29 Effect of ratio of initial feed concentration on membrane surface concentration for constant pressure difference (model estimated) (Reproduced from Pastagia et al. (2003) with permission from Elsevier)

Figures 5.30 and 5.31 shows the variation of permeate flux with pressure difference for the same feed concentration as in Figs. 5.24 and 5.25. The symbols represent the experimental data, and the continuous lines are the model estimations. It is clear from the figure that the flux is higher for higher operating pressure because of an increased driving force (at a fixed feed concentration).

Figure 5.32 presents the permeate flux profile estimated from the model, for a fixed concentration level at different operating pressure differences. This figure reveals two trends. First, the flux profiles are higher in values for higher operating pressures. At higher operating pressure, the driving force across the membrane for solvent transport increases resulting in higher flux values. Second, for same pressure difference, flux decreases marginally with time of operation. As time of operation progresses, the membrane surface concentration also increases (as shown in Fig. 5.28) due to concentration polarization. This leads to an increase

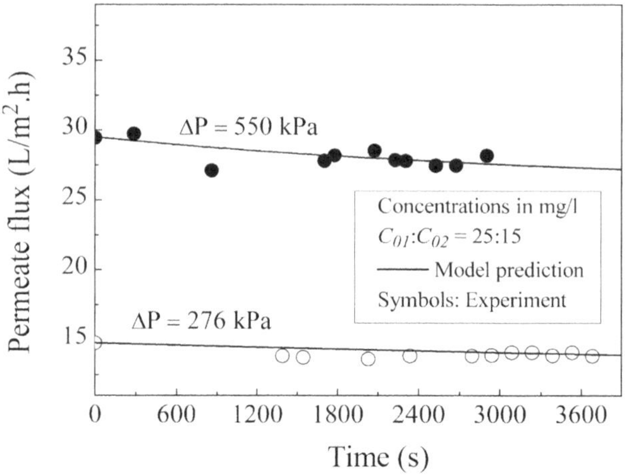

Fig. 5.30 Comparison between experimental and estimated values of permeate flux at a fixed concentration (C_{01}: C_{02} = 25:15) but different pressures in batch cell (Reproduced from Pastagia et al. (2003) with permission from Elsevier)

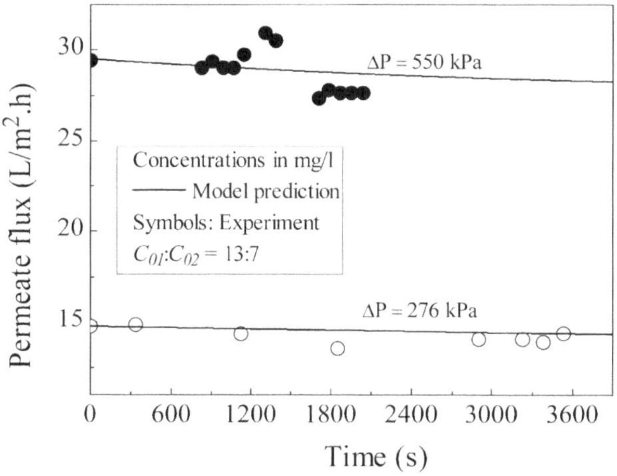

Fig. 5.31 Comparison between experimental and estimated values of permeate flux at a fixed concentration (C_{01}: C_{02} = 13:7) but different pressures in batch cell (Reproduced from Pastagia et al. (2003) with permission from Elsevier)

in osmotic pressure and slight decrease in available driving force across the membrane. Therefore, the permeate flux declines (although marginally) with time of operation.

Figure 5.33 shows the effect of feed concentration on permeate flux. As the total feed concentration increases, there is a decrease in permeate flux. This is due to the fact that as total feed concentration of solutes increases, there is increase in membrane surface concentrations (C_{m1} and C_{m2}) which in turn increases the total solution osmotic pressure, resulting in a decrease in the available driving force and subsequently the permeate flux.

The variations of observed retention (R_0) of both dyes, COD removal, and permeate flux are observed at various conditions of operating pressure and cross flow velocities. Figures 5.34 and 5.35 represent the variation of observed retention of the two dyes with pressure difference for two different feed concentrations

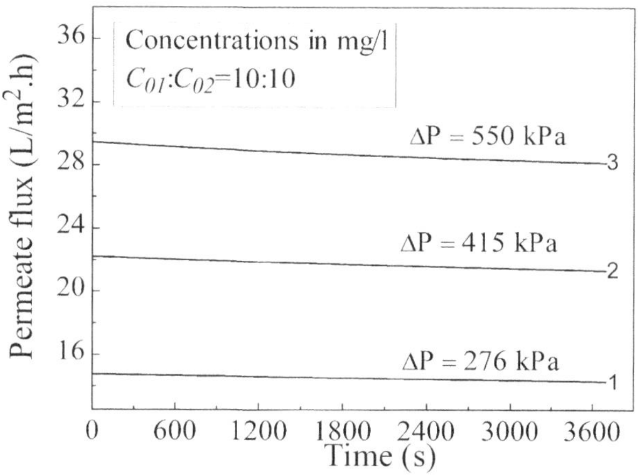

Fig. 5.32 Effect of operating pressure difference on permeate flux for constant initial feed concentration ratio (model estimated) (Reproduced from Pastagia et al. (2003) with permission from Elsevier)

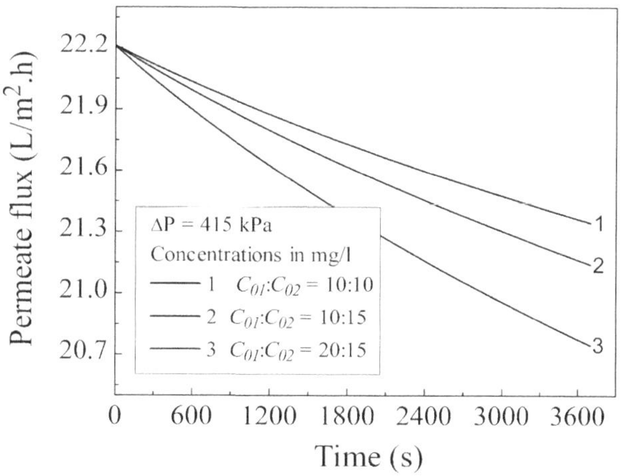

Fig. 5.33 Effect of ratio of initial feed concentration on permeate flux for constant pressure difference (model estimated) (Reproduced from Pastagia et al. (2003) with permission from Elsevier)

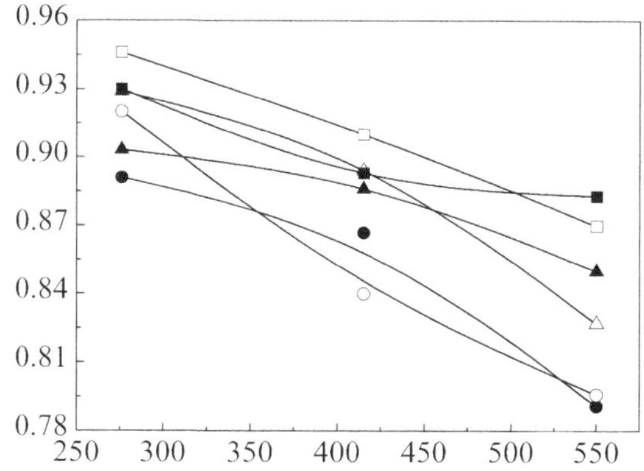

Fig. 5.34 Variation of retention of the two dyes in the effluent mixture with pressure at same feed concentration (25:15) but different cross flow velocities (Reproduced from Chakraborty et al. (2003a) with permission from Elsevier)

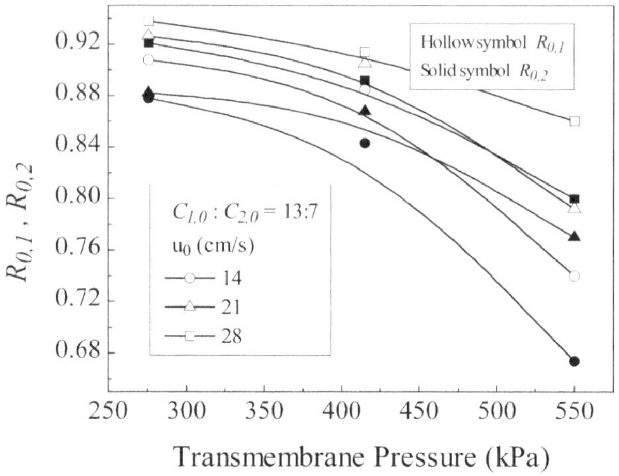

Fig. 5.35 Variation of retention of the two dyes in the effluent mixture with pressure at same feed concentration (13:7) but different cross flow velocities (Reproduced from Chakraborty et al. (2003a) with permission from Elsevier)

(25:15 and 13:7 mg/l, respectively) and three different cross flow velocities (14, 21, and 28 cm/s). For the cross flow cell, the retentions mentioned are the steady-state value since the system reaches steady state in a very short time. In these Figs., hollow and solid symbols represent dye 1 and 2, respectively. From the figures it is observed that R_0 decreases with increase in applied pressure for a particular feed concentration and cross flow velocity. At higher pressure permeate concentration is more (as explained earlier) resulting in lower retention. From the figure it is also clear that as the cross flow velocity increases, R_0 increases, for same feed concentration and applied pressure. As cross flow velocity increases, concentration polarization (CP) on the membrane surface decreases; hence membrane surface concentration decreases leading to a decrease in permeate concentration. It may be observed from Fig. 5.34 that at a pressure of 276 kPa and bulk velocity of 28 cm/s, retention can go up to 95% and 93% for the two dyes, respectively.

Figure 5.36 shows the percentage removal of COD with pressure at three different cross flow velocities for effluent feed mixture at two different feed concentrations. With increase in pressure at constant cross flow velocity, observed retention of dye decreases, and hence, COD of permeate increases, and therefore, percentage removal of COD decreases. For a feed concentration of 13:7 mg/l, a maximum of 94% of COD has been removed at a pressure of 276 kPa and cross flow velocity of 28 cm/s.

The variations of permeate flux with operating pressure for different combinations of feed concentrations and cross flow velocities are presented in Fig. 5.37. It is noted from the figure that the trend in the flux profile of two-component dye mixture is similar with that of the single component system. With increase in operating pressure (at a fixed feed concentration), the driving force across the membrane increases leading to an enhancement of permeate flux. Also, the permeate flux follows the expected trends with variation in feed concentration at a fixed operating pressure, i.e., flux decreases with increase in feed concentration (solid and hollow circle).

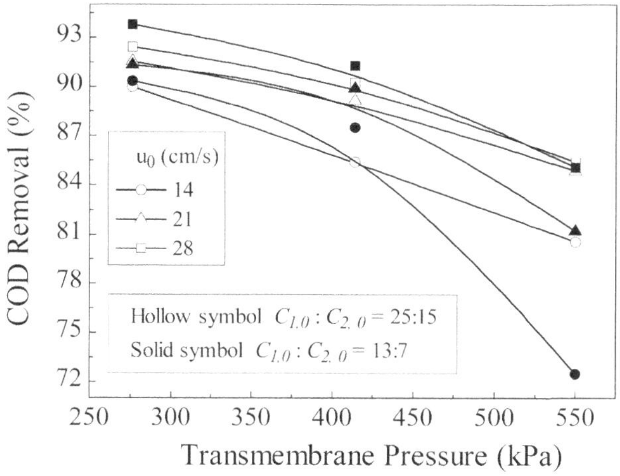

Fig. 5.36 Variation of COD removal with cross flow velocity and transmembrane pressure drop during cross flow NF (Reproduced from Chakraborty et al. (2003a) with permission from Elsevier)

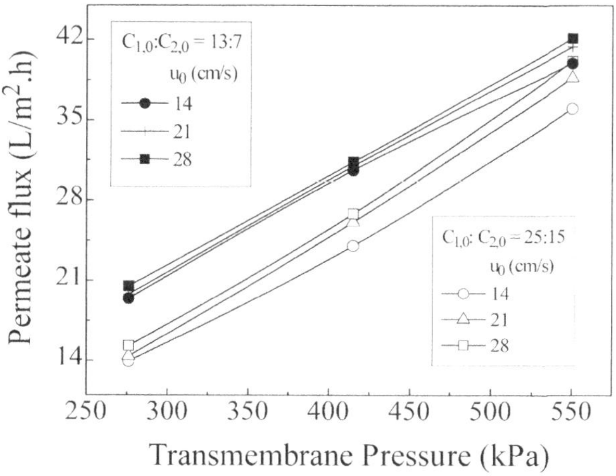

Fig. 5.37 Variation of permeate flux of the industrial effluent with transmembrane pressure drop at two different feed concentrations (Reproduced from Chakraborty et al. (2003a) with permission from Elsevier)

At a fixed feed concentration, flux increases with cross flow velocity due to reduction in concentration polarization, although the difference is marginal as observed from the figure (hollow circle and triangle). A summary of the results of each experiments conducted in cross flow nanofiltration is presented in Table 5.3.

5.4 Nanofiltration of the Textile Effluent in Hollow Fiber Membrane System

A textile effluent was collected from a nearby industry containing four reactive dyes: Cibacron yellow (CIB Y), Cibacron red (CIB R), Cibacron black (CIB B), and basic blue (BB 3). The characterization of the effluent is presented in Table 5.4.

Table 5.3 Properties measured for the textile effluent during NF in the cross flow cell

$C_{0,1}$: $C_{0,2}$	u_0, (cm/s)	ΔP (kPa)	Stream	Conductivity x 10^{-3} mΩ/cm	TS, ppm	% COD removal	% Color removal Dye 1	Dye 2
25:15	14	276	Feed	19.9	1.17×10^4	90	89.1	92.0
			Permeate	13.8	1.02×10^4			
			Retentate	20.5	1.25×10^4			
25:15	21	276	Feed	19.8	1.18×10^4	92.9	90.3	91.45
			Permeate	13.6	1.02×10^4			
			Retentate	20.3	1.25×10^4			
25:15	28	276	Feed	18.9	1.23×10^4	92.63	94.6	93
			Permeate	16.6	1.03×10^4			
			Retentate	19.1	1.27×10^4			
25:15	14	415	Feed	14.48	1.23×10^4	85.4	86.7	84.5
			Permeate	13.23	1.06×10^4			
			Retentate	14.6	1.27×10^4			
25:15	21	415	Feed	9.41	5.16×10^3	89.13	89.4	88.7
			Permeate	7.96	4.06×10^3			
			Retentate	9.53	5.97×10^3			
25:15	28	415	Feed	9.28	4.58×10^3	90.27	91	89.3
			Permeate	7.42	3.85×10^3			
			Retentate	9.56	4.62×10^3			
25:15	14	550	Feed	6.91	3.41×10^3	80.54	79.1	79.7
			Permeate	5.55	3.07×10^3			
			Retentate	6.98	3.57×10^3			
25:15	21	550	Feed	4.37	2.05×10^3	84.86	82.7	85
			Permeate	3.57	1.42×10^3			
			Retentate	4.48	1.98×10^3			
25:15	28	550	Feed	1.49	9.0×10^2	85.4	87	88.3
			Permeate	1.46	6.10×10^2			
			Retentate	1.12	9.80×10^2			
13:7	14	276	Feed	1.25	1.47×10^3	90.37	90.8	87.6
			Permeate	1.018	7.60×10^2			
			Retentate	1.272	1.65×10^3			
13:7	21	276	Feed	1.18	1.78×10^3	91.32	92.6	88.2
			Permeate	0.942	8.10×10^2			
			Retentate	1.185	9.70×10^2			
13:7	28	276	Feed	0.881	3.80×10^2	93.77	93.9	92.1
			Permeate	0.661	1.90×10^2			
			Retentate	0.872	4.20×10^2			
13:7	14	415	Feed	0.614	3.10×10^2	87.5	88.5	84.2
			Permeate	0.497	5.0×10^1			
			Retentate	0.646	3.60×10^2			

(continued)

Table 5.3 (continued)

$C_{0,1}$: $C_{0,2}$	u_0, (cm/s)	ΔP (kPa)	Stream	Conductivity x 10^{-3} mΩ/cm	TS, ppm	% COD removal	% Color removal Dye 1	Dye 2
13:7	21	415	Feed	0.455	1.20×10^2	89.9	90.5	86.8
			Permeate	0.342	8.0×10^1			
			Retentate	0.466	1.15×10^2			
13:7	28	415	Feed	0.332	1.15×10^2	91.3	91.4	89.3
			Permeate	0.244	7.5×10^1			
			Retentate	0.342	1.20×10^2			
13:7	14	550	Feed	0.199	5.40×10^2	72.5	74.1	67.4
			Permeate	0.146	3.40×10^2			
			Retentate	0.204	6.50×10^2			
13:7	21	550	Feed	0.252	1.03×10^3	81.2	79.3	77.1
			Permeate	0.179	8.50×10^2			
			Retentate	0.266	1.15×10^3			
13:7	28	550	Feed	0.162	5.70×10^2	85.1	86.1	80
			Permeate	0.105	3.60×10^2			
			Retentate	0.174	6.20×10^2			

Reproduced from Chakraborty et al. (2003a) with permission from Elsevier

Table 5.4 Effluent characteristics of textile plant

Effluent characteristics	Quantity
pH	11.2 ± 0.3
Conductivity, mS/cm	14 ± 1.2
Total dissolved solids (TDS), g/l	10 ± 0.2
Total solid (TS), g/l	13.4 ± 0.3
Salinity, g/l	8.1 ± 0.2
COD, mg/l	1400 ± 120
CIB yellow (551 Da), concentration, mg/l	67.0 ± 2
Cibacron red (854 Da), concentration, mg/l	240.0 ± 5
Cibacron black (923 Da), concentration, mg/l	204.0 ± 4
Basic blue (360 Da), concentration, mg/l	93.0 ± 2
Salt content (equivalent NaCl), mg/l	9096 ± 20

Three NF hollow fibers (780 Da, 440 Da, and 360 Da MWCO) having polysulfone as the base polymer are used for textile effluent treatment. Experiments are conducted using transmembrane pressure drop of 104 kPa and cross flow rate 20 l/h. The permeate flux profiles are shown in Fig. 5.38. It is observed that in all cases, steady state is reached within 20 min. The steady-state permeate flux is about 4.7, 3.4, and 2.5 l/m^2.h for 780 Da, 440 Da, and 360 Da MWCO membrane, respectively. These are in direct consequence of the pore size of the membrane. Average pore size of these three membranes also decreases in that order. Various physicochemical properties of

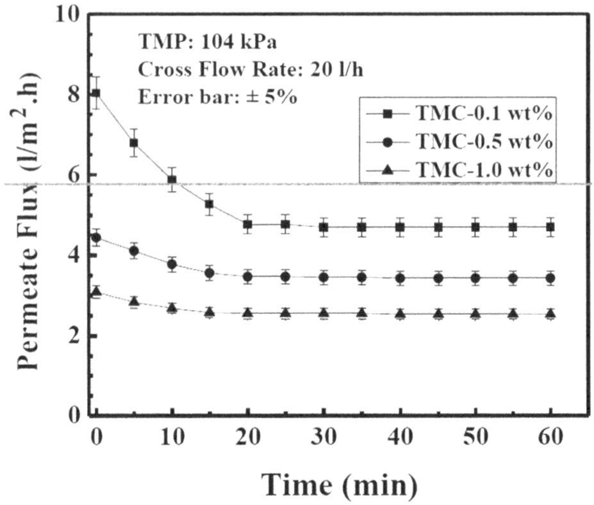

Fig. 5.38 Permeate flux profiles for different hollow fiber NF membranes

the permeate (textile effluent) are also shown in Table 5.5 for these membranes. It is observed from this table that COD values of the filtrate are within permissible limit (250 mg/l) for 440 Da and 360 Da MWCO membrane. Permeate concentration of different dyes is less than 1 mg/l for 440 Da and 260 Da MWCO membranes. In case of 780 Da MWCO membrane, the permeate concentration of dyes is above 3 mg/l. The zeta potential of the 780 Da MWCO membrane is less negative than 440 Da and 360 Da MWCO membrane. Hence, charge-charge repulsion is less (between negatively charged membrane surface and dyes) in 780 Da MWCO membrane than other two NF membranes. In case of TDS and TS, the trend is clear with the MWCO of the membrane. As the pore size of the membrane increases, more solids are permeated. Permeate conductivity and salinity depends upon the surface charge of the membrane. More negative membrane surface shows better percentage of rejection of salinity and conductivity. For example, 360 Da membrane shows 68% and 54% rejection of conductivity and salinity, whereas 780 Da MWCO membrane shows 45% and 36% rejection of conductivity and salinity.

Antifouling capacities of different NF membranes are quantified in terms of FRR and FDR values. A good antifouling membrane shows high FRR and low FDR. From Fig. 5.39, it is observed that 440 Da and 360 Da MWCO membranes have almost same FRR value around 88%. These two membranes show higher FRR value than 780 Da membrane. It is also found that 440 Da and 360 Da membranes show lesser FDR (around 20%) value than 780 Da MWCO membrane. Thus, considering the characterization of permeate quality and antifouling characterization, 440 Da and 360 Da MWCO membranes exhibit comparable performance. Considering throughput of the process, 440 Da membrane performs better and results 40% more steady-state permeate flux compared to 360 Da membrane. Thus, 440 Da MWCO membrane is selected for studying the effects of operating conditions.

Table 5.5 Permeate characteristics of the treated effluent at 20 l/h cross flow rate and 104 kPa transmembrane pressure

	pH	Equivalent NaCl (g/l)	TDS (g/l)	TS (g/l)	Salinity (g/l)	COD (mg/l)	Dye Rejection			
							Blue	Black	Red	Yellow
780 Da	10.6 ± 0.4	4.95 ± 0.05	6.39 ± 0.06	7.12 ± 0.07	5.17 ± 0.04	280 ± 10	94.62 ± 0.08	97.10 ± 0.1	98.42 ± 0.08	92.24 ± 0.1
440 Da	10.6 ± 0.5	3.56 ± 0.04	5.32 ± 0.04	6.25 ± 0.05	4.37 ± 0.06	240 ± 6	99.7 ± 0.06	99.95 ± 0.02	99.91 ± 0.03	99.9 ± 0.02
360 Da	10.6 ± 0.3	2.90 ± 0.06	5.09 ± 0.03	5.85 ± 0.06	3.71 ± 0.05	235 ± 5	99.78 ± 0.04	99.95 ± 0.02	99.91 ± 0.02	99.9 ± 0.02

Fig. 5.39 FRR and FDR of textile effluent for different NF membranes

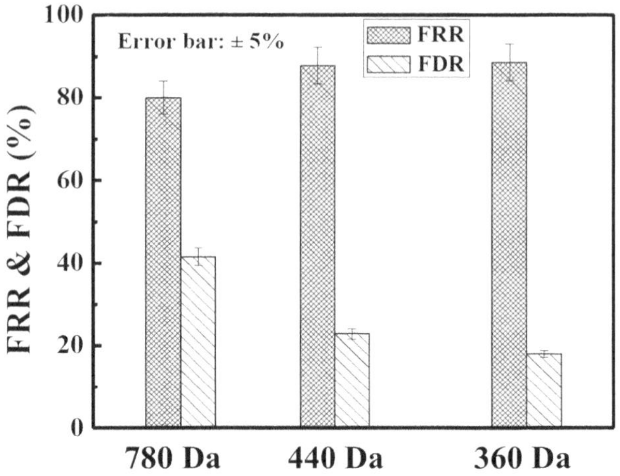

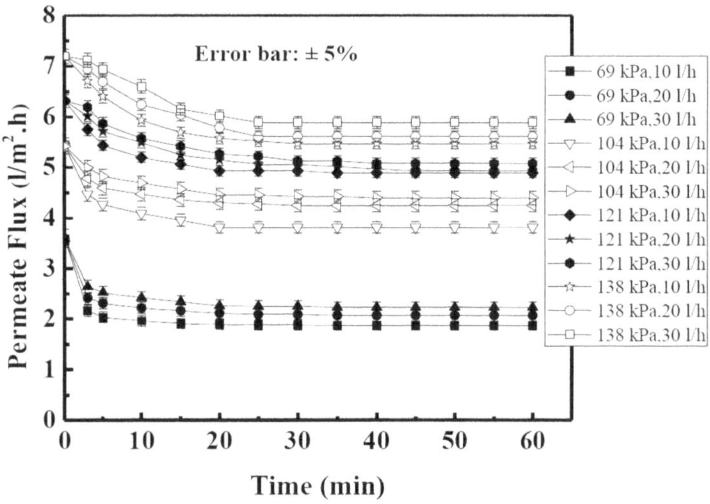

Fig. 5.40 Flux decline profile for NF of textile effluent using 440 Da MWCO hollow fiber membrane

The profiles of permeate flux for various operating conditions are presented in Fig. 5.40. Effects of both transmembrane pressure drop (TMP) and cross flow rate (CFR) are evident from these figures. It is observed from Fig. 5.40 that steady state is reached within 20 min during filtration. Two trends are observed from this figure. For a fixed TMP, the permeate flux increases with the cross flow rate. With increase in cross flow rate, the concentration of solute deposited on the membrane surface decreases due to forced convection, leading to lowering in concentration polarization and increase in permeate flux. For example, at 104 kPa pressure drop, the permeate flux increases from 3.8 to 4.4 l/m^2.h. When cross flow rate increases from

Table 5.6 Permeate stream characteristics of the NF of the textile effluent using 440 Da MWCO hollow fiber membrane

TMP (KPa)	Flow rate (l/h)	pH	Equivalent NaCl (g/l)	TDS (g/l)	TS (g/l)	Salinity (g/l)	COD (mg/l)	Dye rejection			
								Blue	Black	Red	Yellow
69	10	10.6 ± 0.3	3.46 ± 0.02	5.27 ± 0.03	3.95 ± 0.04	5.65 ± 0.04	147 ± 5	98.93 ± 0.03	99.9 ± 0.01	99.88 ± 0.01	99.85 ± 0.04
	20	10.6 ± 0.4	3.48 ± 0.03	5.34 ± 0.04	3.92 ± 0.06	5.71 ± 0.02	153 ± 4	98.93 ± 0.02	99.9 ± 0.02	99.88 ± 0.03	99.85 ± 0.03
	30	10.6 ± 0.5	3.50 ± 0.02	5.45 ± 0.02	4.12 ± 0.03	5.64 ± 0.05	165 ± 6	98.93 ± 0.04	99.9 ± 0.01	99.88 ± 0.02	99.85 ± 0.04
104	10	10.6 ± 0.4	3.56 ± 0.03	5.48 ± 0.02	4.17 ± 0.04	5.81 ± 0.03	220 ± 5	98.93 ± 0.02	99.9 ± 0.03	99.88 ± 0.04	99.84 ± 0.05
	20	10.6 ± 0.5	3.58 ± 0.02	5.51 ± 0.03	4.22 ± 0.03	5.94 ± 0.02	238 ± 4	98.82 ± 0.04	99.87 ± 0.04	99.83 ± 0.05	99.84 ± 0.04
	30	10.6 ± 0.6	3.63 ± 0.04	5.53 ± 0.04	4.19 ± 0.04	5.92 ± 0.04	241 ± 6	98.80 ± 0.05	99.85 ± 0.05	99.79 ± 0.05	99.84 ± 0.03
121	10	10.6 ± 0.3	3.72 ± 0.03	5.71 ± 0.02	4.24 ± 0.05	6.34 ± 0.06	265 ± 5	98.70 ± 0.06	99.85 ± 0.06	99.75 ± 0.06	99.84 ± 0.06
	20	10.6 ± 0.5	3.65 ± 0.04	5.64 ± 0.05	4.31 ± 0.03	6.31 ± 0.07	272 ± 4	98.70 ± 0.07	99.85 ± 0.05	99.7 ± 0.08	99.82 ± 0.04
	30	10.6 ± 0.4	3.71 ± 0.02	5.73 ± 0.04	4.29 ± 0.03	6.37 ± 0.04	281 ± 5	98.66 ± 0.04	99.8 ± 0.06	99.67 ± 0.08	99.82 ± 0.05
138	10	10.6 ± 0.5	3.75 ± 0.03	5.86 ± 0.03	4.35 ± 0.04	6.60 ± 0.05	327 ± 6	98.6 ± 0.04	99.78 ± 0.07	99.65 ± 0.06	99.82 ± 0.03
	20	10.6 ± 0.6	3.73 ± 0.05	5.91 ± 0.01	4.27 ± 0.06	6.57 ± 0.06	332 ± 7	98.55 ± 0.04	99.75 ± 0.08	99.65 ± 0.05	99.8 ± 0.08
	30	10.6 ± 0.5	3.77 ± 0.03	5.87 ± 0.04	4.46 ± 0.02	6.64 ± 0.05	348 ± 4	98.5 ± 0.08	99.7 ± 0.04	99.67 ± 0.04	99.79 ± 0.07

10 to 30 l/h, at 138 kPa, the corresponding increase in permeate flux is from 5.4 to 5.9 l/m^2.h. At a fixed CFR, the permeate flux increases with TMP. This is due to increased driving force. At 30 l/h CFR, the permeate flux increases from 2.2 to 5.9 l/m2.h, when the TMP increases from 69 to 138 kPa.

Quality of permeate after NF at different operating conditions is presented in Table 5.6. Dye rejection (%), conductivity, TDS, TS, and COD are presented in this Table. It is observed from this Table that the rejection of dyes is almost 99% in all the cases due to size exclusion (440 Da MWCO) and charge-charge repulsion. COD of permeate increases with TMP, due to enhanced forced convection. For example, at fixed CFR (30 l/h), the COD value increases from 165 to 348 mg/l, when TMP increases from 69 kPa to 138 kPa. At same TMP, CFR has marginal effect on COD in permeate. Thus, with a feed containing COD of 1400 mg/l, 76–90% COD reduction is obtained for different TMP and CFR in 440 Da MWCO membrane. TS and TDS of permeate increase with TMP due to same reason, and permeation of solids through the membrane pores is increased. At same TMP, CFR has insignificant effect on TS and TDS in permeate. With feed TS of 13.4 g/l, 51–58% TS reduction is achieved with permeate concentration in the range 5.65 g/l–6.64 g/l, for various operating conditions. Conductivity and salinity of permeate increase with TMP, as expected at constant CFR. At higher TMP, convective flux through the NF membrane dominates resulting to reduction in rejection and increase in salt permeation. At higher TMP, highest salt recovery is obtained. It is observed that 48–55% salt recovery is attained under the range of operating TMP and CFR studied herein.

References

Akbari A, Desclause S, Remigy JC, Aptel P (2002) Treatment of textile dye effluents using a new photografted nanofiltration membrane. Desalination 149:101–107

Brandon C, Jernigan D, Gaddis J, Spencer H (1981) Closed cycle textile dyeing: full scale renovation of hot wash water by hyperfiltration. Desalination 39:301–310

Buckley CK, Treffry-Goatley, Simpson M, Bindoff A, Groves G (1985) Pretreatment, fouling and cleaning in the membrane processing of industrial effluents. ACS Symp Ser:261–267

Chakraborty S, Purkait MK, DasGupta S, De S, Basu JK (2003a) Nanofiltration of textile plant for color removal and reduction in COD. Sep Purif Technol 31:141–151

Chakraborty S, Purkait MK, Dasgupta S, De S (2003b) Nanofiltration of textile plant effluent for color removal and reduction in COD. Sep Purif Technol 31:141–151

Chakraborty S, Bag BC, DasGupta S, De S, Basu JK (2003c) Separation and fractionation of dye solution by nanofiltration. Sep Sci Technol 38:219–235

Chakraborty S, Bag BC, Dasgupta S, De S, Basu JK (2003d) Separation and Fractionation of Dye Solution by Nanofiltration. Sep Sci Technol 38:219–235

Chakraborty S, Bag BC, DasGupta S, De S, Basu JK (2004a) Prediction of permeate flux and permeate concentration in nanofiltration of dye solution. Sep Purif Technol 35:141–152

Chakraborty S, Bag BC, Dasgupta S, Basu JK, De S (2004b) Prediction of permeate flux and permeate concentration in nanofiltration of dye solution. Sep Purif Technol 35:141–152

Chen G, Chai X, Yue P, Mi Y (1997) Treatment of textile desizing wastewater by pilot scale NF membrane separation. J Membr Sci 127:93–99

Cheryan M (1986) Ultrafiltration handbook. Technomic publishing, Lancaster

Constantinides A, Mostoufi N (1999) Numerical methods for chemical engineering with MATLAB applications. Prentice Hall, Upper Saddle River

Dhale AD, Mahajani VV (2000) Studies on treatment of disperse dye waste: membrane-wet oxidation process. Waste Managem 20:85–92

El-Nasher A (1980) Energy and water conservation through recycle of dyeing wastewater using dynamic Zr(IV)-PAA membranes. Desalination 33:21–47

Frank MJB, Westerink JB, Schokker A (2002) Recycling of industrial wastewater by using two-step nanofiltration process for the removal of color. Desalination 145:69–74

Kowak KM, Winnicki T, Wisniewski J (1986) Effect of flow conditions on ultrafiltration efficiency of dye solutions and textile effluents. Desalination 71:127–135

Koyuncu I (2002) Reactive dye removal in dye/salt mixtures by nanofiltration membrane containing vinyl sulphone dyes: effects of feed concentration and cross flow velocity. Desalination 143:243–253

Marcucci M, Nosenzo G, Capannelli G, Ciabatti I, Corrieri D, Ciardelli G (2001) Treatment and reuse of textile effluents based on new ultrafiltration and other membrane technologies. Desalination 138:75–82

Opong WS, Zydney AL (1991) Diffusive and convective protein transport through asymmetric membranes. AICHE J 37:1497–1510

Pastagia KM, Chakraborty S, DasGupta S, Basu JK, De S (2003a) Prediction of permeate flux and concentration of two component dye mixture in batch nanofiltration. J Membr Sci 218:195–210

Pastagia KM, Chakraborty S, Dasgupta S, Basu JK, De S (2003b) Prediction of permeate flux and concentration of two-component dye mixture in batch nanofiltration. J Membr Sci 218:195–210

Porter J, Goodman G (1984) Recovery of hot water, dyes and auxiliary chemicals from textile wastewaters. Desalination 49:185–192

Rautenbach R, Vossenkaul K, Linn T, Katz T (1996) Waste water treatment by membrane processes-new development in ultrafiltration, nanofiltration and reverse osmosis. Desalination 108:247–253

Schrig P, Widmer F (1992) Characterisation of nanofiltration membranes for the separation of aqueous dye-salt solutions. Desalination 89:89–107

Treffry-Goatley K, Buckley C, Groves G (1983) Reverse osmosis treatment and reuse of textile dyehouse effluents. Desalination 47:313–320

Treybal RE (1981) Mass transfer operations. McGraw Hill, Singapore

van der Bruggen B, De Vreese I, Vandecasteele C (2001) Water reclamation in the textile industry: nanofiltration of dye baths for wool dyeing. Ind Eng Chem Res 40:3973–3978

Voigt I, Stahn M, Wöhner S, Junghans A, Rost J, Voigt W (2001) Integrated cleaning of coloured waste water by ceramic NF membranes. Sep Purif Technol 25:509–512

Yazhen X, Lebrun RE, Gallo P, Blond P (1999) Treatment of textile dye plant effluent by nanofiltration membrane. Sep Sci Technol 34:2501–2519

Yu S, Gao C, Su H, Liu M (2001) Nanofiltration used for desalination and concentration in dye production. Desalination 140:97–100

Chapter 6
Hybrid Treatment Method of Industrial Effluent

Abstract It can be envisaged that a series system using two technologies can be advantageous in improving the overall efficiency and performance. Two proposed configurations were studied in this chapter: (1) adsorption followed by nanofiltration and (2) advance oxidation followed by nanofiltration. Quantitative comparison of the performance of both of these hybrid methods is elaborated in detail in this chapter.

Keywords Hybrid process • Advance oxidation • Adsorption • Nanofiltration

Depending on the nature of the effluent and the type of solute to be removed, better separation can be achieved by integrating two or multiple processes. Dhale and Mahajani (2000) used an integrated process including nanofiltration (NF) and wet oxidation to treat a disperse dye bath waste. The dye bath waste stream has undergone NF for recovering reusable water from the permeate stream, thus reducing fresh water consumption and minimizing effluent discharge. The concentrate obtained from the membrane unit had a high COD content, which is then treated by wet oxidation (WO) process. In another study, a combination of coagulation, adsorption, and ultrafiltration is used for the treatment of a primary effluent containing organic mineral and dissolved and suspended matter (colloids) (Abdessemed and Nezzal 2002). Microfiltration (MF) or ultrafiltration (UF) was adequate for producing disinfected water suited for different applications. However, direct filtration on membrane was limited by the fouling phenomena, which led to a substantial and continuous decrease of the permeate flux during filtration at constant pressure. In this context, coupled treatment processes such as electrocoagulation-NF (Aouni et al. 2009), MF-NF (Tahri et al. 2012), coagulation-flocculation-NF (Ellouze et al. 2012), ozonation-coagulation (Lin and Lin 1993), chemical-biological oxidation (Rodrigues et al. 2009), and biological treatment-ozonation (Kunz et al. 2001) are explored.

On the other hand, coagulation and adsorption made it possible to remove the colloidal fraction that played a significant role in membrane fouling. It was observed that coagulation significantly improved the ultrafiltration performances. In the study of Lin and Wang (2002), treatment of high-strength phenolic wastewater was investigated by a novel two-step method. The two-step treatment method consisted of chemical coagulation of the wastewater by metal chloride followed by

S. Mondal et al., *Advances in Dye Removal Technologies*, Green Chemistry and Sustainable Technology, DOI 10.1007/978-981-10-6293-3_6

further phenol reduction by resin adsorption. The combined treatment was found to be highly efficient in removing phenol from the aqueous solution and was capable of lowering the phenol concentration considerably. The effectiveness of a combined reduction-biological treatment system for the decolorization of non-biodegradable textile dyeing wastewater was investigated by Ghoreishi and Haghighi (2003). In this treatment system, a bisulfite-catalyzed sodium borohydride reduction followed by activated sludge technique was used in order to remove the color at ambient temperature and pressure.

A hybrid treatment process involving adsorption followed by MF has been reported by the authors (Karmakar et al. 2015). The process involved two steps: in the first step, the dye was adsorbed by activated carbon, and in the second step, a multichannel ceramic MF membrane was used to separate the dye-loaded activated carbon particles, resulting in high throughput. The major advantage of this process was high filtration rate at low operating pressure, and the hazardous reactive dyes were strongly bound to the adsorbent preventing subsequent leaching into the environment under normal conditions (Al-Degs et al. 2008). Baudin et al. (1997) used a combination of adsorption and ultrafiltration for the treatment of surface water in 12 full-scale drinking water treatment plants in Europe. Later on, Meier et al. (2002) used a combination of adsorption and NF for the treatment of severely contaminated wastewater. Powdered adsorbent was injected into the feed of a NF unit and removed from the concentrate subsequently by a thickener. The adsorbent in the feed had a positive effect on permeate quality, permeate flux, and fouling layer in the NF unit. In comparison to reverse osmosis, the combination process had higher maximum recovery rate and lower operating pressure and energy consumption, resulting in lower treatment costs. In another work (Pigmon et al. 2003), the combination of adsorption and ultrafiltration was used in the treatment of colored wastewaters. In this process, adsorption was carried out in batch reactors using activated carbon cloth (ACCs). Adsorption was carried out to uptake the low molecular weight compounds (mainly dyes), while ultrafiltration was used for removing the large-sized compounds (macromolecules, colloids, and turbidity). The membrane filtration step allowed a great removal of turbidity (about 98%), whereas adsorption onto ACC provided the decolorization of the stream with a high adsorption capacity. The continuous process of adsorption onto ACC and ultrafiltration resulted in successful discoloration with a high permeate flow rate.

In this chapter, efficiency of a combined process is demonstrated for removal of dye from textile effluent involving both adsorption and NF. Aqueous synthetic solutions of two typical dyes: crystal violet (CV) and methylene blue (MB) are selected for the primary study. Later on, the process scheme is applied to the treatment of an industrial effluent containing a mixture of two reactive dyes, collected from a textile dyehouse. Experiments are performed to observe the effects of various process parameters in detail. Theoretical models are also developed and used for prediction of the process outputs in each step of the combined process for both single component and two-component systems. The model parameters are estimated, which is important for the industrial design of such a process scheme.

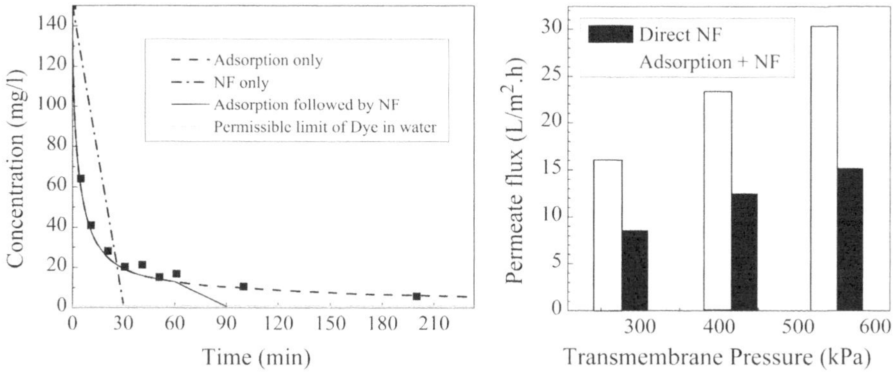

Fig. 6.1 Comparison of removal of crystal violet using two schemes: (*i*) direct NF and (*ii*) adsorption followed by NF

Typical dye concentration in an industrial effluent is usually in the range of 70–200 mg/l. It is observed that if this solution undergoes directly through a membrane-based process like NF, flux declines rapidly due to concentration polarization. The permeate quality also deteriorates. Hence, adsorption is selected to be the first step in order to reduce the bulk dye concentration. An adsorbent prepared from sawdust is used. The aqueous effluent after adsorption is passed through a NF unit, where the remaining dye content is separated and the water from the NF unit can safely be discharged into the environment. Separation of the dyes as well as recovery and reuse of the associated chemicals (mainly salt) and minimization of the process wastewater are the prime objectives of this study.

The efficiency of the proposed method is explained using Fig. 6.1. In this figure a comparison is made among the various methods for dye removal, using a synthetic solution of crystal violet (CV) with initial concentration of about 150 mg/l. These figures are for an overview of the importance of the proposed separation scheme, the details of which are discussed in Chap. 4.

For comparison, only adsorption is carried out for dye separation. In this method, the concentration of dye decreases to about 13 mg/l within the first 1 h, for a typical adsorbent dose of 1 g/l. Thereafter, the rate of concentration decay becomes gradually slower, and even after 3 h of continuous adsorption, the concentration decreases to only about 6 mg/l (dashed line in Fig. 6.1a). In the second method, only NF is performed on the CV solution of same initial concentration. This process is able to bring down the concentration of the feed solution to about 2.5 mg/l in only half an hour (dash-dot line in Fig. 6.1a). However, NF of the concentrated dye solution results in severe concentration polarization, with appreciable decrease of the permeate flux (Fig. 6.1b). Finally, the combination of adsorption and NF is adopted. In this process, after 1 h of adsorption, dye concentration reduces to about 13 mg/l. This dye solution is then charged to the NF unit, where concentration of dye is brought down to about 0.75 mg/l within half an hour. The combination method also results in about 100% gain in the permeate flux, compared to the direct NF (Fig. 6.1b).

Hence, the combination process is able to improve the quality of the treated process water (in terms of the dye concentration) as well as increase the process output (in terms of the permeate flux achieved) considerably. A process involving both adsorption and NF and application in the treatment of a textile dyehouse effluent with the aim of separating the dyes as well as improving the throughput (permeate flux) is envisaged in this chapter, as shown in Fig. 6.1. The recovery and recycling of the associated chemicals (mainly salt content) are also attempted, thus minimizing the waste volume. The effect of various process parameters is also studied using aqueous synthetic solution of the dyes as well as industrial wastewater collected from a textile dyehouse.

6.1 Adsorption Followed by Nanofiltration

Main dye components are reactive red and black in the effluent obtained from M/s, Singhal Brothers, Kolkata, India. Table 6.1 represents the characterization of effluent. Calibration curves for estimating dye concentration are prepared using pure dyes obtained from M/s, Dye Chem Ltd., Kolkata. Dyes are represented as dye 1 (red) and dye 2 (black) for convenience.

6.1.1 Adsorption Equilibrium

Pure dye species are used for equilibrium studies for single component and mixture in 1:2 concentration ratio using the adsorbent charred saw dust. Equilibrium experiments are conducted in a temperature-controlled shaker for 8 h at 298 K.

Equilibrium isotherm data for both dyes (single component) are presented in Fig. 6.2, indicating the adsorption ability of CSD.

Isotherm constants are shown in Table 6.2. Correlation coefficient beyond 0.99 indicates suitability of both Langmuir and Freundlich isotherms for dye 1. On the other hand, Freundlich isotherm is more suitable for dye 2. Values of R_L less than 1 indicate favorable adsorption.

Equilibrium data of dye mixture (Fig. 6.3) show the competitive adsorption of both dyes thereby reducing adsorption of each, represented by the modified Langmuir isotherm using interaction parameter $\eta_{i,j}$. Interaction parameters are optimized as $\eta_{1,1} = 1.69$; $\eta_{2,1} = 4.33$; $\eta_{1,2} = 15.74$; $\eta_{2,2} = 1.70$.

Table 6.1 Characterization of the effluent from the textile plant

Feed	$C_{0,1}$ (mg/l)	$C_{0,2}$ (mg/l)	pH	Conductivity × 10^{-3} (mΩ/cm)	TS (g/l)	Salinity (g/l)	COD (mg/l)
Original	70	148	10.41	117.7	103	98	4661

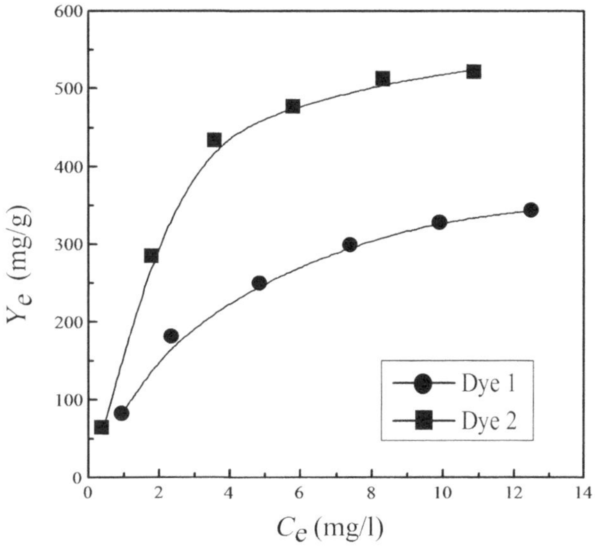

Fig. 6.2 Equilibrium adsorption of the reactive dyes in single component system (*solid lines* are guides for author's eyes) (Reproduced from Chakraborty et al. (2005). With permission from Elsevier)

Table 6.2 Equilibrium analysis from Langmuir and Freundlich model

	Langmuir isotherm			Freundlich isotherm		
Dye	Y_0	K_0	CR	Y_f	n	CR
Reactive red CNN	460.83	0.245	0.99	127.65	2.45	0.99
Reactive black B	689.65	0.338	0.98	261.31	3.15	0.94

Reproduced from Chakraborty et al. (2005). With permission from Elsevier

6.1.2 Adsorption Rate

Rate of adsorption of textile effluent (dye 1, 70 mg/l; dye 2, 148 mg/l) is carried out in 1 l capacity agitated vessel using 1 g of adsorbent at 2000 rpm for 30 min. 6 l solution is collected for subsequent NF. Effects of adsorbent dose are observed by varying the adsorbent concentration from 0.5 to 2.0 g/l. Dye concentrations are measured spectrophotometrically (absorbance peak of dye 1 is 599 nm and that of dye 2 is 535 nm). Figures 6.4 and 6.5 show the effect of mass of adsorbent (M_a) on the concentration decay of dyes 1 and 2, respectively, adding 0.5, 1.0, and 2.0 g of adsorbent per liter of the effluent. From the figures, it is observed that initially increasing the adsorbent amount (from 0.5 to 1.0 g) adsorption increases considerably for both the dyes. The initial rate of adsorption becomes almost independent of

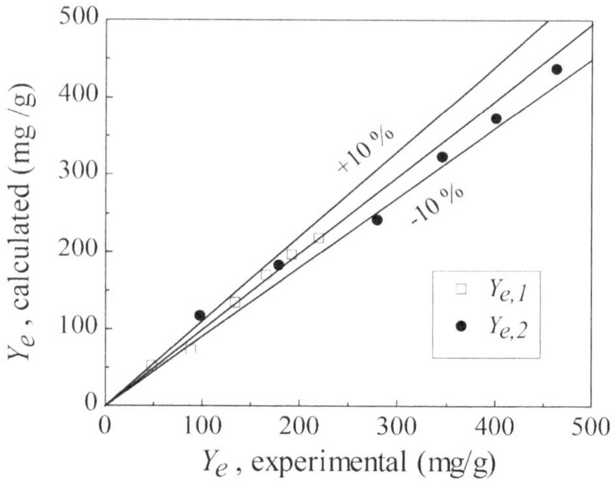

Fig. 6.3 Comparison between the experimental and calculated Y_e values of the dyes in mixture (Reproduced from Chakraborty et al. (2005). With permission from Elsevier)

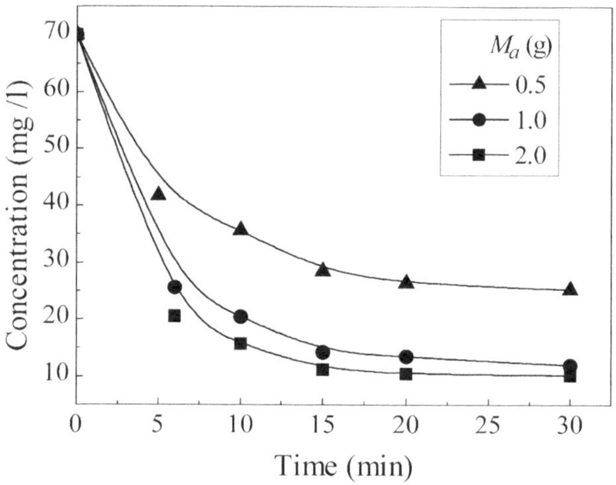

Fig. 6.4 Effect of adsorbent loading on the concentration decay of dye 1 (*solid lines* are guides for author's eyes) (Reproduced from Chakraborty et al. (2005). With permission from Elsevier)

M_a at higher values of M_a. The dye content of the solution decreases drastically and reaches an almost constant value. As the initial rate becomes very high, it is practically independent of M_a at higher values. An adsorbent dose of 1 gm/l is selected for treating the rest of the effluent. Table 6.3 shows the characterization of the effluent after adsorption.

6.1.3 Cross Flow Flat Sheet Membrane Filtration

Figures 6.6 and 6.7 represent the variation of the steady-state permeate concentration of dyes 1 ($C_{p,1}$) and 2 ($C_{p,2}$), respectively, with operating pressure at different

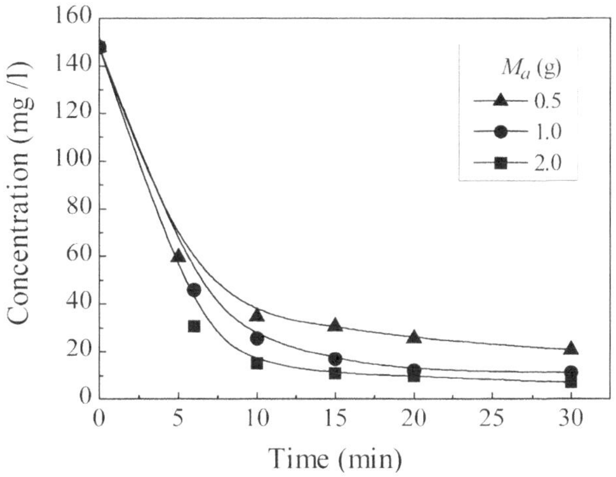

Fig. 6.5 Effect of adsorbent loading on the concentration decay of dye 2 (*solid lines* are guides for author's eyes) (Reproduced from Chakraborty et al. (2005). With permission from Elsevier)

Table 6.3 Characterization of the effluent after adsorption

Feed	$C_{0,1}$ (mg/l)	$C_{0,2}$ (mg/l)	pH	Conductivity × 10^{-3} (mΩ/cm)	TS (g/l)	Salinity (g/l)	COD (mg/l)
After adsorption	11.9	11	10.56	116.5	107	96.93	489

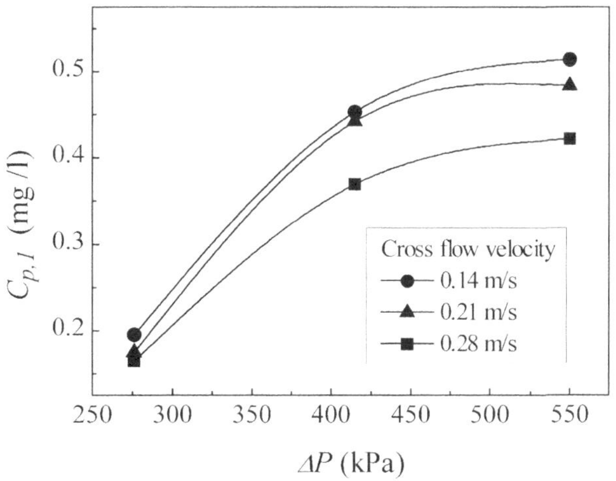

Fig. 6.6 Variation of permeate concentration of dye 1 with pressure and cross flow velocity (*solid lines* are guides for author's eyes)

cross flow velocities. Both the figures show the same trend, i.e., the permeate concentration increases with the operating pressure. Permeate concentration increases with transmembrane pressure drop due to enhanced convection, and it decreases with cross flow velocity due to reduction in concentration polarization.

Fig. 6.7 Variation of permeate concentration of dye 2 with pressure and cross flow velocity (*solid lines* are guides for author's eyes)

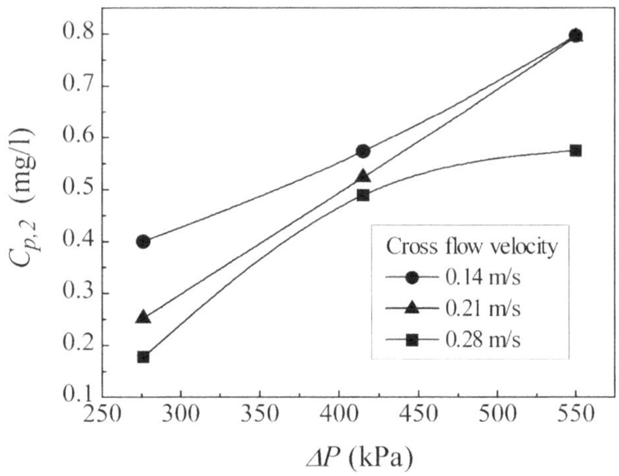

Fig. 6.8 Variation of permeate flux with pressure and cross flow velocity (*solid lines* are guides for author's eyes) (Reproduced from Chakraborty et al. (2005). With permission from Elsevier)

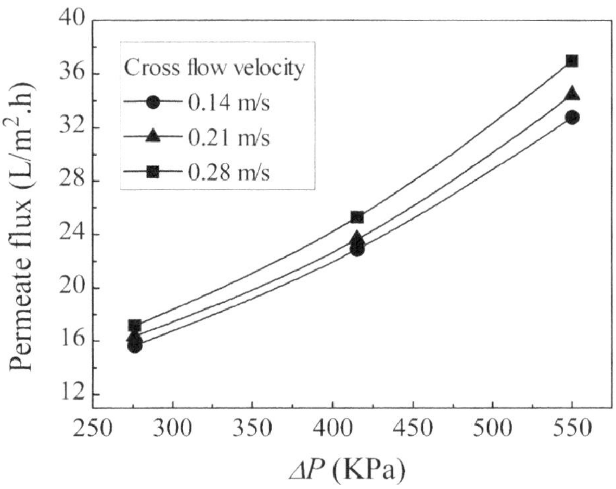

Effects of transmembrane pressure drop and cross flow velocity are apparent from Fig. 6.8. However, these effects are insignificant due to lower osmotic pressure at ppm level of feed concentration.

Table 6.4 presents the results in concise form. The combined treatment results in dye and COD removal beyond 99% and more than 88% of salt recovery for all operating conditions. Dye concentration in permeate is less than 1 mg/l in all cases. There is no color present in permeate with a COD less than 23 mg/l, much less compared to permissible limit, 250 mg/l.

Table 6.4 Properties of the final effluent: NF after adsorption

u_0 cm/s	ΔP kPa	Conductivity $\times 10^{-3}$ (mΩ/cm)	TS mg/l	% COD removal	% Salt recovery	Dye concentration in the permeate (mg/l) Dye 1	Dye 2
14	276	108.2	9.77×10^4	99.70	88.91	0.19	0.40
21	276	91.45	9.79×10^4	99.71	89.26	0.17	0.25
28	276	92.63	9.82×10^4	99.73	89.86	0.16	0.18
14	415	85.4	9.94×10^4	99.59	91.29	0.45	0.57
21	415	89.13	9.93×10^4	99.58	90.86	0.44	0.52
28	415	90.27	10.05×10^4	99.62	92.18	0.37	0.49
14	550	80.54	10.06×10^4	99.51	92.15	0.51	0.80
21	550	84.86	9.85×10^4	99.52	90.23	0.48	0.79
28	550	85.4	10.04×10^4	99.58	90.85	0.42	0.57

Reproduced from Chakraborty et al. (2005). With permission from Elsevier

6.1.4 Direct Nanofiltration

Textile effluent is subjected to direct NF and results are summarized in Table 6.5 indicating more than 98% removal of COD. Average concentration of dyes 1 and 2 in permeate is 2.3 and 1.7 mg/l, respectively.

6.1.5 Comparison of Adsorption Followed by Nanofiltration and Direct Nanofiltration

Performance of direct NF (process 1) and adsorption followed by NF (process 2) is presented in Table 6.6. Dyes 1 and dye 2 are removed by 96% and 98% in process 1, but they are removed by more than 99% by process 2. Dye concentration in permeate is more than 1 mg/l in process 1, but it is less in combined process. Salt recovery and COD reduction are almost at the same extent in both cases. Process 2 leads to significant throughput (permeate flux) than only NF. For example, at 415 kPa and 14 cm/s cross flow velocity, permeate flux in combined method is twice (23 L/m^2 h) that of only NF (11.4 L/m^2 h) due to membrane fouling. Since solids concentration is more in direct NF, membrane fouling (irreversible fouling) is more than the combined one, leading to significant flu decline in long term.

Figures 6.9, 6.10, and 6.11 show a comparative study of dye removal using different methods. It is clear from Fig. 6.9 that dye 1 is degraded quickly to 12 mg/l within 30 min and slowly thereafter and finally to 7 mg/l after 3 h. Using direct NF, concentration of dye is reduced to 2 mg/l. Effect of concentration polarization is

Table 6.5 Properties of the final effluent: Direct NF

u_0 cm/s	ΔP kPa	Conductivity $\times 10^{-3}$ (mΩ/cm)	TS mg/l	% COD removal	% Salt recovery	Dye concentration in the permeate (mg/l) Dye 1	Dye 2
14	415	105.8	9.27×10^4	99.59	86.79	1.98	1.65
21	415	108.9	9.54×10^4	99.58	89.53	1.96	1.35
28	415	109.95	9.64×10^4	99.62	90.46	1.83	1.21
14	550	110.62	9.69×10^4	98.45	91.05	3.63	2.77
21	550	110.3	9.67×10^4	98.81	90.77	2.46	1.96
28	550	109.75	9.62×10^4	99.03	90.28	1.95	1.47

Reproduced from Chakraborty et al. (2005). With permission from Elsevier

Table 6.6 Comparison of the performance of the two processes: Adsorption followed by NF (Proc. 1) and direct NF (Proc. 2)

		% Dye removal				% COD removal		% Salt recovery		Permeate flux (L/m^2.h)	
		Proc. 1		Proc. 2		Proc. 1	Proc. 2	Proc. 1	Proc. 2	Proc. 1	Proc. 2
u_0 cm/s	ΔP kPa	Dye 1	Dye 2	Dye 1	Dye 2						
14	415	99.3	99.6	97.1	98.9	99.6	99.1	91.3	86.8	22.9	11.4
21	415	99.4	99.6	97.2	98.9	99.6	99.0	90.9	89.5	23.6	12.8
28	415	99.5	99.7	97.4	99.4	99.6	99.3	92.2	90.5	25.3	13.2
14	550	99.3	99.5	94.8	98.1	99.5	98.4	92.2	91.0	32.8	13.1
21	550	99.3	99.5	96.5	98.7	99.5	98.8	90.2	90.8	34.4	15.3
28	550	99.4	99.6	97.2	99.0	99.6	99.0	90.8	90.3	37.0	16.9

Reproduced from Chakraborty et al. (2005). With permission from Elsevier

significant in direct NF due to higher concentration of dye leading to decline in permeate flux. In case of adsorption followed by NF, dye 1 is reduced to 12 mg/l after adsorption and thereafter to 0.45 mg/l after NF. Similar behavior for dye 2 is presented in Fig. 6.10. Enhancement in permeate flux is shown in Fig. 6.11 and it is twice that of NF only.

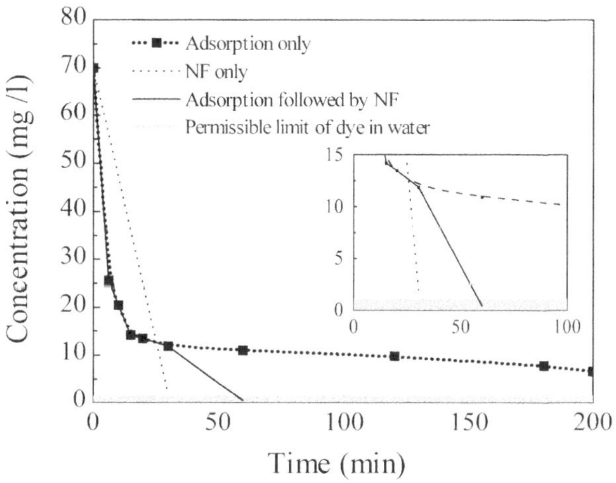

Fig. 6.9 Comparison of the three methods for the removal of dye 1 from industrial effluent

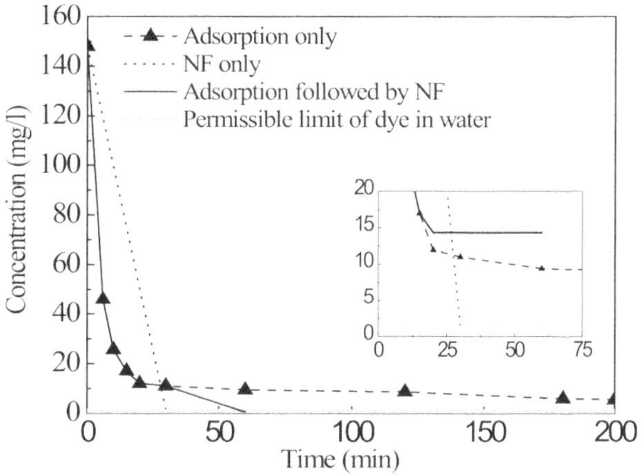

Fig. 6.10 Comparison of the three methods for the removal of dye 2 from industrial effluent (Reproduced from Chakraborty et al. (2005). With permission from Elsevier)

6.2 Advanced Oxidation Processes (AOP) Followed by Nanofiltration

In advanced oxidation process (AOP), hydroxyl radicals are generated by Fenton's reagent, and their power is augmented by ozone and UV radiation degrading dyes into carbon dioxide and water (Weinberg and Glaze 1997; Arana et al. 2001; Rathi et al. 2003; Kuo 1992; Lunar et al. 2000; Robinson et al. 2001; Kang et al. 2002; Pérez et al. 2002). Hydrogen peroxide reacts with ferrous sulfate producing hydrogen radicals and conversion of Fe^{3+} to Fe^{2+} is the limiting step. Hence, AOP is fast initially and becomes slower due to scarcity of Fe^{2+} ions (Neyens and Baeyens, 2003).

A combination of advanced oxidation process (AOP) using Fenton's reagent and NF is proposed to treat a textile effluent containing two dyes (Cibacron red

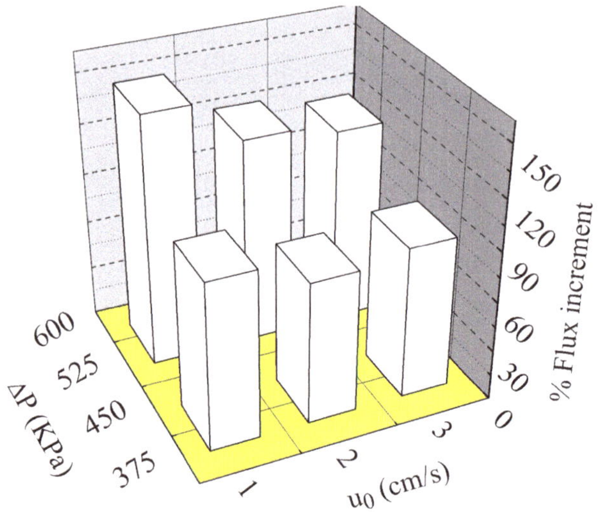

Fig. 6.11 Variation of percentage flux increment with pressure and cross flow velocity in the combination method (cross flow velocities are, 1–14 cm/s, 2–21 cm/s and 3–28 cm/s) (Reproduced from Chakraborty et al. (2005). With permission from Elsevier)

(CR) and Cibacron black (CB)). Effects of concentration of various reagents on degradation of dye were explored during AOP. A detailed study was carried out to observe the effect of the process parameters during NF, namely, applied pressure and cross flow velocity on the process outputs such as dye concentration in permeate, COD removal, permeate flux, and salt recovery. NF operations were carried under laminar flow regime. In case of AOP followed by NF, concentration of CR drops to 15 mg/l and CB falls to 23 mg/l within first 30 min for 556 mg/l of $FeSO_4 \cdot 7H_2O$ and 3240 mg/l of hydrogen peroxide during AOP. After application of NF to the AOP-treated effluent, dye concentrations fell below 1 mg/l at 552 kPa and 689 kPa transmembrane pressure. In scheme 2, after application of NF as the first step, CR concentration varied between 5 and 13 mg/l and CB concentration varied between 6 and 18 mg/l. The permeate in NF in step 2 contained less than 1 mg/l of dyes at 552 and 689 kPa for all cross flow velocities studied herein. The NaCl recovery in permeate is almost 92–99% in both schemes. COD values of finally treated effluent were well within permissible limit, i.e., 250 mg/l, and were in the range of 61–68 mg/l.

The textile effluent used in the present study was collected from a textile dyehouse, located in Kolkata, India. The effluent contained a mixture of two reactive dyes: reactive red (Cibacron red, molecular weight 855.5 g/mol) and reactive black (Cibacron black, molecular weight 924.5 g/mol). The effluent was characterized in terms of dye concentration, conductivity, pH, salt content, COD, and total solid (TS) content and is presented in Table 6.7. In the subsequent sections, the two dyes (reactive red and reactive black) would be denoted by CR and CB, respectively.

AOP is conducted in 500 ml batch at 30° C, 50 rpm, without sunlight for 3 h. Range of concentration of hydrogen peroxide is from 2160 to 4320 mg/l and that for ferrous sulfate is 278 to 556 mg/l. Sludge is separated using a 0.2 micron microfilter.

Table 6.7 Characterization of the effluent from the textile plant

Effluent characteristics	Quantity
CR, concentration	78 mg/l
CB, concentration	165 mg/l
PH	11.1
Conductivity × 10^{-5}	118.0 mS/m
Total solids (TS)	0.85×10^5 mg/l
COD	4840 mg/l
TDS	77.1 mg/l
Salt content (equivalent NaCl)	76.8 mg/l

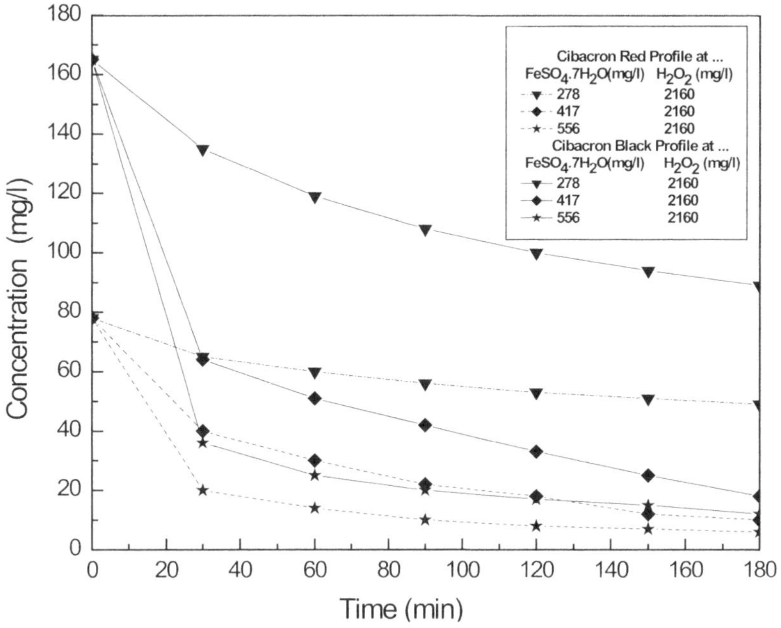

Fig. 6.12 Effect of $FeSO_4.7H_2O$ on degradation of CR and CB in AOP

The clear effluent is treated by 400 MWCO NF membrane (permeability 2×10^{-11} m/Pa.s). The operating pressures of NF are 552, 690, and 828 kPa and cross flow rates corresponding to Reynolds number 688, 1032, and 1362 in laminar regime. Dye concentration profiles during AOP are presented in Fig. 6.12 that shows a sharp reduction initially and a slower decline thereafter. Initially concentration of Fe^{2+} is more leading to fast degradation that slows down due to limited regeneration of Fe^{2+} in long run. Dye 1 is degraded by 37%, 87%, and 92% after 3 h corresponding to ferrous sulfate concentration 278, 417, and 556 mg/l (at H_2O_2 concentration of 2160 mg/l). Corresponding degradation of dye 2 is 18%, 61%, and 78%.

Effect of H_2O_2 is shown in Fig. 6.13. After 3 h, degradation of dye 1 increases from 37 to 53% as H_2O_2 concentration increases from 2160 to 4320 mg/l (at 278 mg/l of $FeSO_4.7H_2O$). Corresponding figures for dye 2 are 46–55%.

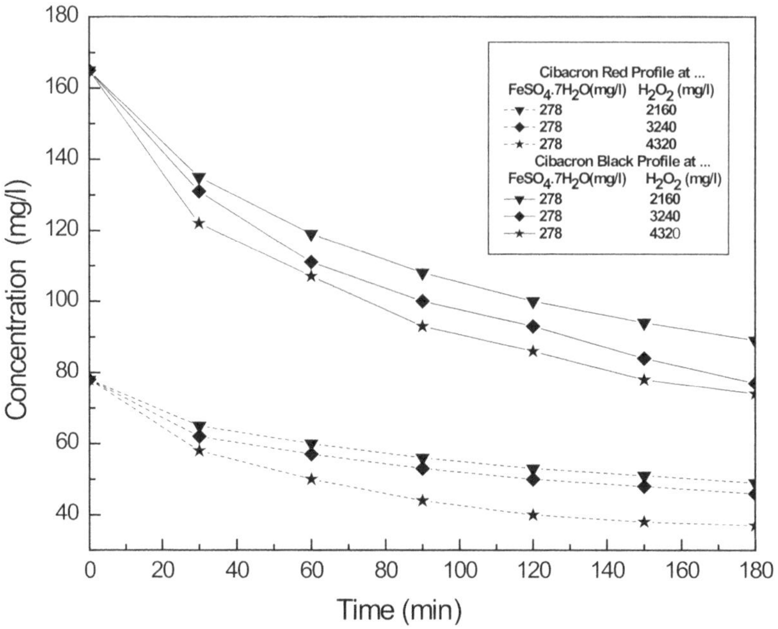

Fig. 6.13 Effect of H_2O_2 on degradation of CR and CB in AOP

Ferrous sulfate reacts with hydrogen peroxide to generate hydroxyl radicals which in turn degrades the dye according to the following reactions (Neyens and Baeyens 2003).

$$Fe^{2+} + H_2O_2 \rightarrow Fe^{3+} + OH^* + OH^- \quad (6.1)$$

$$RH + OH^* \rightarrow H_2O + R^* \quad (6.2)$$

The above equations suggest that the rate of generation of hydroxyl radicals is a function of the ferrous sulfate and hydrogen peroxide concentration. However, the molar concentration of hydrogen peroxide more than that of ferrous sulfate in this study, so the change of ferrous sulfate concentration is more dominant on the dye degradation. This effect is evident from the observations presented in Figs. 6.12 and 6.13.

Concentration of dye 1 and dye 2 is reduced sharply to 15 and 23 mg/l within the first 15 min of AOP using concentration of ferrous sulfate and hydrogen peroxide 556 and 2160 mg/l, producing significant amount of sludge. Reduction of dye concentration is gradual thereafter. Supernatant of AOP is subjected to NF (under laminar flow condition) for half an hour after clarified by MF (refer to Table 6.8 for properties of microfiltered effluent). Steady-state permeate flux for all operating conditions is shown in Fig. 6.15. Permeate flux varies with transmembrane pressure drop almost linearly due to enhanced driving force. However, the effect of transmembrane pressure drop is more dominant compared to cross flow velocity in the range of

Table 6.8 Characterization of effluent after AOP which was used for NF

Effluent characteristics	Quantity
CR concentration (mg/l)	15 mg/l
CB concentration (mg/l)	23 mg/l
PH	8.7
Conductivity $\times 10^{-5}$, mS/m	118.5 mS/m
Total solids (TS), mg/l	0.78×10^5 mg/l
COD (mg/l)	151 mg/l
TDS (mg/l)	76.9 mg/l
Salt content (equivalent NaCl) (mg/l)	76.8 mg/l

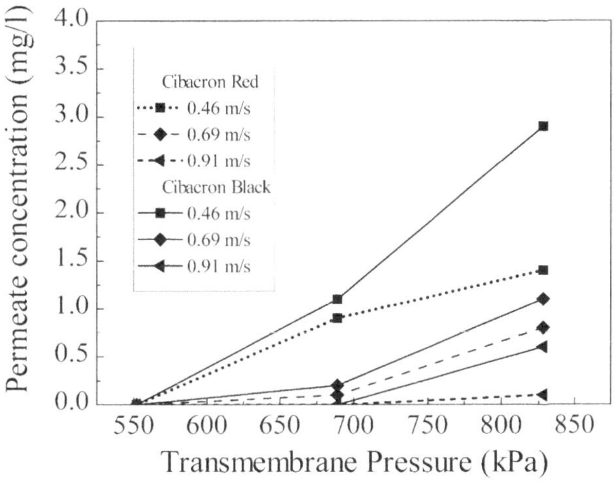

Fig. 6.14 Variation of CR and CB concentrations with operating pressure in permeate in NF

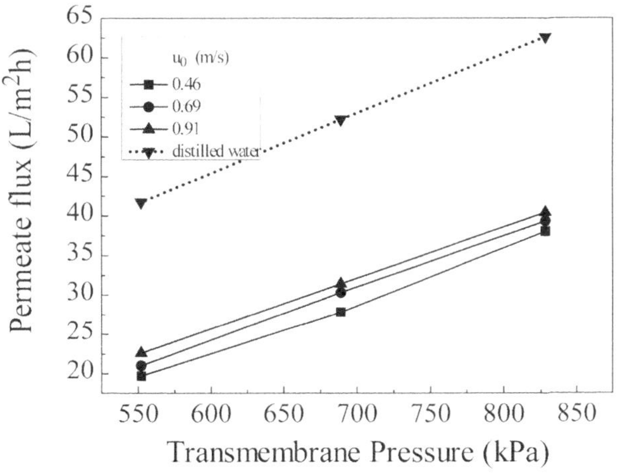

Fig. 6.15 Variation of permeate flux with operating pressure in NF

Table 6.9 Characterization of effluent at the end of AOP followed by NF

S. No.	Pressure (kPa)	Cross flow velocity (m/s)	Conductivity × 10^{-5} (mS/m)	% NaCl recovered	TDS (mg/l)	COD (mg/l)	pH
1	552	0.46	116.1	97.5	75.7	62	8.61
2	-do-	0.69	115.1	96.2	75.0	61	8.58
3	-do-	0.91	113.8	94.7	74.2	60	8.6
4	689	0.46	116.9	98.5	76.2	66	8.63
5	-do-	0.69	116.1	97.5	75.7	63.2	8.62
6	-do-	0.91	115.1	96.3	75.0	63	8.62
7	828	0.46	117.5	99.2	76.6	69	8.65
8	-do-	0.69	116.9	98.5	76.2	67	8.64
9	-do-	0.91	116.0	97.4	75.6	64.5	8.61

operating Reynolds number selected in this study. It may be observed that permeate flux is significantly less compared to pure water flux due to concentration polarization.

Dye concentration of permeate after NF is presented in Fig. 6.14. It is clear that for all operating pressures below 828 kPa, concentration of both dyes is below 1 mg/l. At 828 kPa, dye concentration exceeds only in case of lower cross flow velocity (0.46 m/s). Actually, concentration polarization is favored at higher trans-membrane pressure and lower cross flow velocity, thereby increasing the concentration of solute (dye in this case) at the feed-membrane interface leading to more permeation of solutes. From this study, suitable operating conditions for this stage of NF are 828 kPa pressure and 0.91 cross flow velocity.

Properties of filtrate (using actual effluent) from NF are shown in Table 6.9 corresponding to various operating conditions. Variation of pH, COD, and salt recovery is narrow over different operating conditions. Reduction of COD is about 98% (from 4840 to 62 mg/l) and salt recovery is about 97%.

6.2.1 Two-Stage Nanofiltration

The steady-state permeate flux and permeate concentration after first stage NF are shown in Figs. 6.16 and 6.17 for all operating conditions. Since dye concentrations in feed are more, the permeate flux is almost reduced by half compared to pure water flux. Concentration of dyes varies from 5 to 18 mg/l at 828 kPa. Corresponding figures for second stage NF are shown in Figs. 6.18 and 6.19, respectively. It is observed from Fig. 6.18 that permeate flux increases from 30 to 40 L/m^2 h at 828 kPa and 0.91 m/s in second stage compared to first stage NF. This is due to reduction in dye concentration. On the other hand, concentration of both dyes is less than 1 mg/l except 0.46 m/s after the second stage of NF. 828 kPa pressure and 0.91 m/s cross flow velocity can be selected as operating conditions in both stages of NF so that the dye concentration in permeate is less than 1 mg/l.

Properties of filtrate after the second stage of NF are shown in Table 6.10. The salient features of this table are as follows: (1) pH of the effluent does not need to

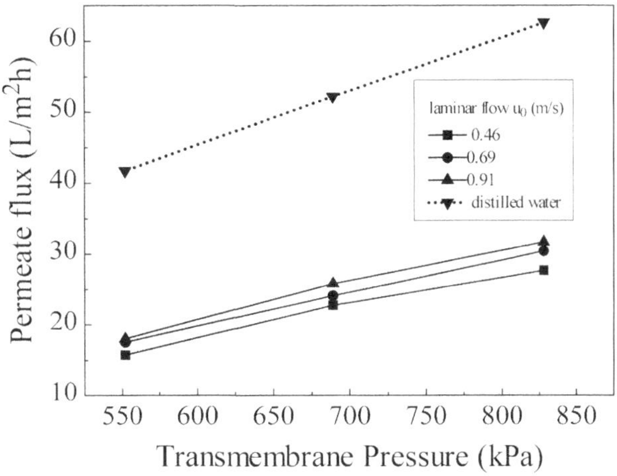

Fig. 6.16 Variation of permeate flux with operating pressure in step 1 (NF)

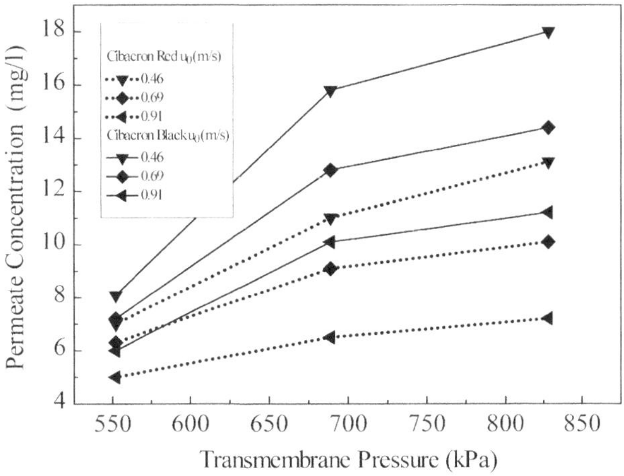

Fig. 6.17 Variation of CR and CB concentrations with operating pressure in permeate after first stage NF

adjust in two-stage NF; (2) final COD is in between 64 and 70 mg/l and hence within the limit of permissible value (250 mg/l); and (3) recovery of NaCl is 92–94% and hence the filtrate can be directly recycled back in upstream processes.

6.2.2 Comparison of AOP Followed by Nanofiltration (Scheme 1) and Two-Stage Nanofiltration (Scheme 2)

As the extent of dye removal is considered, both schemes are comparable. But, in Scheme 2, the pH of the solution does not change and hence can be recycled back. Similarly, salt recovery is also better in Scheme 2, and since the pH of filtrate does not change, it can be directly recycled in upstream process thereby reducing the

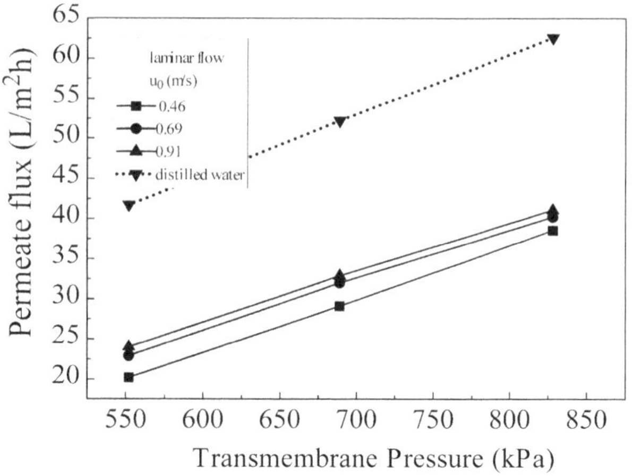

Fig. 6.18 Variation of permeate flux with operating pressure after two-stage NF

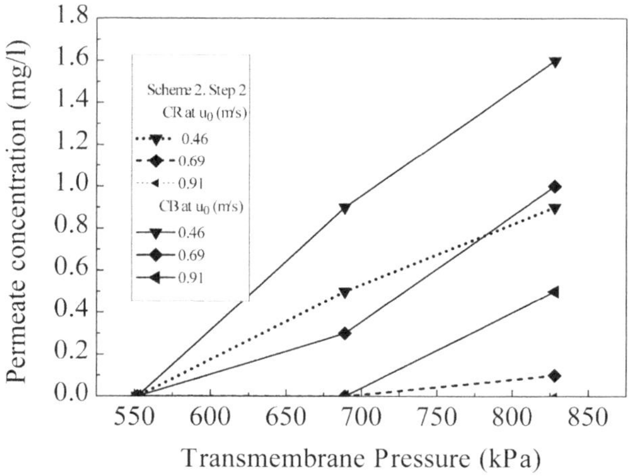

Fig. 6.19 Variation of CR and CB concentrations with operating pressure in permeate after two-stage NF

Table 6.10 Characterization of effluent at the end of double-stage NF

Sl. No.	Pressure (kPa)	Cross flow velocity (m/s)	conductivity × 10^{-5} (mS/m)	% NaCl recovered	TDS	COD	pH
1	552	0.46	112.1	92.6	73.0	64	10.9
2	-do-	0.69	111.2	91.6	72.4	64	10.9
3	-do-	0.91	110.3	90.5	72.0	63	10.8
4	689	0.46	113.6	94.4	74.0	67	10.9
5	-do-	0.69	112.9	93.6	73.4	64.5	10.9
6	-do-	0.91	112.5	93.1	73.3	63.5	10.9
7	828	0.46	114.2	95.2	74.4	71.5	11.0
8	-do-	0.69	113.6	94.4	74.0	66	10.9
9	-do-	0.91	113.0	93.6	73.5	64.5	10.8

operating cost of the plant. This scheme is a step toward achieving zero discharge plant.

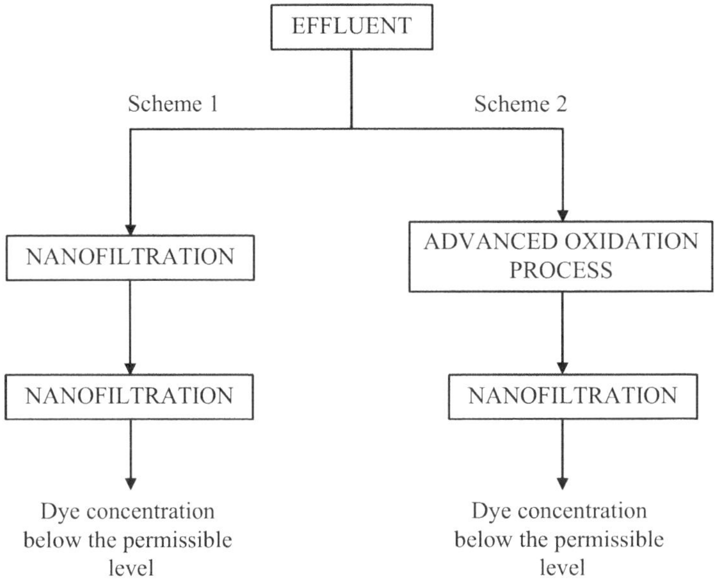

6.3 Adsorption Followed by Microfiltration

A low-energy intensive alternative to two-stage NF is a hybrid process comprising adsorption followed by MF. This was tested using a synthetic solution containing four reactive dyes brown, yellow, red, and black. Basic properties of these dyes are shown in Table 6.11. MF was tested in a tubular 19 channel ceramic membrane (length and diameter of each channel were 0.2 m and 0.03 m) with water permeability 4×10^{-10} m/Pa.s and effective filtration area 4.8×10^{-2} m^2. Concentration of each dye was about 50 mg/l (Pearce et al. 2003; Arami et al. 2005; Selcuk et al. 2005).

Commercial granulated activated carbon was used for adsorption of the dyes in solution phase.

The range of transmembrane pressure drop selected in this work is based on the typical operating range of microfiltration. The cross flow rate varied from 50 to 100 L/h considering the output capacity of the pump used for the filtration experiment.

The adsorption isotherm result for Langmuir and Freundlich isotherm are presented in Table 6.12.

Table 6.11 Chemical structure and physical properties of the reactive dyes

Dye	Chemical structure	λ_{max} (nm)	Molecular weight (g/mol)	Molecular volume (Å^3)
Reactive yellow 15	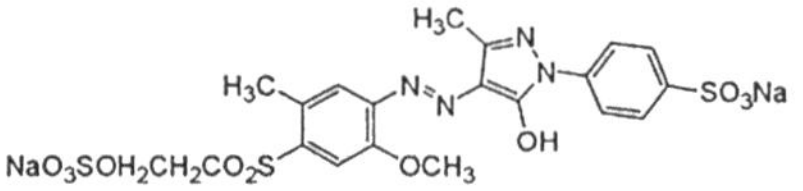	416	634.57	498
Reactive black 5	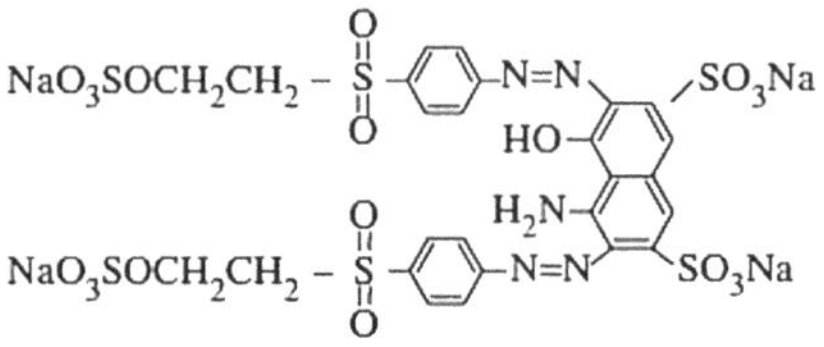	596	991.82	578
Reactive red 24	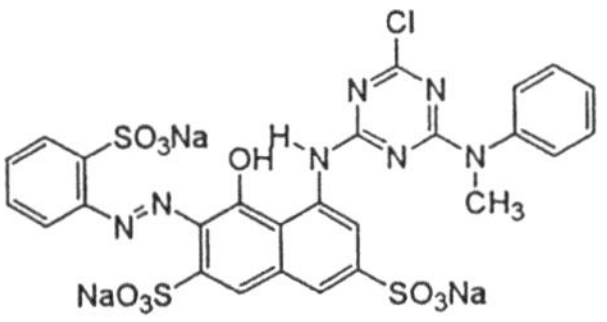	531	808.48	667
Reactive brown 10	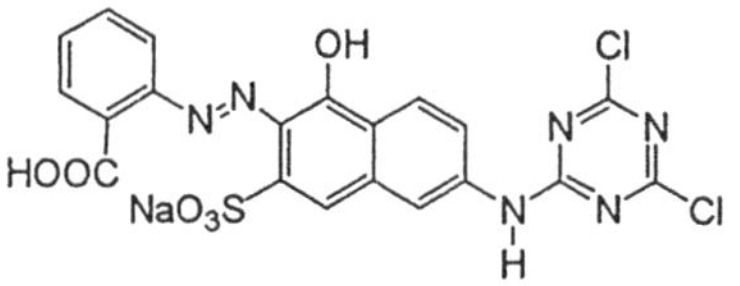	405	557.30	798

Adsorption of reactive dyes with variation in pH and salt concentration are presented in Fig. 6.20, respectively. It is evident from Fig. 6.20 that maximum adsorption occurs at lower pH in the range 2–5. Extent of adsorption in this pH range is also uniform. This result is in accordance with the observations reported by Al-Degs et al. (2008). The value of pH_{ZPC} of commercial activated carbon is 9.0, and pKa values of the reactive dyes are in the range of 4.4–5.5 (Al-Degs et al. 2008). At lower pH the sulfonate groups of the dyes were protonated making it -SO_3H, which makes the dye solution almost neutral or slightly positive. Therefore, it is likely that electrostatic interaction dictates repulsion or marginal interaction between the dye molecules and positively charged adsorbent surface, resulting to lower adsorption. On the other hand, in this pH range, the adsorption is maximum and close to 100%. This clearly indicates that electrostatic interaction is not the sole mechanism of facilitation of adsorption of reactive dyes on the activated carbon surface. Strong hydrogen bonding between the dyes and carbon surface and hydrophobic interaction between them may be the reason of enhanced adsorption at lower pH range (Newcombe et al. 1996; Newcombe and Drikas 1997; Al-Degs et al. 2008). At higher pH above 9, both dye molecules and the adsorbent possess negative charge and electrostatic repulsion plays an important role, leading to lower adsorption. At pH 11, the adsorption of various dyes is in the range of

Table 6.12 Values of Langmuir and Freundlich isotherm constants

Dye	Langmuir constants K (l/mg)	Q_m (mg/g)	R^2	Freundlich constants K_f (mg$^{1-1/n}$.l$^{1/n}$.g^{-1})	n	R^2	Separation factor (R_L)
Reactive yellow dye	0.92	152	0.99	50.7	4.5	0.68	0.001–0.1
Reactive brown dye	0.85	147	1.0	47.2	4.4	0.69	0.001–0.11
Reactive red dye	0.77	135	1.0	39.8	4.1	0.68	0.001–0.12
Reactive black dye	0.67	130	0.99	39.1	4.1	0.61	0.0012–0.13

Reproduced from Chakraborty et al. (2005). With permission from Elsevier

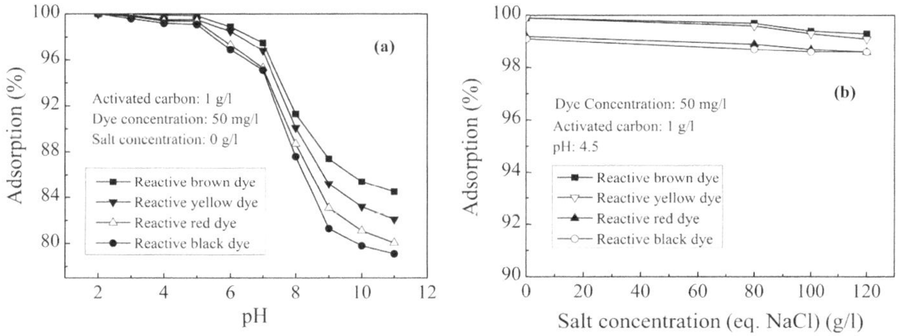

Fig. 6.20 Effect of (**a**) pH and (**b**) salt concentration on adsorption of reactive dyes on activated carbon (Reproduced from Karmakar et al. (2015). With permission from Taylor & Francis Ltd)

80–85%. This also confirms that electrostatic interaction is not the sole mechanism of adsorption. In between pH 4 and 6, the dye molecules are practically neutral, and hence their adsorption onto the carbon surface (positively charged) is not facilitated by electrostatic interactions, thereby decreasing the extent of adsorption.

From Fig. 6.20b, it can be seen that with increase in salt concentration in the feed, the adsorption of reactive dye decreases marginally at pH 4.5. At this pH, the adsorbent is positively charged and the dye molecules are almost neutral. Therefore, there is a competitive adsorption between the negatively charged chloride ions and almost neutral dye molecules. However, chloride ions being much smaller in size, this phenomenon leads to slight decrease in dye adsorption. With the increase in salt concentration from 80 to 120 g/l, the percentage decrease in adsorption of dyes by activated carbon is 0.5–0.8%, which is negligible.

Figure 6.21a–d represent the permeate flux profiles corresponding to different dye solution having initial concentration of 50 mg/l (before adsorption) with adsorbent dose 1 g/l at pH 4.5. Activated carbon particles, being larger in size (average diameter is 14 μm) compared to pore size of the membrane (0.2 μm),

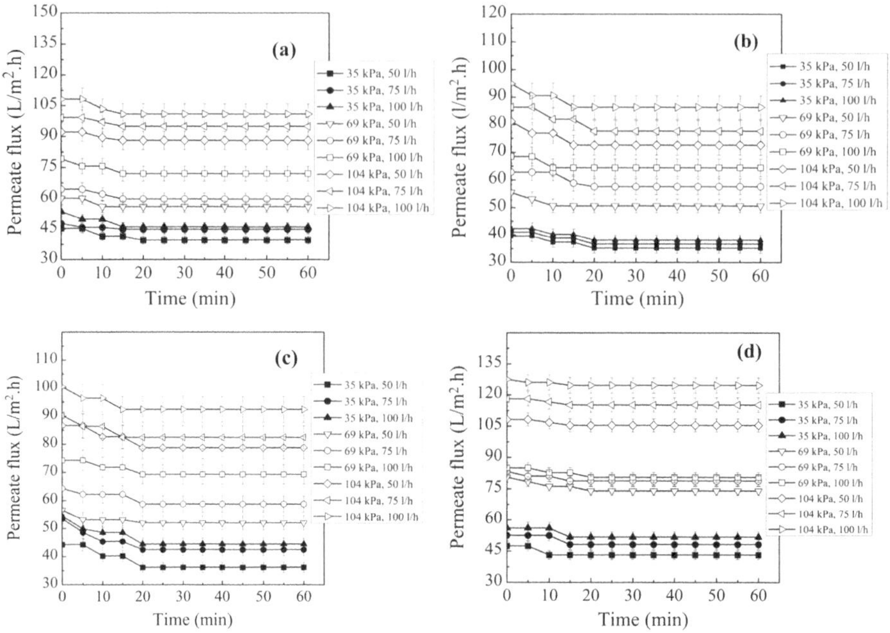

Fig. 6.21 Profiles of permeate flux under cross flow recycle mode (dye concentration, 50 mg/l). (**a**) Reactive yellow dye, (**b**) reactive black dye (**c**) reactive red dye, (**d**) reactive brown dye (Reproduced from Karmakar et al. (2015). With permission from Taylor & Francis Ltd)

deposit on the membrane surface forming a cake type of layer almost at the start of the experiment. For example, at TMP at 104 kPa, the pure water flux was 150 L/m^2.h. At the same TMP, the measured flux at the beginning is 110 L/m^2.h for yellow dye, 95 L/m^2.h for black, 100 L/m^2.h for red, and 128 L/m^2.h for brown reactive dye. These data confirm the instantaneous growth of cake layer of carbon particles on the membrane surface. These figures exhibit the expected trends that the permeate flux increases with TMP due to increase in driving force and with cross flow velocity due to increase in the shearing effect imposed by forced convection induced by feed cross flow. For example, in case of reactive yellow dye, at 35 kPa TMP, the steady-state flux is 40 L/m^2.h at a flow rate of 50 l/h and that for 69 kPa and 104 kPa is 56 L/m^2.h and 88 L/m^2.h, respectively, at the same cross flow rate. Thus, the increment in steady-state flux is around 40% for 69 kPa and 120% for 104 kPa (compared to TMP 35 kPa). Similar trends are observed for other dyes. These data clearly confirm that within the range of TMP used in this study, the filtration is not pressure independent. Pressure-independent cake filtration may occur at higher TMP. Also with the increase in cross flow rate from 50 l/h to 100 l/h, the increment in permeate flux is around 15%, 11%, and 15% for 35, 69, and 104 kPa, respectively. However, the axial pressure drop was in the range of 7–10 kPa only for the cross flow rates considered herein. This shows improvement in mass transfer with cross flow and reduction in the cake layer resistance against the solvent flux.

From Fig. 6.21b, it is also observed that the permeate flux for reactive black dye is the lowest among all the dyes corresponding to identical set of operating conditions. This is because, the molecular size of reactive black dye is the highest among the dyes, as evident from their molecular weight and volume as shown in Table 6.11. Thus, the effective size of the adsorbent particles loaded with the adsorbed dyes is more for reactive black, leading to enhanced thickness of the cake layer, thereby lowering the permeate flux. For example, at 104 kPa TMP and 100 l/h cross flow rate, the steady-state permeate flux is 90 L/m^2.h for reactive black. Under the same operating conditions, the permeate flux is 105, 95, and 125 L/m^2.h for reactive yellow, red, and brown dyes, respectively. The trend is similar for other operating conditions.

As observed in Fig. 6.21c, the steady-state flux for reactive red under all operating conditions is less than that of reactive yellow but more than that of reactive black. The variation of flux values is marginal. This is due to the difference in molecular sizes of the dyes (refer to Table 6.5 for the molecular weight and volume of dyes). Increasing cross flow rate from 50 to 100 l/h, the permeate flux increases from 36 to 44 L/m^2.h (22%) at 35 kPa, 52 to 69 L/m^2.h (33%) at 69 kPa, and 79 to 92 L/m^2.h (16%) at 104 kPa. The enhancement of permeate flux with cross flow rate is due to the reduction of thickness of the cake layer at higher cross flow rate, imparting more shear force on the cake layer as described earlier. At higher TMP, the enhancement due to increase in cross flow rate of identical magnitude leads to reduced increment (16% compared to 33%) indicating that the cake layer becomes more compact imparting extra resistance to the permeation of the solvent at higher TMP. In Fig. 6.21d, the flux profiles of reactive brown are presented. The permeate flux for reactive brown is the highest among four dyes. The molecular size of this dye is the lowest (refer to Table 6.5).

The permeate flux profiles of various dyes are shown in Fig. 6.22 at dye concentration of 150 mg/l (before adsorption). At enhanced feed concentration, additional dye particles are adhered to the surface of activated carbon making it more bulky, offering more resistance to the solvent flux during its passage through cake layer formed by adsorbents. This leads to decrease in permeate flux compared to that at 50 mg/l dye concentration before adsorption under all operating conditions as shown in Fig. 6.21. However, since the particle size of the dyes is quite small compared to that of activated carbon, the decrease in flux is marginal. For example, at 104 kPa TMP and 100 l/h cross flow rate, the steady-state permeate flux of reactive yellow is 100 L/m^2.h at 50 mg/l dye concentration before adsorption. Under the same operating conditions, the permeate flux is reduced to 94 L/m^2.h at 150 mg/l dye concentration before adsorption. This trend is similar for all other operating conditions for all dyes.

In the adsorption-MF hybrid process, the dyes are removed in the adsorption step and dye-loaded adsorbents are removed by MF. Thus, the removal of dyes occurs in the first step itself, and hence their removal is independent of the operating conditions of MF.

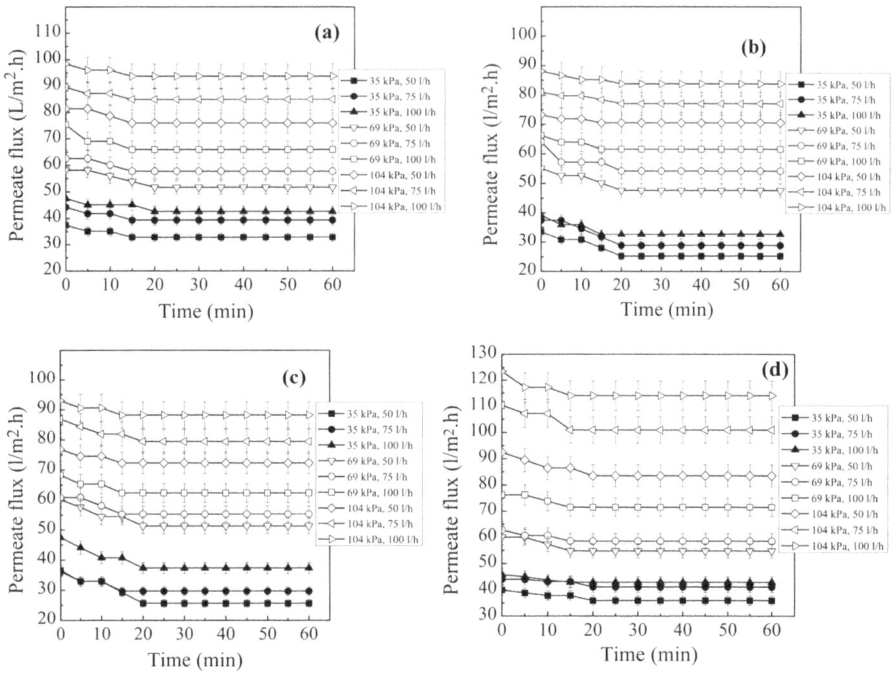

Fig. 6.22 Profiles of permeate flux under cross flow under cross flow recycle mode (dye concentration, 150 mg/l). (**a**) Reactive yellow dye, (**b**) reactive black dye, (**c**) reactive red dye, (**d**) reactive brown dye (Reproduced from Karmakar et al. (2015). With permission from Taylor & Francis Ltd)

However, adsorbent dose plays an important role in dye removal. Removal of all dyes is more than 99% at 50 mg/l feed concentration at 1 g/l dose of adsorbent as shown in Fig. 6.23a. Under the same loading of adsorbent, the removal of dyes at feed concentration 150 mg/l (before adsorption) is more than 99% except reactive black (79–80%) and red (85–86%). The molecular weight and hence the size of reactive black and red are the highest (991 and 808 g/mol, respectively, in Table 6.5) among the four dyes. Thus, less number of dye particles at higher concentration (150 mg/l) is adsorbed on the adsorbent compared to those corresponding to 50 mg/l as shown in Fig. 6.23b. On the other hand, for other dyes with molecular weight less than that of reactive yellow (molecular weight 634 g/mol), the size of the dye molecules is such that even at 150 mg/l feed concentration, the adsorption is complete. However, the trial runs indicate that slight increase in adsorbent dose to 1.5 g/l leads to complete removal of these two reactive dyes even at concentration 150 mg/l as evident from Fig. 6.23b. The other quality parameters of the feed are pH 4.2, conductivity 106.3 mS/cm, and total dissolved solids 72.5 g/l. These values for the permeate are 4.2, 44 mS/cm and 31 g/l, respectively.

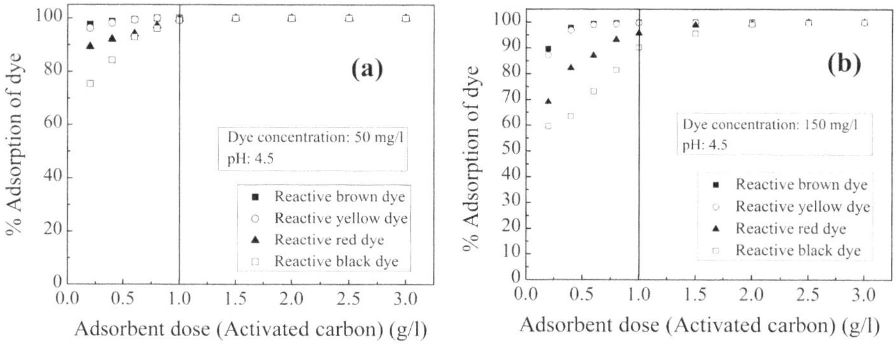

Fig. 6.23 Percentage adsorption of (**a**) 50 mg/l and (**b**) 150 mg/l of reactive dyes with adsorbent dose (Reproduced from Karmakar et al. (2015). With permission from Taylor & Francis Ltd)

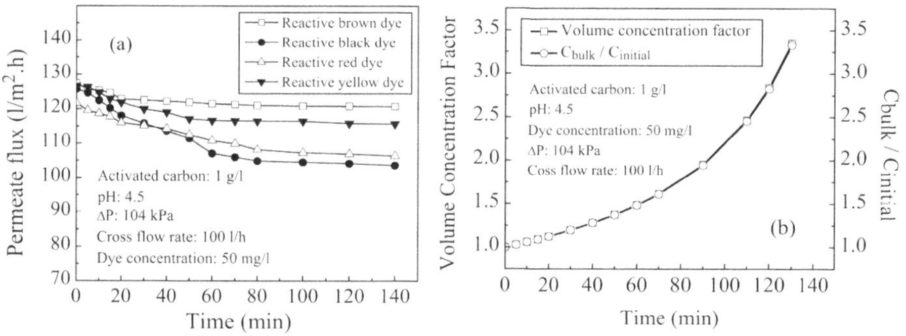

Fig. 6.24 Profiles of (**a**) permeate flux and (**b**) VCF under batch mode of cross flow filtration (Reproduced from Karmakar et al. (2015). With permission from Taylor & Francis Ltd)

In the batch mode, the permeate is not recycled back to the feed tank. As a result, the feed concentration increases due to decrease in feed volume during filtration. The volume concentration factor, VCF ($\frac{V_0}{V^t}$ where V_0 is defined as the initial feed volume and V^t is defined as the volume of the feed after a certain time "t"), increases with time. Since the feed concentration increases in time, the permeate flux never attains a steady state under the batch mode. The profiles of permeate flux for all the dyes at 50 mg/l feed concentration and at 1 g/l adsorbent dose are shown in Fig. 6.24a. The striking feature of this figure compared to that under total recycle mode (Fig. 6.21) is that the flux decline is more prominent in this case. The decline is sharp up to 50 min for reactive black and it is up to 20 min for reactive brown. As the filtration progresses, the concentration of solutes (adsorbents) increases in the feed, resulting to more deposition of them on the membrane surface, leading to a thicker cake layer. Since the cake layer grows in time, the resistance against the solvent flux also increases and finally permeate flux declines to a substantial extent. For example, in case of reactive black, the permeate flux declines by about 20%

(127 to 103 L/m^2.h) within 140 min. However, the profiles of flux decline for different dyes reveal an interesting trend. Permeate flux decline is steeper in case of dyes of higher molecular weight. Flux decline for reactive brown having the lowest molecular weight (557) has the most gradual decline (from 127 to 120 L/m^2.h in 120 min). Bulkier dyes (higher molecular weight) make the adsorbent-dye agglomerate larger in size, thereby forming a thicker cake layer, and hence the flux decline is more. Flux decline is greater for reactive black > red > yellow > black in that order, and the molecular weight and volume of the dyes also decrease in the same order (refer to Table 6.5). A profile of VCF and bulk concentration of activated carbon is presented in Fig. 6.24b. It was observed from this figure that VCF attains the value 3.3 within 2.5 h of operation in batch mode. At the same time, the bulk concentration also increases up to three times the initial feed concentration. Similar trends (including numerical values) are observed for other dyes. This is due to the fact that the adsorbent particles are mainly filtered in MF.

References

Abdessemed D, Nezzal G (2002) Treatment of primary effluent by coagulation- adsorption-ultrafiltration for reuse. Desalination 152:367–373

Al-Degs YS, El-Barghouthi MI, El-Sheikh AH, Walker GM (2008) Effect of solution pH, ionic strength, and temperature on adsorption behavior of reactive dyes on activated carbon. Dyes Pigments 77:16–23

Aouni A, Fersi C, Ali MBS, Dhahbi M (2009) Treatment of textile wastewater by a hybrid electrocoagulation/nanofiltration process. J Hazard Mater 168:868–874

Arami M, Limaee NY, Mahmoodi NM, Tabrizi NS (2005) Removal of dyes from colored textile wastewater by orange peel adsorbent: equilibrium and kinetic studies. J Colloid Interf Sci 288:371–376

Arana J, Rendón ET, Rodříyguez JMD, Melián JAH, Ďíyaz OG, Pena JP (2001) Highly concentrated phenolic wastewater treatment by the photo-Fenton reaction, mechanism study by FTIR-ATR. Chemosphere 44:1017–1023

Baudin I, Chevalier MR, Anselme C, Cornu S, Laine JM (1997) L'Apie and Vigneux case studies: first months of operation. Desalination 113:273–275

Chakraborty S, De S, Basu JK, Dasgupta S (2005) Treatment of a textile effluent: application of a combination method involving adsorption and nanofiltration. Desalination 174:73–85

Dhale AD, Mahajani VV (2000) Studies on treatment of disperse dye waste: membrane-wet oxidation process. Waste Manage 20:85–92

Ellouze E, Tahri N, Amar RB (2012) Enhancement of textile wastewater treatment process using nanofiltration. Desalination 286:16–23

Ghoreishi SM, Haghighi R (2003) Chemical catalytic reaction and biological oxidation for treatment of non-biodegradable textile effluent. Chem Eng J 95:163–169

Kang SF, Liao CH, Chen MC (2002) Pre-oxidation and coagulation of textile wastewater by the Fenton process. Chemosphere 46:923–928

Karmakar S, Mondal M, Ghosh S, Bandopadhaya S, Majumder S, De S (2015) Removal of reactive dyes using a high throughput-hybrid separation process. Desalination Wat Treat. https://doi.org/10.1080/19443994.2015.1033762

Kitis M, Adams CD, Daigger GT (1999) The effects of Fenton's reagent pretreatment on the biodegradability of nonionic surfactants. Water Res 33:2561–2568

Kunz A, Reginatto V, Duran N (2001) Combined treatment of textile effluent using the sequence Phanerochaete chrysosporium–ozone. Chemosphere 44:281–287

Kuo WG (1992) Decolourising dye wastewater with Fenton's reagent. Water Res 26:881–886

Lin SH, Lin CM (1993) Treatment of textile waste effluents by ozonation and chemical coagulation. Water Res 27:1743–1748

Lin SH, Wang CS (2002) Treatment of high-strength phenolic wastewater by a new two-step method. J Hazard Mater B90:205–216

Lunar L, Sicilia D, Rubio S, Perez-Bendito D, Nickel U (2000) Degradation of photographic developers by Fenton's reagent: condition optimization and kinetics for metal oxidation. Water Res 34:1791–1802

Meier J, Melin T, Eilers LH (2002) Nanofiltration and adsorption on powdered adsorbent as process combination for the treatment of severely contaminated wastewater. Desalination 146:361–366

Newcombe G, Drikas M (1997) Adsorption of NOM activated carbon: electro-static and non-electrostatic effects. Carbon 35:1239–1250

Newcombe G, Donati C, Drikas M, Hayes R (1996) Adsorption onto activated carbon: electrostatic and non-electrostatic interactions. Water Supply 14:129–144

Neyens E, Baeyens J (2003) A review of classic Fenton's peroxidation as an advanced oxidation technique. J Hazard Mater B 98:33–50

Pearce CI, Lloyd JR, Guthrie JT (2003) The removal of colour from textile wastewater using whole bacterial cells: a review. Dyes Pigment 58:179–196

Pérez M, Torrades F, Dome'nech X, Peral J (2002) Fenton and photo-Fenton oxidation of textile effluents. Water Res 36:2703–2710

Pigmon HM, Brasquet CF, Cloiree PL (2003) Adsorption of dyes onto activated carbon cloths: approach of adsorption mechanisms and coupling of ACC with ultrafiltration to treat colored wastewaters. Sep Purif Technol 31:3–11

Rathi A, Rajor HK, Sharma RK (2003) Photodegradation of direct yellow-12 using UV/H_2O_2/Fe^{2+}. J Hazard Mater B 102:231–241

Robinson T, McMullan G, Marchant R, Nigam P (2001) Remediation of dyes in textile effluent: a critical review on current treatment technologies with a proposed alternative. Bioresour Technol 77:247–255

Rodrigues CSD, Madeira LM, Boaventura RAR (2009) Treatment of textile effluent by chemical (Fenton's reagent) and biological (sequencing batch reactor) oxidation. J Hazard Mater 172:1551–1559

Selcuk H (2005) Decolorization and detoxification of textile wastewater by ozonation and coagulation processes. Dyes Pigment 64:217–222

Tahri N, Masmoudi G, Ellouze E, Jrad A, Drogui P, Amar RB (2012) Coupling microfiltration and nanofiltration processes for the treatment at source of dyeing-containing effluent. J Clean Prod 33:226–235

Weinberg HS, Glaze WH (1997) A unified approach to the analysis of polar organic by-products of oxidation in aqueous matrices. Water Res 31:1555–1572

Chapter 7
Micellar-Enhanced Ultrafiltration (MEUF)

Abstract Typically, the micelles are charged globular particles formed due to agglomeration of surfactant molecules when present above a critical concentration in water. Due to the charge interactions, micelles act as excellent binding agent to the dye molecules, thus solubilizing it. Even in case of uncharged dyes, nonionic surfactant micelles can solubilize it inside the micelle core. Once the low molecular weight dyes are solubilized by the micelles, the micelle-dye complex can be effectively separated by an open membrane (large pore size, large molecular weight cutoff), thus producing high throughput of clean water per unit pressure drop. This chapter deals with various types of dyes and surfactant combination in different membrane separation systems, analyzing its performance and relative separation efficiencies.

Keywords Micellar-enhanced ultrafiltration • Micelle • Ultrafiltration • Gel layer • Surfactant

Micellar-enhanced ultrafiltration employs surfactant micelles to solubilize inorganic and organic pollutants from the effluent stream and subsequently filtered using an open membrane to restrict the micelle-pollutant complex in the permeate stream. Micelles being large in size can be removed along with the solubilized organic contaminants using a relatively porous membrane at lower operating pressure. More than 90% removal efficiency along with high throughput can be attained by using pollutant-specific surfactant (or mixed surfactant system) and high permeability membrane, depending on the charge and other physical properties of the contaminants. Since more open-sized membranes are used, the process is energy efficient. After the separation of the waste from the effluent/process stream, recovery of surfactant is also possible. Therefore, MEUF may be an alternative to overcome the limitations of the RO and NF processes (De and Mondal 2012).

S. Mondal et al., *Advances in Dye Removal Technologies*, Green Chemistry and Sustainable Technology, DOI 10.1007/978-981-10-6293-3_7

7.1 Micelle Formation and Solubilization

Surfactant molecules are amphiphilic in nature having a hydrophilic (head) and hydrophobic (tail) part. The hydrophobic part is generally a long-chain hydrocarbon or aromatic ring. The head part of the surfactant is hydrophilic due to presence of some electrophilic or nucleophilic groups. In aqueous solution, and at low concentration, surfactant molecules are present in monomer form. The concentration above which surfactant molecules are dynamically associated (~50–150 monomers) to form molecular aggregates (micelles) is called critical micellar concentration (CMC). Depending on the nature of the surfactant and solvent, the aggregates may form micelles, reverse micelles, microemulsion, vesicles, monolayers, and bilayers (Scamehorn and Harwell 1989). This configuration allows the removal of the hydrocarbon chains from the aqueous environment, while the hydrophilic groups are still exposed to the polar solvent.

When surfactants are present in aqueous solutions at concentration above the critical micellar concentration (CMC), they are in two possible states; one is monomer and the other is aggregated form of monomers called micelles. As the total concentration of surfactant is increased beyond the CMC, almost all of the added surfactants increase the micelle concentration. In MEUF, the surfactant concentration is designed to be far above the CMC, so that free monomer constitutes only a very small fraction of the total surfactant in solution.

Organic or inorganic compounds may dissolve or solubilize within micelles at four different locations: (1) in the hydrocarbon core, (2) in the palisade layer or region between the hydrocarbon core, (3) adsorbed on the micelle surface, and (4) in the polyoxyethylene shell of micelles composed of nonionic surfactants (Elworthy et al. 1968; Rosen 2004; Mukherjee 1979). Organic solute species tend to solubilize preferentially in the region within the micelle which are similar chemically and polarity to these molecules. Micelles composed of ionic surfactants have a high charge, and counterions (opposite charge to the surfactant, generally metal ions, oxyanions, phosphates, ferrocyanides, etc.) will bind to the surface of the charged micelle. Multivalent counterions will bind in higher proportion to the surfactant compared to the monovalent counter ions. Higher is the valence state, greater is the binding (Purkait et al. 2005a).

Solubilization into micelles can be viewed conceptually as being similar to extraction of a dissolved organic from an aqueous phase into an immiscible extraction phase. In MEUF, the extractant phase (micelles) is dispersed in the aqueous phase. As the surfactant concentration is increased (micelle concentration is increased) in a stream containing organic pollutant, the fraction of the organics solubilized in micelles increases and the fraction present as unsolubilized species in the water decreases. As an approximation, the ratio of the concentration of pollutant in solubilized form to that in unsolubilized form is proportional to the surfactant concentration constituting micelles. If the surfactant is well above the CMC and the vast majority of the organic solute is solubilized in micelles, the concentration of the unsolubilized organic is approximately inversely proportional to the surfactant concentration (Dasan et al. 1988).

Solubilization, in a broad sense, encompasses both co-micellization and hydrotropy, as it involves the preparation of a thermodynamically stable isotropic solution of a solute. Several practical applications are encountered related to solubilization system, such as drug formulation (Attwood and Florence 1983), paints and cosmetics (Gillberg 1984), catalysis of chemical reactions (Fendler and Fendler 1975), separation of close boiling compounds (Armstrong 1985), modified fuels (Gillberg and Friberg 1978), etc.

7.2 Selection of Surfactant

For the selection of a surfactant in MEUF, some of the desirable characteristics are: (1) high solubilization capacity for the solute; (2) formation of large micelles, such that large membrane pore size resulting high throughput can be used; (3) low monomer concentration (so little surfactant is wasted); (4) minimal phase-separation problems (macroemulsion formation, precipitation, gelling, etc.); (5) The Krafft point temperature for ionic surfactant should be very low. Nonionic surfactants do not have a specific temperature effect on property curve; and (6) CMC of the surfactant should be as low as possible.

One desirable characteristic of a surfactant is a long hydrocarbon chain, since it results in large micelles, high solubilization, and low monomer concentrations. Anionic surfactants are restricted to hydrocarbon chain lengths of about 12 carbons or less if the process is to be applied at room temperature. The Krafft temperature is above room temperature for longer hydrophobic groups, so the surfactant precipitates and MEUF is ineffective (Rosen 2004).

Nonionic surfactants forming micelles can have high solubilization capacities (per mole of surfactant) and have low monomer concentration in micellar solutions. These surfactants would appear to be good candidates for use in MEUF. The solubilization capacities of the nonionic surfactants are not very high compared to that of anionic or cationic surfactants because of their high molecular weights. Since surfactants are sold on a weight basis rather than on a molar basis, nonionic surfactants are not economically attractive for use in MEUF (Dunn et al. 1985).

Ionic surfactants generally have much lower Krafft temperatures than those of nonionic surfactants of corresponding hydrophobic group size. Therefore, ionic surfactants with large hydrophobic groups can be used in MEUF, resulting in large micelles, high solubilization capacities, and low monomer concentrations. Ionic surfactants do not pose significant environmental risks (Boethling 1984). In general, ionic surfactants are the surfactants of choice in MEUF.

7.3 Applications of MEUF

MEUF is a nondestructive separation technique used to separate a dissolved material from a stream. This can be applied to a situation where the purified stream produced is the main goal or to a situation where the aim is the concentration of some valuable product for further downstream purification. Wastewater cleanup would be an example of the need to purify an aqueous stream for emission or reuse.

The pollutants those can be removed most effectively from wastewater using MEUF are, in general, are least soluble in water Saraf and Bhagwat (1995). However, if the solubility is low enough, not enough pollutant could dissolve in the stream to reach toxic levels and cause a problem. Therefore, there is a solubility range where MEUF can be useful in pollution control.

So far, MEUF is used to separate different organics (Boethling 1984; Saraf and Bhagwat 1995; Alesson et al. 2001a, b; Sabaté et al. 2002; Syamal et al. 1997; Adamczak et al. 1999), chromate (Gzara and Dhahbi 2001), and dye (Purkait et al. 2003, 2004) using different surfactants. A review of literature related to MEUF also reveals that a number of studies have been carried out to remove multivalent metal ions from aqueous medium (Akita et al. 1999; Tung et al. 2002; Juang et al. 2003). However, most of the researches are focused on the MEUF of single component, and few researchers reported the MEUF of multicomponent system (Baek et al. 2003a, b; Jadhav et al. 2001). Competition due to difference in binding power of different compounds plays an important role during the MEUF of multicomponent system.

Untreated polluted water contains a large number of metal ions (Cu^{2+}, Cr^{3+}, Zn^{2+}, Sr^{2+}, Ca^{2+}, Pb^{2+}, Ni^{2+}, Mn^{2+}, Co^{2+}, As^{3+}, Fe^{2+}, etc.), anion (oxyanions, phosphates, ferrocyanide, etc.), and organic compounds like phenol, β-naphthol, p-nitrophenol, m-nitrophenol, catechol, o-chlorophenol, aniline etc. MEUF is particularly effective for removal of single components, such as Cd^{2+} (Ke et al. 2007), Mn^{2+} (Juang et al. 2003), Zn^{2+} (Rahmanian et al. 2010), Cu^{2+} (Liu and Li 2005), Cr^{3+} (Witek et al. 2006), Pd^{2+} (Ghezzi et al. 2008), Au^{3+} (Akita et al. 1997), etc., for simultaneous removal of Ni^{2+} and Co^{2+} (Karate and Marathe 2008) and Ni^{2+} and Zn^{2+} (Channarong et al. 2010).

Removal of very small amounts of organic substances in water is removed effectively by MEUF (Dunn et al. 1985, 1987; Gibbs et al. 1987; Adamczak et al. 1999; Sabaté et al. 2002; Syamal et al. 1997). Generally, dissolved organics are present along with heavy metals from industrial wastewater. Simultaneous removal of phenol or orthocresol and zinc or nickel ions using MEUF is reported by Dunn et al. (1989) and Witek et al. (2006).

Removal of more than one anionic pollutant such as chromate, nitrate, permanganate, etc., using MEUF has been performed (Baek et al. 2003a, Baek and Yang 2004; Purkait et al. 2005a, b). Separation of a mixture of anionic and cationic pollutants can be treated by mixed micelle system. Micelles of cationic surfactants solubilize the anionic pollutants, while the anionic surfactants solubilize the cationic pollutants.

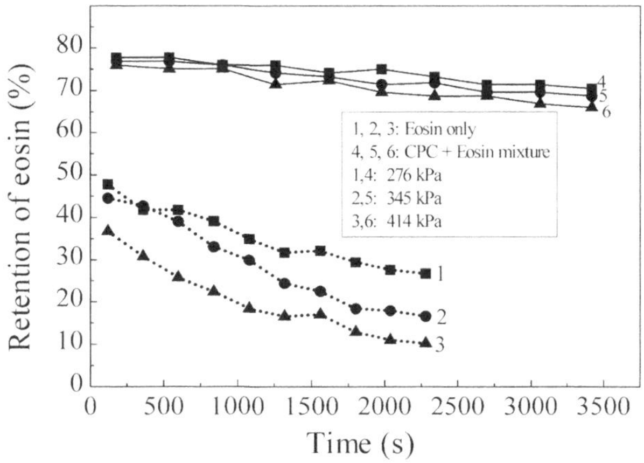

Fig. 7.1 Variation of observed retention of eosin with time at different operating pressure differences. Feed eosin concentration is 10×10^{-3} kg/m^3 and CPC concentration is 10 kg/m^3 (Reproduced from Purkait et al. (2004). With permission from Elsevier)

7.4 Micellar-Enhanced Ultrafiltration of Dye

Removal of eosin dye by cationic micelles is reported in the subsequent sections. The experiments were conducted in a typical batch filtration unit using a 1 kDa MWCO polyamide with an effective membrane area of 13.4 m^2. The permeability of the membrane is estimated using distilled water and they are 3.57×10^{-11} m/Pa. s. Variation of eosin retention is conducted with and without surfactant is presented in Fig. 7.1 for different transmembrane pressures varying from 276 to 552 kPa and a typical dye concentration of 0.01 kg/m^3.

In both the cases, it is observed that the retention decreases with time. As the filtration progresses, more solutes will be deposited on the membrane surface leading to an increase in the membrane surface concentration (concentration polarization). This results in an increase in the convective transport of the solutes to the permeate side, thereby increasing the permeate concentration and subsequently decreasing the observed retention with time. For higher operating pressure, at the same time instant, retention is low. This is due to the fact that at higher operating pressure, the convective transport of the solutes through the membrane is high leading to higher value of the permeate concentration, lowering observed retention. It can be observed from the figure that for dye solution without surfactant, the observed retention varies from 10 to 26% at the end of the operation under different operating pressures. When surfactant is used, the retention of the dye has been significantly increased, as shown in the figure to about of 68–74%, at the end of the experiment. This clearly indicates that the dyes are solubilized within the surfactant micelles, which are subsequently rejected by the ultrafiltration membrane.

Variation of the permeate flux with and without surfactants is shown in Fig. 7.2. Two trends can be observed from this figure. First, in both cases, the permeate flux declines over the time of operation, and, second, the permeate flux is more at higher operating pressure. As discussed earlier, due to concentration polarization, the

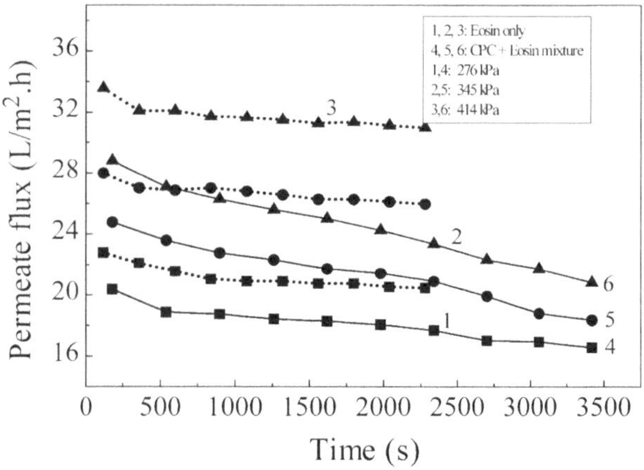

Fig. 7.2 Variation of the permeate flux with time at different operating pressure differences. Feed eosin concentration is 10×10^{-3} kg/m^3 and CPC concentration is 10 kg/m^3 (Reproduced from Purkait et al. (2004). With permission from Elsevier)

membrane surface concentration increases with time. This increases the osmotic pressure at the membrane solution interface and therefore reduces the driving force for the permeating solution. This leads to a decline of the permeate flux with the time of operation. For example, in the case of filtration of eosin only, the permeate flux decreases from about 33.5 L/m^2.h (at 180 s) to 31 L/m^2.h (at 2200 s) at 414 kPa. This indicates flux drop of about 7.5% during 40 min of operation. For the case of MEUF of eosin and CPC mixture, the permeate flux drops from 28.8 L/m^2.h (at 180 s) to 20.9 L/m^2.h (at 3400 s.) at 414 kPa. Therefore, the flux of the micelle containing mixture drops about 27% during 57 min of operation. At the same instance of time, increase in operating pressure simply increases the driving force across the membrane; thus, flux is more at higher pressure. For example, in case of filtration of only eosin, at the end of the experiment, the permeate flux is 20.5 L/m^2.h at 276 kPa pressure and it is 31 L/m^2.h at 414 kPa. This indicates about 51% increase in flux when pressure increases from 276 to 414 kPa. On the other hand, in case of MEUF of eosin and CPC mixture, the permeate flux increases from 16.6 L/m^2.h to 20.9 L/m^2.h at the end of the experiment, while pressure increases from 276 to 414 kPa, indicating about 26% increase in flux.

It is also evident from the figure, that the permeate flux is more for only dye solution compared to that with the surfactants. By addition of surfactants above the critical micellar concentration, the surfactant micelles form aggregates, generating a deposited layer over the membrane surface. This increases the resistance against the solvent flux through the membrane. This results in a decrease in the permeate flux compared to that of the dye solution alone. For example, at 276 kPa pressure, the permeate flux is about 16.6 L/m^2.h at the end of the experiment for eosin and surfactant mixture, whereas, for eosin only it is about 20.5 L/m^2.h.

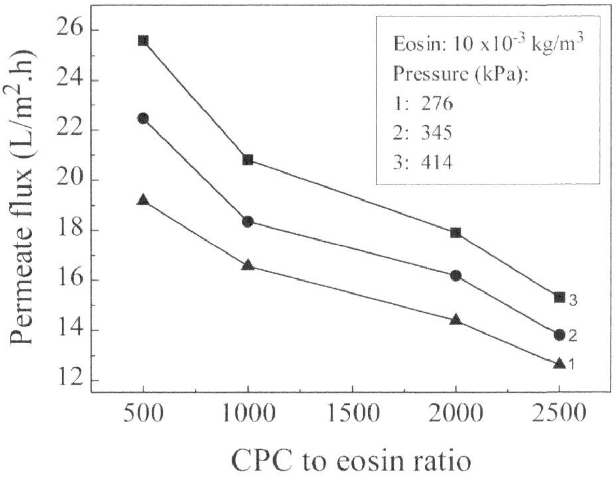

Fig. 7.3 Variation of permeate flux with CPC to eosin ratio at different pressures at the end of experiment. Feed eosin concentration is 10.0×10^{-3} kg/m^3 (Reproduced from Purkait et al. (2004). With permission from Elsevier)

7.4.1 Effects of Operating Pressure and Feed CPC Concentration on the Permeate Flux and Observed Retention of Eosin

Effects of the operating pressure and different dosage of feed CPC concentration for a fixed eosin concentration of 10×10^{-3} kg/m^3 on the permeate flux are presented in Fig. 7.3. Figure 7.3 represents the flux data at the end of experiment. The figure shows that the permeate flux decreases with feed CPC concentration at a fixed operating pressure. For example, at 276 kPa pressure, flux decreases from 20.9 L/m^2.h to about 12.6 L/m^2.h when CPC concentration increases from 5 to 25 kg/m^3. This trend may be explained by the increase in resistance against the solvent flux due to micellar aggregates as described in the preceding section. At the constant CPC concentration, flux increases with pressure. For example, at CPC to eosin ration 2500 (i.e., feed CPC concentration 25 kg/m^3), flux increases from 12.6 L/m^2.h to 15.3 L/m^2.h, while pressure increases from 276 to 414 kPa. This occurs due to increase in the effective driving force.

The effect of operating pressure on the observed retention as a function of CPC to eosin concentration ratio at the end of the experimental duration is shown in Fig. 7.4. From the figure, it can be observed that for a fixed pressure, eosin retention increases with CPC concentration. With increase in surfactant concentration, the micelle concentration in the solution increases. This results in more solubilization of eosin in CPC micelles, thereby increasing the retention of eosin. For a fixed CPC to eosin ratio, the retention of eosin increases with decrease in operating pressure. This may be due to the fact that at higher operating pressure, micelles may become compact and therefore solubilization capability of the micelles decreases (Dunn et al. 1987; Syamal et al. 1997). Therefore, less amount of dye is solubilized in the micelles at higher operating pressure, and the permeate concentration is more leading to a decrease in observed retention of eosin at higher pressure. It can also

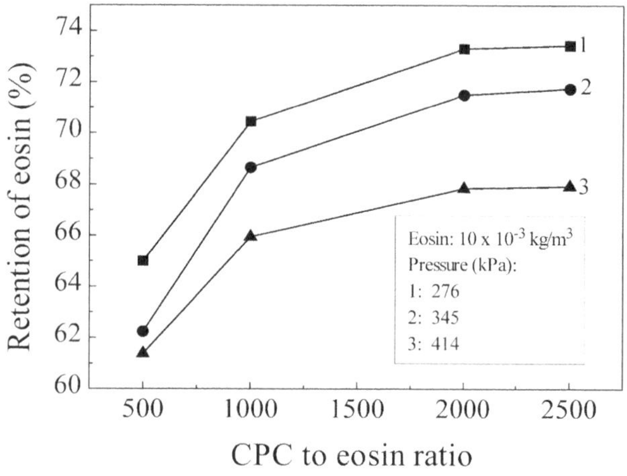

Fig. 7.4 Variation of eosin retention with CPC to eosin ratio at different pressures after 1 h of experiment. Feed concentrations are eosin 10×10^{-3} kg/m^3 and CPC 5, 10, 20, and 25 kg/m^3 (Reproduced from Purkait et al. (2004). With permission from Elsevier)

be observed from Fig. 7.4 that at a fixed pressure, the increase of eosin retention is fast for lower CPC to eosin concentration ratio and is gradual as this ratio increases. Beyond a ratio of 2000, the increase in eosin retention is marginal. Therefore, lower operating pressure and a ratio of feed concentration of CPC to eosin of 2000 may be considered as optimum for maximum removal of eosin.

7.4.2 *Effects of Feed Eosin Concentration on the Observed Retention of Dye at Fixed CPC Concentration*

In this case, the surfactant concentration is kept fixed at 10 kg/m^3, and eosin concentration is varied as 5×10^{-3}, 10×10^{-3}, 20×10^{-3}, 30×10^{-3}, and 40×10^{-3} kg/m^3. The experiments are conducted at a relatively low pressure of 276 kPa as lower pressure is a favorable operating condition for higher retention of eosin as shown in Fig. 7.4. In Fig. 7.5, the variation of retention of eosin with CPC to eosin ratio at the end of 1 h of operation at 276 kPa pressure is presented for both the combinations of surfactant and dye mixture. This figure reveals important information regarding the ratio of CPC to dye to obtain maximum solubilization of the dye in the surfactant micelles. Curve 1 in the figure indicates that the dye concentration is fixed and CPC concentration is gradually increased, whereas the curve 2 indicates that the CPC concentration is fixed and the ratio is varied by varying the concentration of dyes. Although the CPC to eosin ratio is in the same range in both the cases, the retention of dye is more in the first case compared to that in the second case. In the first case, since CPC concentration is increased, concentration of micelles increases and more dye will be solubilized in the micelle. In the latter case, the micelle concentration is fixed (as the surfactant concentration is kept constant) and dye concentration is gradually increased. Since the concentration of micelle is constant in this case, the solubilization capacity of dye is also constant.

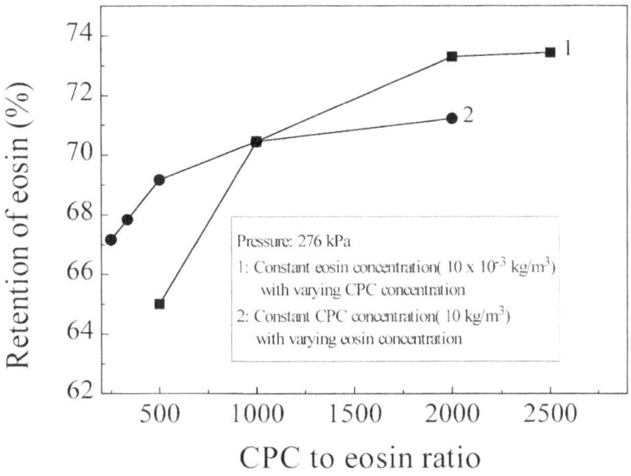

Fig. 7.5 Variation of eosin retention with CPC to eosin ratio at 276 kPa (Reproduced from Purkait et al. (2004). With permission from Elsevier)

Hence further increase in dye concentration only results in an increase of unsolubilized dye concentration which will be increasing the permeate concentration and hence reduces the observed retention of dyes. In essence, it can be concluded from this figure that in order to determine an optimum surfactant-to-dye ratio to get maximum removal of the dye, the ratio should be changed by varying the surfactant concentration (keeping dye concentration fixed) to calculate the concentration of CPC required for a given dye concentration.

Retention of CPC at the end of the experimental run is presented in Fig. 7.6 for both the combinations of CPC and dye mixture at different pressures. It is observed from the figure that for the case of constant eosin concentration (10×10^{-3} kg/m^3), retention of CPC is in the range of 94–98% with varying CPC to surfactant ratio. There is an increasing trend of observed retention of CPC with CPC to eosin ratio. Since all the feed concentrations of surfactants are much above CMC, the micelles formed are retained by the membrane, and the free surfactants at the concentration of CMC pass to the permeate stream. Hence, with the increase in feed CPC concentration (C_0), permeate CPC concentration (C_p) remains around the CMC (0.322 kg/m^3) value, and therefore, the ratio C_p/C_0 decreases and observed retention increases with CPC to eosin ratio. In case of constant CPC and varying dye concentrations, a constant CPC retention of about 97.5% is obtained. This is because the permeate surfactant concentration remains at CMC and the observed retention of CPC remains unaltered although CPC to eosin ratio has been increased by addition of more dyes at constant feed CPC concentration.

7.4.3 Flux Decline Mechanism

To analyze the flux decline data of the dye-containing micellar solution, $1/v_w^2$ is plotted against time for different pressures. Figure 7.7 shows this plot for initial

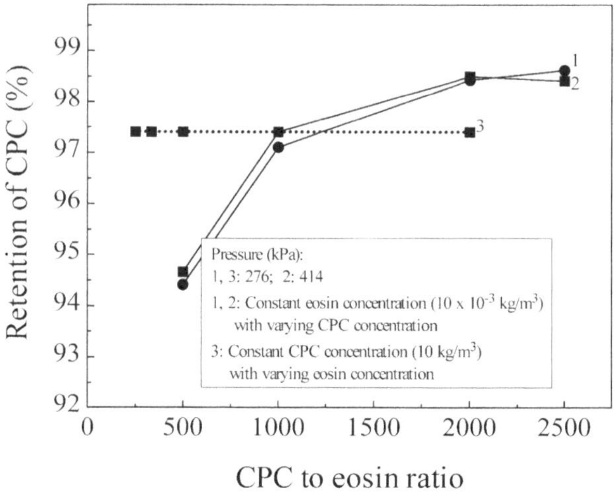

Fig. 7.6 Variation of CPC retention with CPC to eosin ratio (Reproduced from Purkait et al. (2004). With permission from Elsevier)

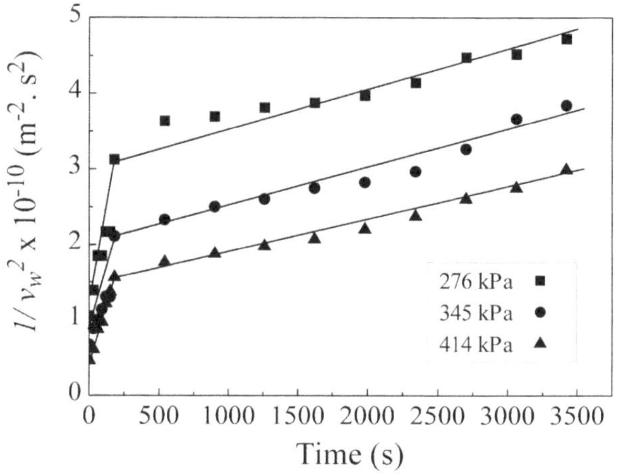

Fig. 7.7 Variation of $1/v_w^2$ with time at different pressure. Feed eosin concentration is 10×10^{-3} kg/m^3 and CPC concentration is 10 kg/m^3 (Reproduced from Purkait et al. (2004). With permission from Elsevier)

CPC concentration of 10 kg/m^3 and Fig. 7.7 shows for initial CPC concentration of 20 kg/m^3. In both the figures, there is a distinct time zone, beyond which the increase in the curves is linear. This time point is 180 s in both the figures, and the same trends are observed for the experiments with CPC concentrations, 5 and 25 kg/m^3, under various operating conditions. The linear increase of $1/v_w^2$ with time confirms the growth of a gel-type layer over the membrane surface. Below 180 s, the curve increases nonlinearly and changes its slope after 180 s. This behavior of flux decline data is therefore assumed to be due to pore blocking. The short-term flux decline data for a typical operating condition are shown in Fig. 7.9. The figure shows that the flux drops sharply and gradually thereafter. Once the membrane is cleaned after the experiment, the permeability of it is again checked and it is found that the permeability remains almost unchanged. This prompts to the assumption that the initial flux decline is due to a pore blocking mechanism which is reversible in nature.

7.5 Theoretical Analysis

Flux decline is the major problem in UF. Various models have been developed to analyze and predict UF flux behavior during separation/fractionation of macromolecular solute. All of them can be classified into three broad categories: (a) osmotic pressure controlled model, (b) gel polarization model, and (c) resistance in series model. According to resistance in series model, the flux decline is due to the combined effects of (1) fouling of the membrane (reversible and/or irreversible pore blocking) (Boerlage et al. 1997; Wiesner and Chellam 1999) and (2) concentration polarization over the membrane surface (Madsen 1973; Kishihara et al. 1989; De et al. 1997). But in gel polarization model, the decrease in flux is due to the hydraulic resistance of deposited gel layer over the membrane surface (De et al. 1997; Karode 2000). On the other hand in osmotic pressure controlled model, flux decline is due to the decrease in effective transmembrane pressure drop which occurs as the osmotic pressure of the retained solution increases (Wijmans et al. 1984; Sreenivas et al. 2002). In a typical membrane filtration process, the flux drop due to osmotic pressure, pore blocking, and gel layer growth may coexist. The model for the quantification of flux decline due to these is available in the literature (Song 1998). That model assumes the solutes as hard spheres and pore blocking is an irreversible process. However, the micelles may not behave as hard spheres and the pore blocking may be reversible leading to a departure from the applicability of the referenced model.

As described earlier (refer to Figs. 7.7 and 7.8), the flux decline can be divided into two distinct regions: first, the short-term flux decline due to the reversible pore

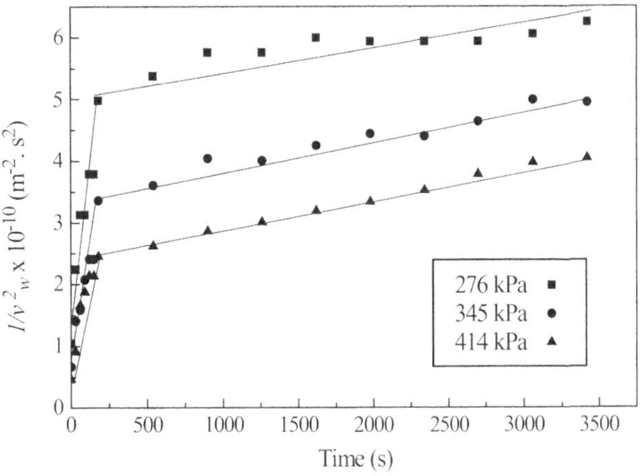

Fig. 7.8 Variation of $1/v^2_w$ with time at different pressure. Feed eosin concentration is 10×10^{-3} kg/m^3 and CPC concentration is 20 kg/m^3 (Reproduced from Purkait et al. (2004). With permission from Elsevier)

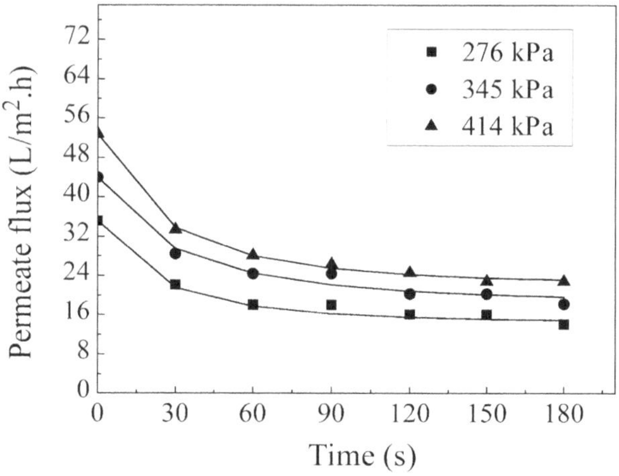

Fig. 7.9 Variation of short-term flux decline with time at different pressure. Feed eosin concentration is 10×10^{-3} kg/m^3 and CPC concentration is 25 kg/m^3 (Reproduced from Purkait et al. (2004). With permission from Elsevier)

blocking for a period $t < 180$ s and long-term flux decline due to the growth of a gel type of layer over the membrane surface for a period of $t > 180$ s. In both the time domains, the theoretical development of flux decline is presented below.

It may be noted that there may be an additional resistance due to the osmotic pressure buildup from the bulk concentration to the gel layer concentration leading to a decrease in the operating pressure difference (Elimelech and Bhattacharjee 1998). The osmotic pressure of the micellar solution increases marginally beyond the critical micellar concentration, unlike the solutions of non-aggregating solutes (Rosen 2004; Clint 1992). The osmotic pressure corresponding to the critical micellar concentration for CPC is about 2 kPa. The osmotic pressure difference corresponding to the concentration difference between the bulk and the gel layer will be very small (compared to the operating pressures in the range of 276–414 kPa for the experiments reported in this work). Moreover, the calculated value of the resistance due to osmotic pressure turns out to be about 2–4% of the overall resistance after pore blocking for all the experiments. Therefore, the effect of osmotic pressure difference in the subsequent equations is insignificant with reference to flux calculations and is not included.

7.5.1 Short-Term Flux Decline: Reversible Pore Blocking (t < 180 s)

The initial flux decline data for various operating conditions have been observed to be best fitted in a logistic type of curve as follows:

$$v_w = \frac{v_w^0(1-b)}{1-be^{-ct}}, \tag{7.1}$$

where v_w^0 $(= L_P \Delta P)$ is pure water flux and b and c are the parameters to be estimated. Equation (7.1) can be expressed in terms of resistances as

$$v_w = \frac{\Delta P}{\mu[R_m + R_{PB}(t)]} = \frac{v_w^0}{1 + R_{PB}^*(t)}, \tag{7.2}$$

where the membrane permeability is related to the hydraulic resistance of the membrane as $L_P = \frac{1}{\mu R_m}$ and the dimensionless resistance due to pore blocking is given as

$$R_{PB}^* = \frac{b}{1-b}(1 - e^{-c\ t}) \tag{7.3}$$

The significances of the parameters "b" and "c" in Eqs. (7.1 and 7.3) can be stated as follows. The flux value at the end of pore blocking or the magnitude of the resistance due to pore blocking depends on "b." The rate of flux decline or the rate of the growth of the resistance due to the pore blocking is dictated by the value of "c." The forms of Eqs. (7.1 and 7.3) also predict zero slope at the end of the pore blocking cycle (about 3 min) as supported by the experimental evidence presented in the next section.

7.5.2 *Long-Term Flux Decline: Growth of Gel-Type Layer (t > 180 s)*

$$v_w = \frac{1}{A}\frac{dV}{d\tau} = \frac{\Delta P}{\mu\left(R_m + R_{PB,180} + R_g(\tau)\right)}, \tag{7.4}$$

where $\tau = t - 180$

Equation (7.4) can be arranged as

$$v_w = \frac{\Delta P}{\mu R_m\left(1 + R^*{}_{PB,180} + R_g^*\right)} = \frac{v_{w,180}}{1 + \left(\frac{R_g^*}{1 + R_{PB,180}^*}\right)} \tag{7.5}$$

The gel layer resistance is expressed as (Geankoplis 1997)

$$R_g = \frac{\alpha\left(c_0 - c_{cmc}\right)V}{A}, \tag{7.6}$$

where α is the specific gel resistance. The specific cake resistance (assuming gel characteristics to be the same of a cake) α is obtained from the Kozeny-Karman equation (Zihao et al. 1996) in terms of gel forming particle properties,

$$\alpha = 180\frac{\left(1 - \varepsilon_g\right)}{\varepsilon_g^3 d_p^2 \rho_g}, \tag{7.7}$$

where ε_g, ρ_g, and d_p are the gel porosity, density, and diameter of the gel forming particles, respectively.

In nondimensional term, the gel layer resistance can be written as

$$R_g^* = \frac{\alpha\left(c_0 - c_{cmc}\right)V}{AR_m} \tag{7.8}$$

Using Eqs. (7.5 and 7.8), Eq. (7.4) can be integrated to obtain the expression of cumulative volume as

$$V = \frac{\sqrt{1 + 2Av_{w,180}\beta\tau} - 1}{\beta}, \tag{7.9}$$

where $\beta = \dfrac{\alpha(c_0 - c_{cmc})}{AR_m\left(1 + R_{PB,180}^*\right)}$

From Eqs. (7.8 and 7.9), the nondimensional gel layer resistance can be expressed as a function of time as

$$R_g^* = \left(1 + R_{PB,180}^*\right)\left[\left(\sqrt{1 + 2Av_{w,180}\beta\tau}\right)\right] \tag{7.10}$$

Using Eqs. (7.4 and 7.9), the flux decline due to gel layer growth can be obtained after simplification:

$$\frac{1}{v_w^2} = \frac{1}{v_{w,180}^2} + \alpha\varphi\tau, \tag{7.11}$$

where $\phi = \frac{2\mu(c_0 - c_{cmc})}{\Delta P}$. From the slope of $1/v_w^2$ versus τ for a given set of operating conditions, namely, ΔP and $c_{0,}$ the specific cake resistance α can be determined as shown in the subsequent sections.

The growth of the gel-type layer can be obtained by doing a simple mass balance in the gel layer, and the expression of gel layer thickness is given as

$$L(\tau) = \frac{(c_0 - c_{cmc})V(t)}{A(1 - \varepsilon_g)\rho_g} \tag{7.12}$$

7.5.3 Determination of the Constants in Short-Term Flux Decline

The constants b and c in Eq. (7.1) are evaluated by matching the flux and derivative of the flux between the short- and long-term flux decline at 180 s.

By matching the flux values at $t = 180$ or $\tau = 0$, the following equation is obtained:

$$\frac{v_w^0(1 - b)}{1 - be^{-180c}} = \frac{1}{v_{w,180}} \tag{7.13}$$

By matching the derivatives of the flux at $t = 180$ or $\tau = 0$, the following equation is obtained:

$$\frac{\alpha\phi v_{w,180}^3}{2} = \frac{bc(1 - b)v_w^0 e^{-180c}}{\left(1 - be^{-180c}\right)^2} \tag{7.14}$$

For a given operating conditions (ΔP and c_0), using the experimental value of flux at 180 s ($v_{w,180}$) and the determined value of α, Eqs. (7.13 and 7.14) are solved iteratively and the constants b and c are estimated.

7.5.4 Total Resistance

At any point of time, the total resistance (R_t) is defined as,

$$\text{for } t < 180, \quad R_t^*(t) = 1 + R_{PB}^*(t) \tag{7.15}$$

$$\text{for } t > 180 \text{ or } \tau > 0, R_t^*(t) = 1 + R_{PB,180}^* + R_g^*(\tau) \tag{7.16}$$

7.5.5 Determination of the Specific Resistance of the Gel-Type Layer

It has already been discussed that the straight-line portion is $1/v_w^2$ versus time plot due to the formation of gel-type layer. From the each set of experiments (i.e., the operating pressure and feed CPC concentration), the slope of $1/v_w^2$ versus t (Eq. 7.11) is calculated. The specific gel-type layer resistance α is obtained as $\alpha = slope/\phi$. Next, the specific gel-type layer resistance is correlated with the operating conditions as

$$\alpha = \alpha_0(\Delta P)^{n1}(c_0)^{n2}, \quad (7.17)$$

where $\alpha_0 = 1.897 \times 10^6$, $n_1 = 1.464$, and $n_2 = -0.553$.

7.5.6 Determination of the Constants in the Reversible Pore Blocking Model

The constants b and c are evaluated for every operating condition by matching the flux and its time derivative at $t = 180$ s. For each operating condition, the specific cake resistance is evaluated from Eq. (7.17), and using the experimental flux value at $t = 180$ s, Eqs. (7.13 and 7.14) are solved iteratively to estimate the constants b and c. The correlations of these two parameters with the operating conditions are presented below:

$$b = b_0(\Delta P)^{n3}(c_0)^{n4} \quad (7.18)$$

$$\text{and } c = c_1(\Delta P)^{n5}(c_0)^{n6}, \quad (7.19)$$

where $b_0 = 0.030$, $n_3 = 0.161$, and $n_4 = 0.283$ and $c_1 = 18.523$, $n_5 = -0.524$, and $n_6 = 0.080$.

7.5.7 Determination of the Gel Concentration of CPC Micelles

UF runs are conducted at a constant stirring speed of 500 rpm and various feed CPC concentrations of 3.125, 6.25, 12.5, 25, 50, 100, and 150 kg/m^3 keeping feed dye concentration constant (10×10^{-3} kg/m^3) for all the cases. The steady-state permeate flux has been plotted against the feed concentration and is shown in Fig. 7.10. From the figure, it can be concluded that the gel concentration is about 340 kg/m^3 which is in corroboration to the literature value (Wiesner and Chellam 1999).

Fig. 7.10 Variation of steady-state permeate flux with feed CPC concentration (eosin concentration is constant at 10×10^{-3} kg/m^3 for all the cases) at 276 kPa pressure and 500 rpm (Reproduced from Purkait et al. (2004). With permission from Elsevier)

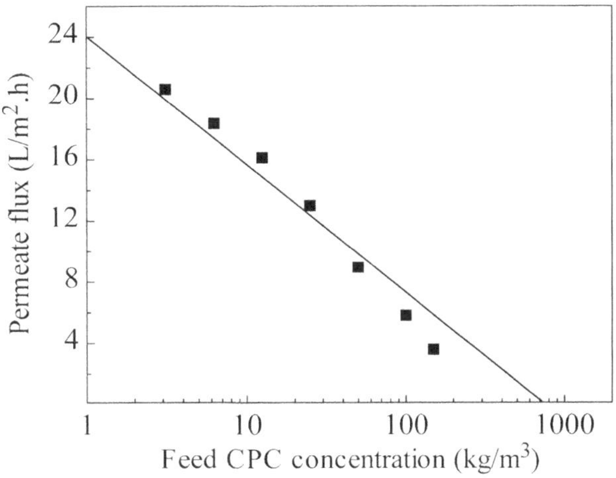

7.5.8 Determination of the Gel Porosity

Typical spherical diameters of CPC micelles are 5 nm (Markels et al. 1994). The gel porosity is obtained in terms of particle diameter, gel concentration, etc., from the following expression (De and Bhattacharya 1997):

$$\varepsilon_g = 1.0 - 10^3 \frac{\pi}{6} c_g \frac{N_A d_p^3}{M_w}, \tag{7.20}$$

where N_A is the Avogadro number, M_w is the molecular weight of the micelles, c_g is the gel concentration, and d_p is the micelle diameter. Assuming aggregation number to be 136 (Markels et al. 1994), the molecular weight of the CPC micelle is 48,690. The porosity of the gel layer is found to be 0.725.

7.5.9 Analysis of Various Resistances

Having determined the model parameters, the overall flux decline has been predicted for both short- and long-term flux decline from Eqs. (7.2 and 7.4). The matching of the calculated and experimental flux decline profiles for various pressures, and two typical surfactant concentrations of 10 and 20 kg/m^3 are shown in Figs. 7.11 and 7.12, respectively. Similar match is found for two other concentrations, namely, 5 and 25 kg/m^3. It is evident from the figures that at any point of operating time, the permeate flux is more for higher pressure because of larger driving force. Also at the same pressure, the permeate flux is lower for higher feed concentration. This is due to the fact that at higher feed surfactant

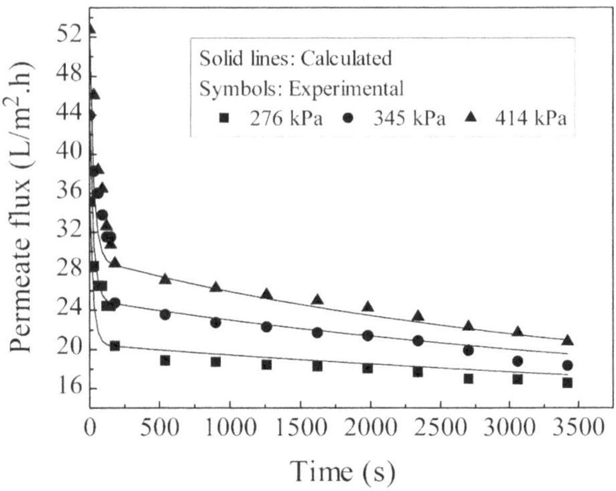

Fig. 7.11 Variation of the permeate flux with time at different operating pressure differences. Feed eosin concentration is 10×10^{-3} kg/m^3 and CPC concentration is 10 kg/m^3 (Reproduced from Purkait et al. (2004). With permission from Elsevier)

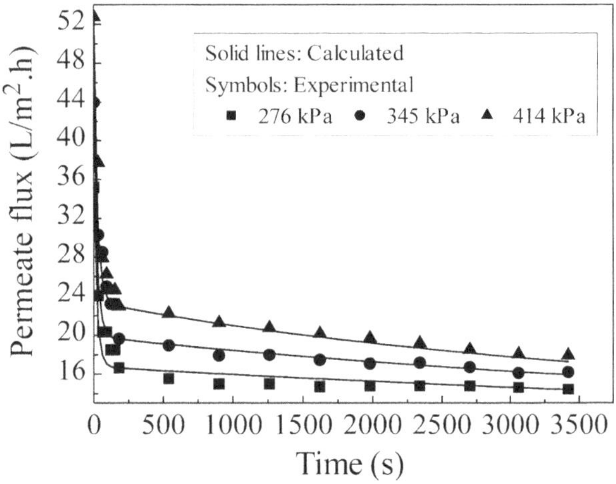

Fig. 7.12 Variation of the permeate flux with time at different operating pressure differences. Feed eosin concentration is 10×10^{-3} kg/m^3 and CPC concentration is 20 kg/m^3 (Reproduced from Purkait et al. (2004). With permission from Elsevier)

concentration, the number of micelles is more leading to the increase in gel-type resistance over the membrane surface. The various resistances for different operating conditions are tabulated in Table 7.1 at the end of 1 h of operation. It has been observed that for the lower concentration of surfactants (up to 10 kg/m^3), the membrane hydraulic resistance is the maximum of the total resistance, whereas the resistance due to the pore blocking is larger than that of the gel-type resistance. For feed surfactant concentration for 20 kg/m^3, the magnitude of the resistance due to pore blocking becomes competitive to that of the hydraulic resistance. Feed concentration at 25 kg/m^3, the pore blocking resistance is more than that of the membrane hydraulic resistance. This clearly indicates that at higher surfactant concentrations, the number of the micelles increases, and, consequently, the

Table 7.1 Comparison of different resistances with respect to the total resistance. Gel layer and the total resistances are at the end of the experiment

Pressure (kPa)	Feed (kg/m^3) CPC	Eosin	$R_t \times 10^{-13}$ (m^{-1})	R_m/R_t (%)	$R_{PB,180}/R_t$ (%)	R_g/R_t (%)
276	5	10×10^{-3}	5.795	54.512	29.200	16.288
	10	10×10^{-3}	6.709	47.086	34.229	18.685
	20	10×10^{-3}	7.717	40.935	45.611	13.454
	25	10×10^{-3}	8.813	35.844	53.576	10.580
345	5	10×10^{-3}	6.185	51.075	30.054	18.871
	10	10×10^{-3}	7.568	41.741	32.357	25.902
	20	10×10^{-3}	8.579	36.822	45.685	17.493
	25	10×10^{-3}	10.063	31.392	44.123	24.485
414	5	10×10^{-3}	6.513	48.503	29.634	21.863
	10	10×10^{-3}	8.007	39.453	32.828	27.719
	20	10×10^{-3}	9.308	33.938	43.988	22.074
	25	10×10^{-3}	10.898	28.986	39.359	31.655

Reproduced from Purkait et al. (2004). With permission from Elsevier

reversible pore blocking is more severe and is the dominant resistance responsible for the flux decline. On the other hand, the resistance of the gel-type layer ranges from 10 to 30% of the total resistance for various operating conditions. Although this resistance is the smallest among the three, it is not negligible and therefore contributes significantly to the overall resistance. The gel layer resistance also increases with feed concentration as can be observed from the table. This signifies that the number concentration of the micelles increases with feed concentration and, therefore, offers more resistance against the solvent flux.

The profiles of reversible pore blocking resistance with time for a fixed feed composition (surfactant and dye concentrations of 10 kg/m^3 and 10×10^{-3} kg/m^3, respectively) at various operating pressures are shown in Fig. 7.13. In this figure, it can be observed that for a fixed pressure, the resistance for pore blocking increases initially and levels off thereafter. The pore blocking resistance is more for higher pressure. This indicates that at higher pressure, smaller-sized micelles are transported by forced convection toward the membrane and block the pores to a great extent. The pore blocking resistance at the end of 3 min for CPC to dye ratio for various operating pressures is plotted in Fig. 7.14. It is evident that the pore blocking resistance increases with feed CPC concentration. This indicates that the number concentration and size distribution become wider with the initial feed surfactant concentrations leading to more severe pore blocking.

The profiles of gel layer resistance with operating time have been plotted in Fig. 7.15 for a fixed feed composition (surfactant and dye concentrations of 10 kg/m^3 and 10×10^{-3} kg/m^3, respectively) at different operating pressure. It is evident from the figure that the gel layer resistance increases with time almost linearly. The gel layer resistance is more at higher pressure. This is because the specific cake resistance is an increasing function of pressure (Eq. 2.17), leading to

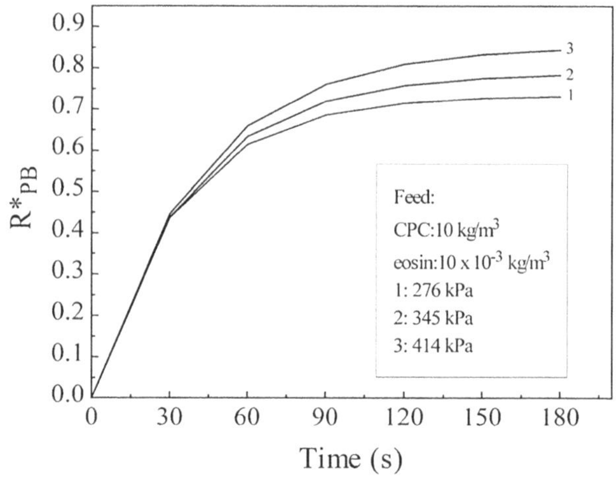

Fig. 7.13 Variation of dimensionless pore blocking resistance with time at different operating pressures (Reproduced from Purkait et al. (2004). With permission from Elsevier)

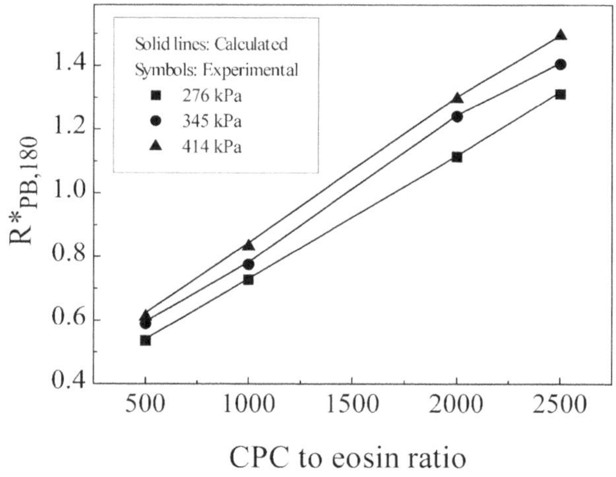

Fig. 7.14 Variation of dimensionless pore blocking resistance with CPC to eosin ratio at different pressures after 180 s of experiment. Feed eosin concentration is 10×10^{-3} kg/m^3 (Reproduced from Purkait et al. (2004). With permission from Elsevier)

more resistance to the solvent flux. Figure 7.16 shows the variation of gel layer resistance with CPC to eosin ratio at the end of the experiment for different operating pressures. The figure shows that the gel layer resistance increases with the feed surfactant concentrations. It is noted that the experimental results of the variation of R_g with feed concentration are less conclusive (compared to R_{PB} and R_t), but an overall increasing trend may be observed.

The profile of the total resistance (R_t^*) is plotted for a fixed feed composition (surfactant and dye concentrations of 20 kg/m^3 and 10×10^{-3} kg/m^3, respectively) at different pressures in Fig. 7.17. It can be observed from the figure that the total resistance increases with pressure. Figure 7.18 shows the increasing trend of the total resistance with CPC to surfactant ratio at various pressures.

Fig. 7.15 Variation of dimensionless gel-type layer resistance with time at different operating pressure (Reproduced from Purkait et al. (2004). With permission from Elsevier)

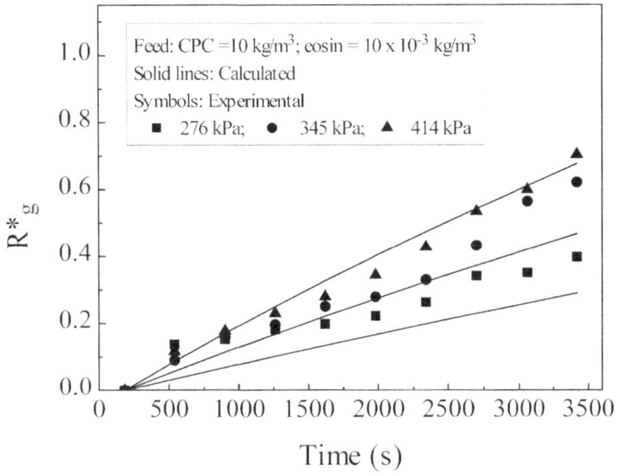

Fig. 7.16 Variation of dimensionless gel-type layer resistance with CPC to eosin ratio at different pressures after 1 h of experiment. Feed eosin concentration is 10×10^{-3} kg/m^3 (Reproduced from Purkait et al. (2004). With permission from Elsevier)

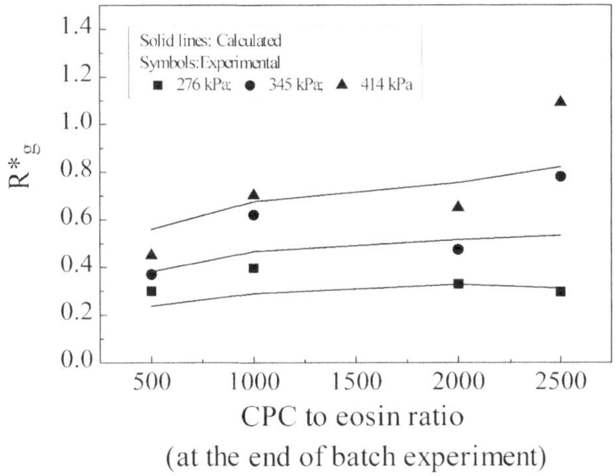

Having determined system parameters, Eq. (7.12) is solved to obtain the profiles of the gel layer thickness with time. The growth of gel layer with time is shown in Fig. 7.19 for a fixed feed composition (surfactant and dye concentrations of 20 kg/m^3 and 10×10^{-3} kg/m^3, respectively) and different operating pressures. As expected, the gel layer thickness is more at higher pressure. At a fixed pressure of 414 kPa, the gel layer thickness is plotted against the operating time for various feed CPC concentrations and fixed eosin concentration (10×10^{-3} kg/m^3) and is shown in Fig. 7.20. As discussed earlier, the gel layer grows with time almost linearly, and at a fixed time, the gel layer thickness is more at higher feed surfactant concentrations.

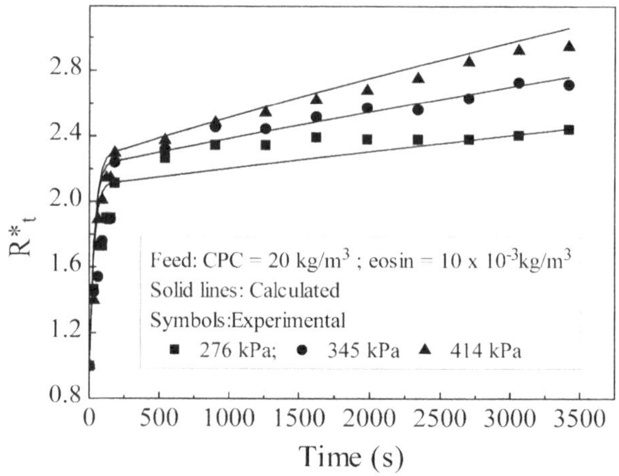

Fig. 7.17 Variation of dimensionless total resistance with time at different operating pressure (Reproduced from Purkait et al. (2004). With permission from Elsevier)

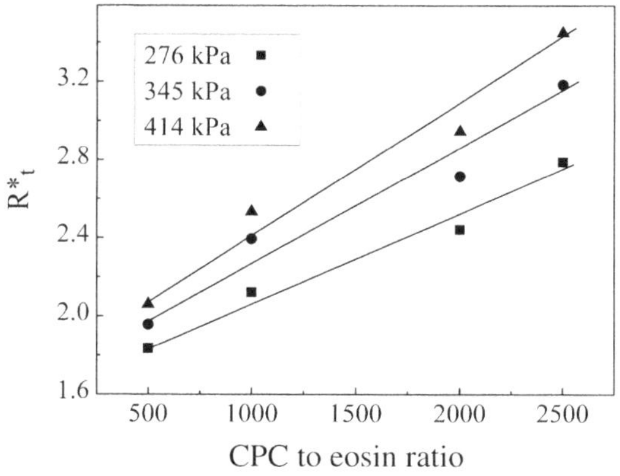

Fig. 7.18 Variation of dimensionless total resistance with CPC to eosin ratio at different pressures after 1 h of experiment. Feed eosin concentration is 10×10^{-3} kg/m^3 (Reproduced from Purkait et al. (2004). With permission from Elsevier)

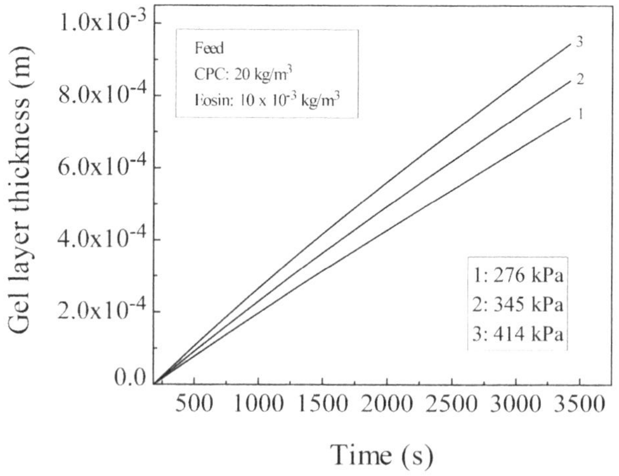

Fig. 7.19 Variation of the thickness of the gel layer with time at different operating pressure (Reproduced from Purkait et al. (2004). With permission from Elsevier)

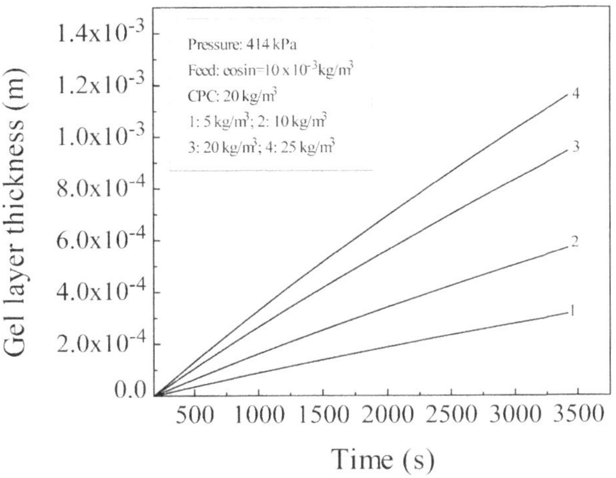

Fig. 7.20 Variation of the thickness of the gel layer with time at different feed CPC concentration (Reproduced from Purkait et al. (2004). With permission from Elsevier)

7.6 MEUF in Continuous Cross Flow Cell

Once the concept of the separation of dye in an unstirred batch cell using MEUF is successfully validated, the mixture is subjected to a steady state cross flow ultrafiltration as this should substantially improve flux and retention. A detailed parametric study is also conducted to observe the effects of the operating conditions on the permeate flux and observed retention.

7.6.1 Effect of the Feed CPC Concentration on Permeate Flux and the Retention of Both Dye and CPC

The effect of feed CPC to dye concentration ratio on the retention and permeate flux is presented in Fig. 7.21. The concentration ratio of CPC to dye is varied by changing the CPC concentration at a fixed dye concentration of 10×10^{-3} kg/m^3. The figure shows that the retention of both CPC and eosin increases with feed CPC concentration. But beyond the ratio of 2000, the increase in retention becomes gradual. For example, when CPC to eosin ratio increases from about 500 to 2000, the retention increases from 69.6 to 77.8%, but retention becomes 78.2% when the ratio increases to 2500. Figure 7.21 also shows that the permeate flux decreases with feed CPC concentration at constant dye concentration. For example, flux decreases from 23 L/m^2.h to about 16.2 L/m^2.h when CPC concentration is increased from 5 to 25 kg/m^3 (CPC to eosin ratio increases from 500 to 2500). This is due to the resistance offered by the deposited layer of the micellar aggregates over the membrane surface.

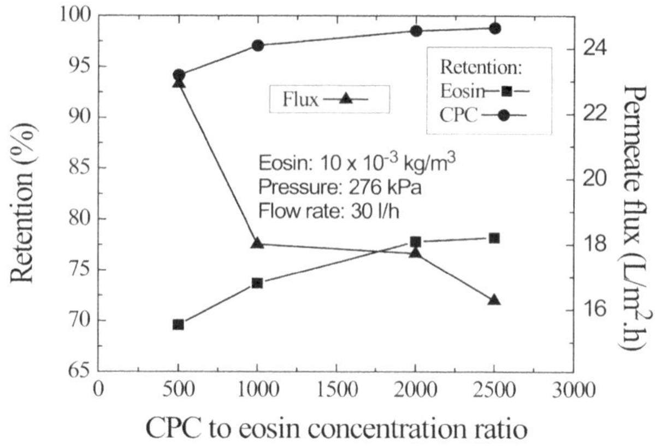

Fig. 7.21 Effect of the feed CPC concentration on permeate flux and the retention of both dye and CPC

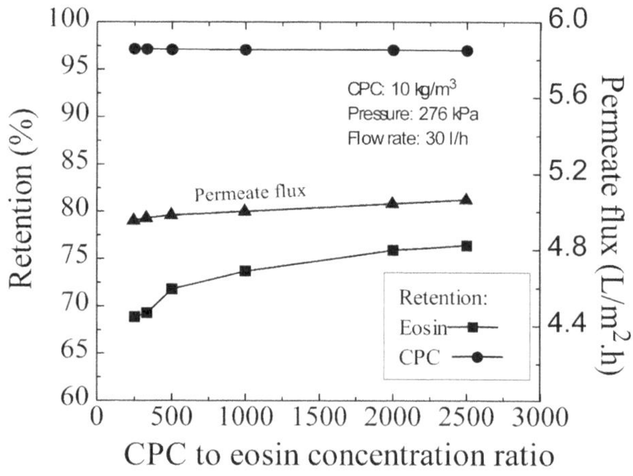

Fig. 7.22 Effect of the feed dye concentration on the permeate flux and retention of both dye and CPC

7.6.2 *Effect of the Feed Dye Concentration on the Permeate Flux and Retention of Both Dye and CPC*

The effect of dye concentration in feed on the retention of dye and permeate flux is shown in Fig. 7.22. In this case, CPC to dye ratio is varied by changing the dye concentration keeping CPC concentration constant at 10 kg/m^3. It may be observed from Fig. 7.22 that the flux and retention increase marginally with CPC to dye ratio. Beyond the ratio of 2000, the retention remains almost unchanged. On the other hand, the variation of the permeate flux is marginal with CPC to dye concentration ratio. It may be noted that the major contribution of resistance against the solvent flux comes from the CPC micelles. Since CPC concentration remains constant and the solute concentration varies over a range between 4 and 40 $\times$ 10^{-3} kg/m^3, the permeate flux remains almost unaltered.

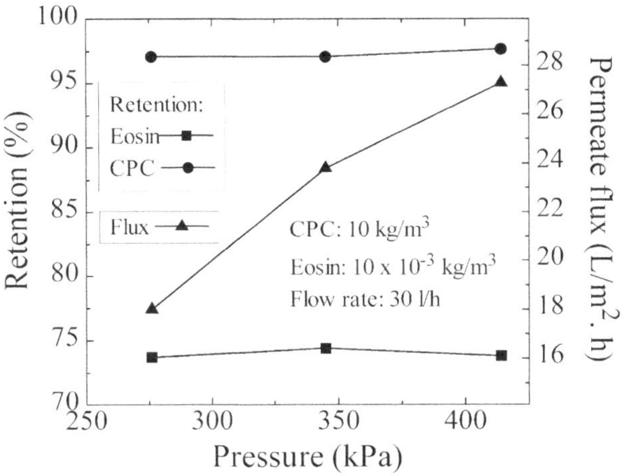

Fig. 7.23 Effect of pressure on the observed retention and the permeate flux

7.6.3 *Effect of Pressure Drop on the Observed Retention of Dye and Permeate Flux*

Variation of the dye retention with the applied pressure is shown in Fig. 7.23. It may be observed from the figure that the retention of dye and CPC remains almost independent of pressure. The effect of the operating pressure on the permeate flux has also been presented in Fig. 7.23 using 10 kg/m^3 of CPC and 10×10^{-3} kg/m^3 of eosin. The figure shows that the flux increases with pressure almost linearly within the pressure range. This occurs due to an increase in the effective driving force.

7.6.4 *Effect of Cross Flow Rate on the Observed Retention of Dye and Permeate Flux*

The effects of cross flow rate on permeate flux and observed retention of eosin dye are presented in Fig. 7.24. It may be observed from the figure that (1) the retention of both dye and CPC remains almost unchanged with cross flow rate which shows the independency of the extent of solubilization with cross flow rate and (2) marginal increase in flux value with cross flow rate. This is due to the fact that the resistance offered by the deposited layer of the micellar aggregates is slightly lowered with the increase in cross flow rate considered herein.

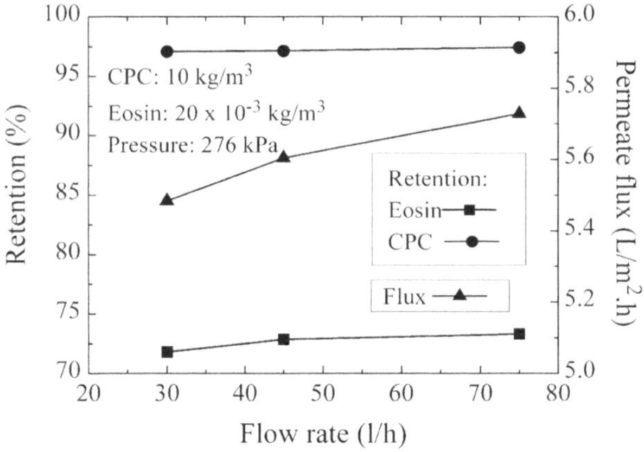

Fig. 7.24 Effect of cross flow rate on the observed retention and the permeate flux

7.7 Regeneration of Surfactant from the Permeate and Retentate Stream

To make the MEUF system more economical, it is necessary to recycle the surfactant present both in the permeate and retentate stream before the final disposal. Several authors have studied the precipitation method for the recycling of sodium dodecyl sulfate (SDS) using multivalent counter ions (Scamehorn and Harwell 1989). Air stripping method for the separation of surfactant and contaminant and surfactant reuse has also been reported (Lipe et al. 1996). Many researchers performed MEUF experiments using CPC as the cationic surfactant (Jadhav et al. 2001), but experiments for the recovery of CPC are still scant. The present work includes the recovery of CPC by the precipitation method (Purkait et al. 2003). A two-step chemical treatment process has been adopted to recover the surfactant from the permeate stream. The main reactions of chemical treatment I and II as shown in Fig. 7.25 are as follows.

7.7.1 *Procedure*

Recovery of surfactant from the permeate and retentate stream before final disposal is carried out by two successive chemical treatments. A series of experiments have been carried out to obtain the optimum ratio of KI to CPC for maximum precipitation of the surfactant in chemical treatment I and the ratio of $CuCl_2$ to CPI (cetylpyridinium iodide) in chemical treatment II for maximum regeneration of CPC. Keeping a constant CPC concentration (same as the permeate and retentate concentration), KI concentrations are varied. The solution is filtered with Whatman filter papers after complete precipitation. In chemical treatment II, $CuCl_2$ solutions are added in different concentrations to the precipitate that is obtained from

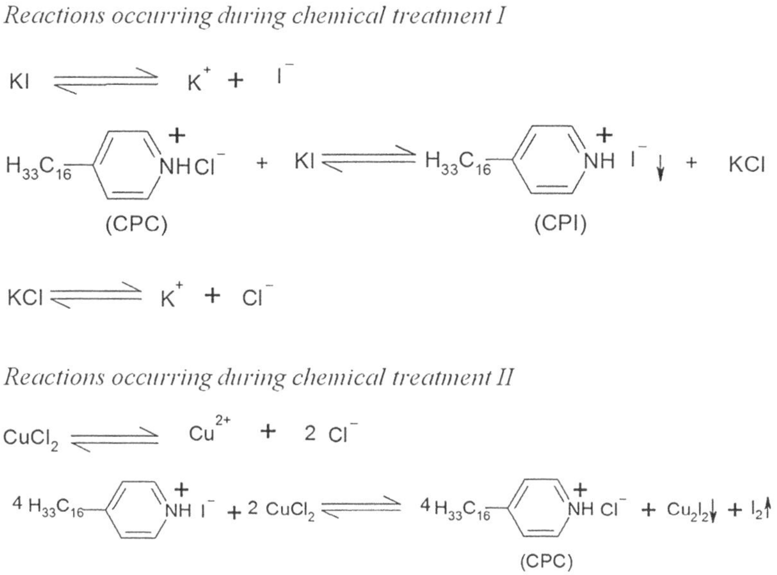

Fig. 7.25 Schematic of the chemical reactions involved during regeneration

chemical treatment I. Percent redissolution of CPC are calculated in the same way as in chemical treatment I.

7.7.2 *Regeneration of Surfactant from Permeate Stream*

7.7.2.1 Chemical Treatment I

With addition of KI, in the permeate, simultaneous production of CPI, KCl, and HCl will start. Extent of precipitation reaction depends on the degree of dissociation of KI. Again, degree of dissociation of KI depends on the concentration of KCl produced by the precipitation reaction. Since KCl is more ionic than KI, dissociation of KI is restricted due to common ion effect. At KI to CPC ratio of 1.5, the percentage CPC precipitation is maximum, and above that ratio the percentage CPC precipitation gradually decreases due to common ion effects as mentioned above.

7.7.2.2 Chemical Treatment II

The effects of concentrations of $CuCl_2$ on the percentage redissolution of CPI have been studied as a function of $CuCl_2$ to CPI ratio. With increase of $CuCl_2$ to CPI ratio, the percentage dissolution of CPC increases for both the cases, and beyond 3.5 the percentage dissolution of CPC becomes gradual with $CuCl_2$ concentration.

Table 7.2 Performance of chemical treatment I and II for a typical retentate

Solute	Retentate CPC conc. (kg/m^3)	Retentate solute conc. (kg/m^3)	KI/CPC (wt/wt)	CPC precipitated (%)	Solute in filtrate (%)	$CuCl_2$/CPI	Recovered with respect to retentate (%)	
							CPC	Solute
P	10.0	0.20	1.5	94.8	90.9	3.50	90.87	<1.0
PD	10.0	0.20	1.5	97.7	99.3	3.75	94.08	<1.0

7.7.3 Regeneration of Surfactant from Retentate Stream

The two-step treatment process is also applied to recover CPC from the retentate stream of phenol and potassium dichromate. Typical retentate streams of phenol and potassium dichromate are treated with the optimum amount of KI and $CuCl_2$ (as obtained in the last section), and the results are presented in Table 7.2. It is observed from the table that more than 90 percent CPC can be recovered from the retentate stream which can be reused.

References

Adamczak H, Materna K, Urban'ski R, Szymanowski J (1999) Ultrafiltration of micellar solutions containing phenols. J Colloid Interf Sci 218:359–368

Akita S, Yang L, Takeuchi H (1997) Micellar-enhanced ultrafiltration of gold(III) with nonionic surfactant. J Membr Sci 133:189–194

Akita S, Sastillo LP, Nii S, Takahashi K, Takeuchi H (1999) Separation of Co(II)/Ni(II) via micellar enhanced ultrafiltration using organo phosphorous acid extractant solubilized by nonionic surfactant. J Membr Sci 162:111–117

Alesson FIT, Urbanski R, Szymanowsk J (2001a) Evolution of resistance to permeation during micellar enhanced ultrafiltration of phenol and 4-nitro phenol. Colloids Surf A Physicochem Eng Asp 178:71–77

Alesson FIT, Adamczak H, Szymanowski J (2001b) Micellar-enhanced ultrafiltration of phenol by means of oxyethylated fatty acid methyl esters. J Membr Sci 192(1):155–163

Armstrong DW (1985) Micelles in separation: a practical and theoretical review. Sep Purif Method 14:213–304

Attwood D, Florence AT (1983) Surfactant systems: their chemistry, pharmacy and biology. Chapman and Hall, London

Baek K, Yang JW (2004) Cross-flow micellar-enhanced ultrafiltration for removal of nitrate and chromate: competitive binding. J Hazard Mater B108:119–123

Baek K, Kim BK, Yang JW (2003a) Application of micellar-enhanced ultrafiltration for nutrients removal. Desalination 156:137–144

Baek K, Lee HH, Yang JW (2003b) Micellar-enhanced ultrafiltration for simultaneous removal of ferricyanide and nitrate. Desalination 158:157–166

Boerlage SFE, Kennedy MD, Bonne MAC, Galjaard G, Schippers JC (1997) Workshop on membranes in drinking water production-technical innovation and health aspects. L'Aquila, Italy

Boethling RS (1984) Environmental fate and toxicity in wastewater treatment of quaternary ammonium surfactants. EPA, Washington, DC

Channarong B, Lee SH, Bade R, Shipin OV (2010) Simultaneous removal of nickel and zinc from aqueous solution by micellar-enhanced ultrafiltration and activated carbon fiber hybrid process. Desalination 262:221–227

Clint JH (1992) Surfactant aggregation. Blackie, Glasgow

Dasan DT, Ginn ME, Shah DO (eds) (1988) Surfactants in chemical/process engineering. Surfactant science series, vol 28. Marcel Dekker, Inc., New York

De S, Bhattacharya PK (1997) Modeling of ultrafiltration process for a two component aqueous solution of low and high (gel-forming) molecular weight solutes. J Membr Sci 136:57–69

De S, Mondal S (2012) Micellar enhanced ultrafiltration: fundamentals and applications. Taylor and Francis, Boca Raton

De S, Dias JM, Bhattacharya PK (1997) Short and long term flux decline analysis in ultrafiltration. Chem Eng Comm 159:67–89

Dunn RO Jr, Scamehorn JF, Christian SD (1985) Use of micellar-enhanced ultrafiltration to remove dissolved organics from aqueous stream. Sep Sci Technol 20:257–284

Dunn RO Jr, Scamehorn JF, Christian SD (1987) Concentration polarization effects in the use of micellar-enhanced ultrafiltration to remove dissolved organic pollutants. Sep Sci Technol 22:763–789

Dunn RO Jr, Scamehorn JF, Christian SD (1989) Simultaneous removal of dissolved organics and divalent metal cations from water using micellar-enhanced ultrafiltration. Colloids Surfaces 35:49–56

Elimelech M, Bhattacharjee S (1998) A novel approach for modeling concentration polarization in cross flow membrane filtration based on the equivalence of osmotic pressure model and filtration theory. J Membr Sci 145:223–241

Elworthy PH, Florence AT, MacFarlane CB (1968) Solubilization by surface active agents and its application in chemistry and biological sciences. Chapman and Hall, London

Fendler JH, Fendler EJ (1975) Catalysis in micellar and macromolecular systems. Academic Press, New York

Geankoplis CJ (1997) Transport processes and unit operations. Prentice Hall of India, New Delhi

Ghezzi L, Robinson BH, Secco F, Tiné MR, Venturini M (2008) Removal and recovery of palladium(II) ions from water using micellar-enhanced ultrafiltration with a cationic surfactant. Colloids Surf A Physicochem Eng Asp 329:12–17

Gibbs LL, Scamehorn JF, Christian SD (1987) Removal of n-alcohols from aqueous streams using micellar-enhanced ultrafiltration. J Membr Sci 30:67–74

Gillberg G (1984) Practical use of microemulsions. In: Lissant KJ (ed) Emulsion and emulsion technology-part III. Surfactant science series, vol 6. Marcel Dekker, New York, Ch. 1, 1984

Gillberg G, Friberg S (1978) Microemulsion as diesel fuels. In: Zung JT (ed) Evaporation-combustion of fuels. Advances in chemistry series, vol 166. American Chemical Society, Washington

Gzara L, Dhahbi M (2001) Removal of chromate anions by micellar enhanced ultrafiltration using cationic surfactants. Desalination 137:241–250

Jadhav SR, Verma N, Sharma A, Bhattacharya PK (2001) Flux and retention analysis during micellar enhanced ultrafiltration for the removal of phenol and aniline. Sep Purif Technol 24:541–557

Juang RS, Xu YY, Chen CL (2003) Separation and removal of metal ions from dilute solutions using micellar enhanced ultrafiltration. J Membr Sci 218:257–267

Karate VD, Marathe KV (2008) Simultaneous removal of nickel and cobalt from aqueous stream by cross flow micellar enhanced ultrafiltration. J Hazard Mater 157:464–471

Karode SK (2000) A method for prediction of the gel concentration in macromolecular ultrafiltration. J Membr Sci 171:131–139

Ke X, Guang-ming Z, Jin-hui H et al (2007) Removal of Cd^{2+} from synthetic wastewater using micellar-enhanced ultrafiltration with hollow fiber membrane. Colloids Surf A: Physicochem Eng Asp 294:140–146

Kishihara S, Fujii S, Komoto M (1989) Clarification of technical sugar solution through dynamic layer formed on porous ceramic membrane. J Membr Sci 41:103–114

Lipe KM, Sabatini DA, Hasegawa MA, Harwell JH (1996) Micellar-enhanced ultrafiltration and air stripping for surfactant contaminant separation and surfactant reuse. Ground Water Monit Remidiat 16:85–92

Liu CK, Li CW (2005) Combined electrolysis and micellar enhanced ultrafiltration (MEUF) process for metal removal. Sep Purif Technol 43:25–31

Madsen RF (1973) Application of ultrafiltration and reverse osmosis to cane juice. Int Sugar J 75:163–167

Markels JH, Lynn S, Radke CJ (1994) Micellar ultrafiltration in an unstirred batch cell at constant flux. J Membr Sci 86:241–261

Mukherjee P (1979) Solubilization in aqueous Micellar systems. In: Mittal KL (ed) Solution chemistry of surfactants. Plenum Press, New York, pp 153–174

Purkait MK, DasGupta S, De S (2003) Removal of dye from wastewater using micellar enhanced ultrafiltration and recovery of surfactant. Sep Purif Technol 37:81–92

Purkait MK, DasGupta S, De S (2004) Resistance in series model for micellar enhanced ultrafiltration of eosin dye. J Colloid Interface Sci 270:496–506

Purkait MK, DasGupta S, De S (2005a) Micellar enhanced ultrafiltration of phenolic derivatives from their mixtures. J Colloid Interf Sci 285:395–402

Purkait MK, DasGupta S, De S (2005b) Simultaneous separation of two oxyanions from their mixture using micellar enhanced ultrafiltration. Sep Sci Technol 40:1439–1460

Rahmanian B, Pakizeh M, Maskooki A (2010) Micellar-enhanced ultrafiltration of zinc in synthetic wastewater using spiral-wound membrane. J Hazard Mater 184:261–267

Rosen MJ (2004) Surfactants and interfacial phenomena. Wiley, Hoboken

Sabaté J, Pujolà M, Llorens J (2002) Comparison of polysulfone and ceramic membranes for the separation of phenol in micellar-enhanced ultrafiltration. J Colloid Interf Sci 246:157–163

Saraf YP, Bhagwat SS (1995) Interfacial effects in the solubilization of o-, p-substituted phenols. Sep Technol 5:207–212

Scamehorn JF, Harwell JH (eds) (1989) Surfactant based separation processes. Surfactant science series, vol 33. Marcel Dekker, New York

Song L (1998) Flux decline in cross flow microfiltration and ultrafiltration: mechanisms and modeling of membrane fouling. J Membr Sci 139:183–200

Sreenivas K, Ragesh P, DasGupta S, De S (2002) Modeling of cross-flow osmotic pressure controlled membrane separation processes under turbulent flow conditions. J Membr Sci 201:203–212

Syamal M, De S, Bhattacharya PK (1997) Phenol solubilization by cetyl pyridinium chloride micelles in micellar-enhanced ultrafiltration. J Membr Sci 137:99–107

Tung CC, Yang YM, Chang CH, Maa JR (2002) Removal of copper ions and dissolved phenol from water using micellar-enhanced ultrafiltration with mixed surfactants. Waste Manag 22:695–701

Wiesner MR, Chellam S (1999) The promise of membrane technology. Environ Sci Technol 33:360A–366A

Wijmans JG, Nakao S, Smolders CA (1984) Flux limitations in ultrafiltration: osmotic pressure model and gel layer model. J Membr Sci 20:115–124

Witek A, Koltuniewicz A, Kurczewski B, Radziejowska M, Hatalski M (2006) Simultaneous removal of phenols and Cr^{3+} using micellar-enhanced ultrafiltration process. Desalination 191:111–116

Zihao W, Yuanli J, Jufu F (1996) The entrainment swelling of emulsion during lactic acid extraction by LSMs. J Membr Sci 109:25–34

Chapter 8
Cloud Point Extraction

Abstract Phase separation of a surfactant-loaded solution happens beyond a certain critical thermodynamic state (known as the cloud point), separating the hydrophobic-rich phase in a nonpolar microenvironment from the aqueous supernatant. The dye molecules are bounded to the surfactant and subsequently separated by changing the environmental factor (temperature is commonly altered in cloud point extraction) beyond the cloud point. This is a popular extraction method in bioseparations but is also applicable for purification of dye solution. The chapter discusses the effect of the various operating conditions and different surfactants on extracting clear water from the dye solution with this technology.

Keywords Cloud point extraction • Congo red • Eosin dye • Chrysoidine • Surfactant recovery

Surfactant aggregates orient their hydrophobic tail toward their center of the micelle core, creating a hydrophobic region in aqueous solution. Hydrophobic and covalent organic compounds present in the aqueous solution are favorably partitioned in the nonpolar hydrophobic microenvironment. When the solution condition changes such as altering temperature or pressure, phase separation occurs for the micellar solution into two isotropic phases. A coacervate phase (surfactant rich) is loaded with the hydrophobic burden of the initial solution, while the aqueous supernatant has a concentration of surfactant, close to surfactant critical micellar concentration. The two phases, aqueous and surfactant rich, are usually separated by centrifugation. The temperature above which the phases separate is known as the cloud point temperature, and it is a function of the surfactant concentration. The solute present in aqueous solution of nonionic surfactant is distributed between the two phases at the cloud point temperature (Wang et al. 2003). The ability of this process to concentrate and separate nonpolar target species from aqueous stream has been demonstrated in analytical chemistry and separation science.

S. Mondal et al., *Advances in Dye Removal Technologies*, Green Chemistry and Sustainable Technology, DOI 10.1007/978-981-10-6293-3_8

8.1 Mechanism of Phase Separation

Although the exact mechanism via which this phenomenon occurs is unknown, possible reports suggest:

1. The dielectric constant of water decreases with increase in temperature, reducing the interaction between the hydrophilic portion of surfactant and water. Thus, above the CPT, dehydration occurs in the external layer of micelles of nonionic surfactant due to rise in temperature (Lindman and Wennerstrom 1991).
2. The phase separation above CPT may be due to micellar attraction. This is because at lower temperature (below CPT), intermicellar repulsive force is predominant that becomes attractive when temperature exceeds the CPT (Lindman and Wennerstrom 1991).
3. Phase separations occur due to the competition between enthalpy (which favors separation) and entropy (which favors miscibility of micelles in water) (Liu et al. 1996), which infers to the fact that the clouding and phase separation process is thermodynamically reversible.

A homogeneous system is produced by driving the micelles to combine with the aqueous phase, possibly by reestablishing the initial solution configuration. Kjellander and his group have developed a model describing the phase separation in nonionic surfactants – water system (Silva et al. 1998; Qiao and Easteal 1998; Toerne et al. 2003; Inoue et al. 2003). At low temperatures, each surfactant monomer is solvated with water molecules resulting in the hydrogen bond with the polyoxyethylene units and the polar head. In increasing the temperature, the system entropy is raised, which dehydrates the oxyethylene chains and sabotages the molecular layer of water. Subsequently, the weak intra- as well as intermolecular van der Waals forces play a significant role in the formation of the micellar aggregate, responsible for the phase separation.

The insoluble, sparingly soluble, or highly soluble solute in water dissolves extensively in/on the micelles of surfactant. The extent of solubilization and the location of solubilization in the micelles of nonionic surfactant are related to one another, but not clearly understood. Some authors have proposed that for nonionic surfactant, the core is surrounded by a mantle of aqueous hydrophilic chains and solubilization may occur in both the core and the mantle (Hiemenz and Rajagopalan 1997). The relative amount of solubilization in these two regions of nonionic micelles depends on the polarity of solubilizate. Nonionic surfactants appear relatively more hydrophobic at higher temperature, due to an equilibrium shift that favors dehydration of the ether oxygens. As the cloud point is approached, the solubilization of nonpolar solubilizates increases, probably due to an increase in the aggregation number of the micelles. For polar solubilizates, solubilization decreases owing to dehydration of the polyoxyethylene chains coiling more tightly. These observations demonstrate that nonpolar compounds are solubilized in the core of micelles, while polar solubilizates are located on the mantle. Both the temperature effects cited here are consistent with variations in the space available for the solubilized molecules in the micelles (Hiemenz and Rajagopalan 1997).

8.2 Applications of Cloud Point Separation

The first applications of phase separation based on the physics of cloud point phenomena was reported by Watanabe and Tanaka (1978) for the preconcentration of Zn(II) using 1-(2-pyridylazo) naphthol as a ligand and PONPE as extractant. Later, this methodology was also applied to the determination of different metal ions in different types of samples. Another application of the CPE focuses on the isolation and purification of species of biological interest, mainly proteins. It is in this field of bioseparations that CPE currently finds one of its main use, as evident by the considerable volume of literature related to the extraction and purification of membrane proteins and other biomaterials (Saitoh and Hinze 1995; Heegaard et al. 1997). The use of CPE for the extraction of organic compounds other than biomolecules is relatively recent (Saitoh and Hinze 1995). CPE has been evaluated for the extraction of a series of chlorinated phenols from water (Fernandez et al. 1998). Cloud point technique has been successfully employed for the preconcentration of polycyclic aromatic hydrocarbons (Pinto et al. 1994; Garcia et al. 1992; Ferrer et al. 1996), polychlorinated compound (Fernandez et al. 1998), and vitamins (Saitoh and Hinze 1995; Casero et al. 1999; Sirimanne et al. 1998).

8.3 Effects of Surfactant Concentration on Extraction

Surfactant solutions and different dyes are prepared by dissolving accurately weighed amount of surfactant and dye in distilled water at different concentrations. Each experiment is conducted in a fixed volume of different concentrations of surfactants containing dye and salt solution at a constant temperature bath for 20 min. After complete phase separation, the experimental solution is removed from the temperature bath and cooled for 2 min. The volumes of the coacervate phase and concentration of dilute phase have been measured. The concentration of the surfactant as well as the dyes is measured in UV spectrophotometer at the maximum absorbance wavelength of 226 nm (TX-100), 223 nm (TX-114), 457 nm (chrysoidine), 517 nm (eosin), and 499 nm (Congo red). The concentration of chrysoidine dye is varied from 12 to 200 ppm, while it is varied from 1 to 200 ppm for eosin dye and from 1 to 555 ppm for Congo red. The experiments are conducted with varying temperature from 70 to 95 °C for TX-100 and from 40 to 44 °C for TX-114.

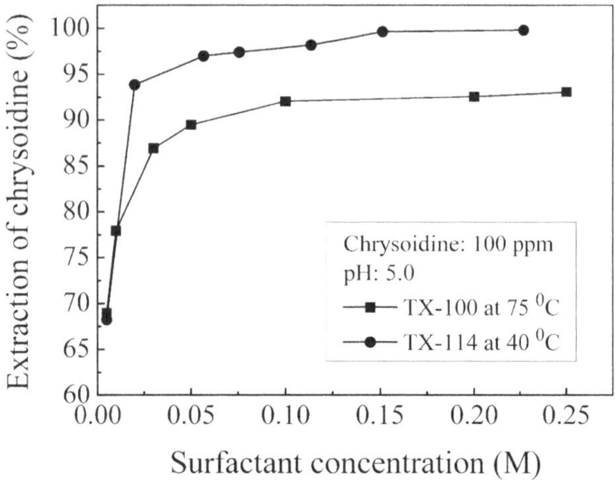

Fig. 8.1 Effect of surfactant concentration on extraction of dye (Reproduced from Purkait et al. (2006). With permission from Elsevier)

8.3.1 Chrysoidine

Figure 8.1 shows the effect of concentration of TX-100 and TX-114 on the extraction of chrysoidine at 75 °C and 40 °C, respectively, at a feed chrysoidine concentration of 100 ppm. It has been observed from the figure that for a chrysoidine concentration of 100 ppm, extraction of the dye increases sharply when TX-100 concentration increases from 0.005 to 0.1(M). Beyond 0.1(M), increase in extraction efficiency becomes gradual and is constant beyond 0.15 (M). It may also be observed from Fig. 8.1 that at a surfactant concentration of 0.15(M), dye extraction is about 92% when TX-100 is used. The extraction increases to about 99.6% when TX-114 is used at the same concentration level of dye and surfactant.

At constant temperature and feed chrysoidine concentration, the fractional coacervate phase volume increases (as shown in Table 8.1) with feed surfactant concentration. This is because of the fact that the concentration of surfactant in the coacervate phase remains nearly constant at constant temperature (Kimchuwanit et al. 2000), and hence to maintain material balance, the coacervate phase volume increases. Therefore, there is more surfactant in micellar form present in the micellar-rich phase (as volume of surfactant-rich phase is more). This increases the extent of solubilization and also the extraction efficiency. However, beyond a surfactant concentration of 0.1(M) for TX-100 and 0.075(M) for TX-114), the extraction becomes nearly constant. The surfactant partition coefficient (defined as the ratio of the concentration of surfactant in the coacervate phase to that in the dilute phase) remains almost constant with increasing surfactant concentration as shown in Fig. 8.2. As the partition coefficient remains unchanged, the distribution of the dye in these two phases remains almost constant. This is reflected in the near constant values of extraction in Fig. 8.1.

Table 8.1 Fractional coacervate phase volume and surfactant extraction data for some selective conditions for CPE of chrysoidine

Dye (ppm)	TX-100 (M)	TX-114 (M)	NaCl (M)	$CaCl_2$ (M)	Temperature (°C)	pH	Fractional coacervate volume	Surfactant extraction (%)
100	0.005	....	....	....	75	5.0	0.004	94.34
100	0.010	....	....	....	75	5.0	0.10	97.11
100	0.030	....	....	....	75	5.0	0.06	99.04
100	0.050	....	....	....	75	5.0	0.10	99.41
100	0.100	....	....	....	75	5.0	0.12	99.69
100	0.200	....	....	....	75	5.0	0.20	99.85
100	0.250	....	....	....	75	5.0	0.23	99.88
100	....	0.075	....	....	40	5.0	0.15	99.62
12.5	0.1	....	....	....	75	5.0	0.05	99.66
25	0.1	....	....	....	75	5.0	0.09	99.67
50	0.1	....	....	....	75	5.0	0.10	99.69
200	0.1	....	....	....	75	5.0	0.16	99.71
200	....	0.075	....	....	40	5.0	0.19	99.81
100	0.05	....	....	....	80	5.0	0.09	99.42
100	0.05	....	....	....	85	5.0	0.08	99.45
100	0.05	....	....	....	90	5.0	0.07	99.49
100	....	0.075	....	....	40	5.0	0.15	99.44
100	0.10	....	....	....	75	2.0	0.18	99.98
100	0.10	....	....	....	75	5.0	0.12	99.98
100	0.10	....	....	....	75	9.0	0.11	99.98
100	0.10	....	....	....	75	12.0	0.09	99.99
100	....	0.075	....	....	40	12.0	0.08	99.03
100	0.10	....	0.05	....	75	5.0	0.12	99.97
100	0.10	....	0.10	....	75	5.0	0.11	99.97
100	0.10	....	0.30	....	75	5.0	0.09	99.97
100	0.10	....	0.50	....	75	5.0	0.08	99.97
100	0.10	....	....	0.05	75	5.0	0.11	99.97
100	0.10	....	....	0.10	75	5.0	0.10	99.97
100	0.10	....	....	0.30	75	5.0	0.09	99.97
100	0.10	....	....	0.50	75	5.0	0.08	99.97

Reproduced from Purkait et al. (2006). With permission Elsevier

8.3.2 *Eosin*

For successful CPE of dye, it is desirable to use minimum amount of surfactant for maximum extraction of dye. Figures 8.3 and 8.4 show the effect of concentration of TX-100 on the extraction of eosin for different initial concentrations at 80 and 95 °C. It has been observed from both figures that for eosin concentration of more than 1 ppm, extraction of dye increases sharply when TX-100 concentration

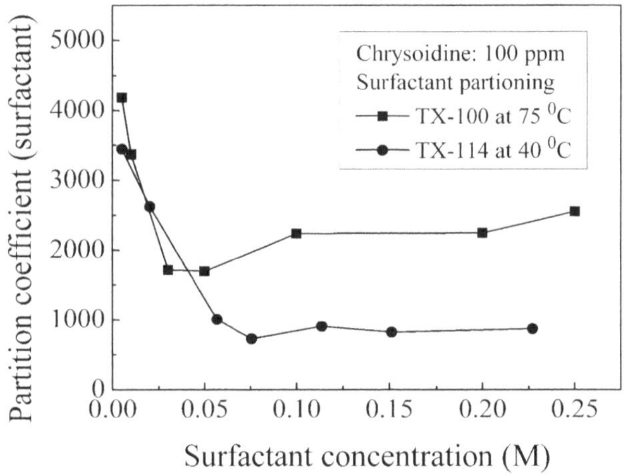

Fig. 8.2 Effect of surfactant concentration on the partition coefficient of surfactant (Reproduced from Purkait et al. (2006). With permission from Elsevier)

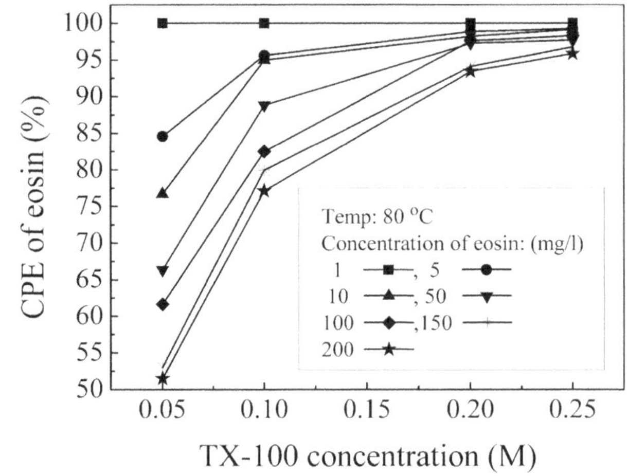

Fig. 8.3 Effect of concentrations of TX-100 and eosin on the efficiency of CPE of dye at 80 °C (Reproduced from Purkait et al. (2005). With permission from Elsevier)

increases from 0.05(M) to 0.1(M). Beyond 0.1(M), increase in extraction efficiency becomes gradual. For eosin concentration of 1 ppm, close to 100% dye extraction is possible with a TX-100 concentration of 0.05(M). For 5, 10, 50, and 100 ppm of dye concentration, 0.2(M) TX-100 may be an optimum concentration to achieve about 98% extraction of dye. For higher concentrations, e.g., 150 and 200 ppm, about 96% extraction of eosin is achieved with 0.25(M) TX-100. It has also been observed that for a particular TX-100 concentration, percent extraction decreases with increase in the dye concentration. With increase in surfactant concentration, the volume of coacervate phase increases, as the concentration of surfactant in coacervate phase remains almost constant. This increase in coacervate phase volume renders higher solubilization of dye, which explains the higher extraction of dye at higher TX-100 concentration at constant temperature and initial dye concentration.

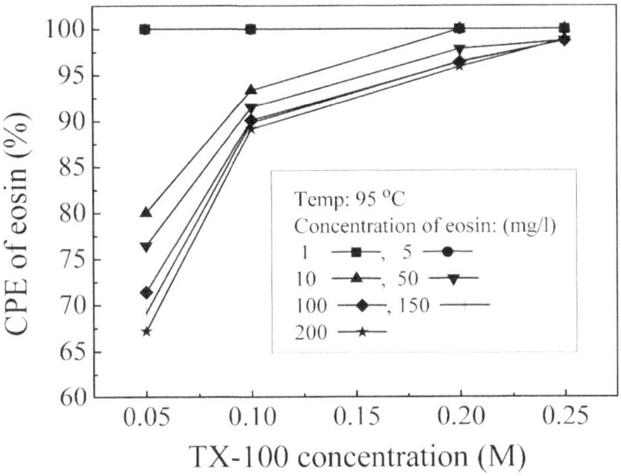

Fig. 8.4 Effect of concentrations of TX-100 and eosin on the efficiency of CPE of dye at 95 °C (Reproduced from Purkait et al. (2005). With permission from Elsevier)

Figure 8.4 presents the efficiency of dye extraction with surfactant concentration at 95 °C. This figure clearly indicates that the extraction efficiency increases with temperature. For example, extraction efficiency of concentrated dye solution at 200 ppm is about 51.5% at 0.05(M) surfactant concentration at 80 °C (Fig. 8.3), whereas it is about 67% at the same concentration level at 95 °C (Fig. 8.4). At surfactant concentration of 0.25(M), the extraction efficiency of all the eosin concentration used in the present experiments increases from a range of 96 to 100% (Fig. 8.3) to about 98.8 to 100% (Fig. 8.4) when temperature increases from 80 to 95 °C. Increase in temperature leads to an increase in aggregation number of micelles thereby increasing the extent of solubilization of dyes at higher temperature. It may be noted from Figs. 8.3 and 8.4 that beyond a surfactant concentration of 0.2(M), the increase of extraction efficiency of eosin is gradual for all the feed eosin concentrations used in this study. Therefore, 0.2 (M) TX-100 may be considered as the optimum surfactant dose for efficient CPE of eosin dye.

8.3.3 *Congo Red*

Similar observation is found in case of CPE of Congo red using TX-100. The results in Figs. 8.5 and 8.6 show for different initial Congo red concentrations at 70 °C and 85 °C, respectively. It has been observed from both figures that for a Congo red concentration of more than 35 ppm, the extraction of dye increases sharply when TX-100 concentration increases from 0.02(M) to 0.1(M). Beyond 0.1(M), increase in extraction efficiency becomes gradual.

It may be observed from Fig. 8.5 that for a dye concentration of 35 ppm, 100% Congo red extraction is possible with a TX-100 concentration of 0.02(M). For

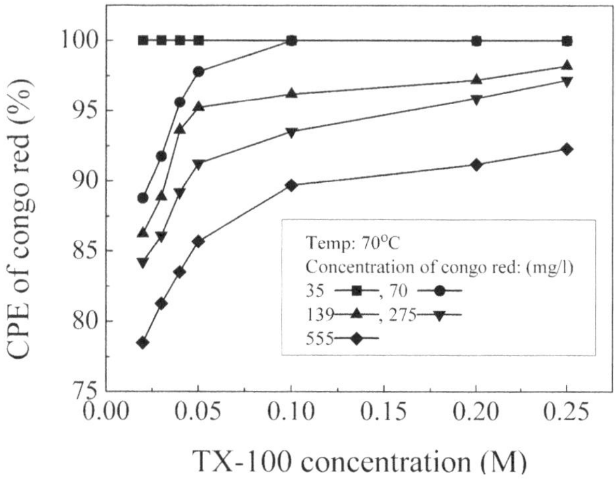

Fig. 8.5 Effect of concentrations of TX-100 and Congo red on the CPE of dye at 70 °C (Reproduced from Purkait et al. (2004). With permission from Elsevier)

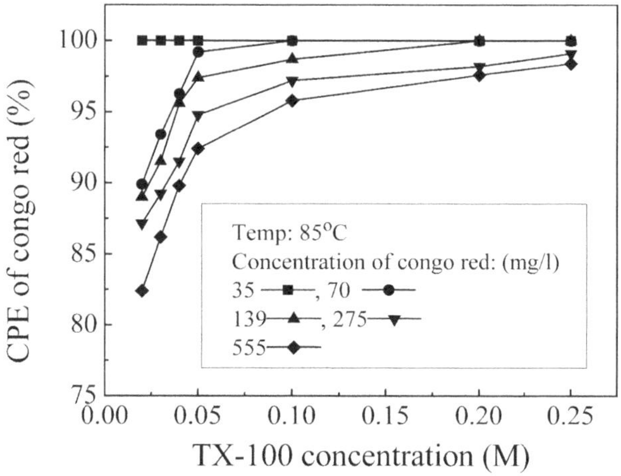

Fig. 8.6 Effect of concentrations of TX-100 and Congo red on the CPE of dye at 85 °C (Reproduced from Purkait et al. (2004). With permission from Elsevier)

70 ppm of Congo red concentration, 0.1(M) TX-100 may be an optimum concentration to achieve about 100% extraction of dye. For higher dye concentrations, e.g., 139, 275, and 555 ppm, with 0.25(M) TX-100, about 98.2, 97.2, and 92.3% extractions of Congo red are achieved. The extraction efficiency of concentrated Congo red solution of 555 ppm is about 92.3% at 0.25(M) surfactant concentration, at 70 °C (Fig. 8.5), whereas it is about 98.4% at the same concentration level at 85 °C (Fig. 8.6). At a surfactant concentration of 0.25(M), the extraction efficiency of all the dye concentrations used in the present experiments increases from 92.3 to 100% (Fig. 8.5) and from 98.4 to 100% (Fig. 8.6) when temperature increases from 70 to 85 °C. It may be noted from Figs. 8.5 and 8.6 that beyond a surfactant

concentration of 0.2(M), the increase of extraction efficiency for a dye concentration more than 70 ppm is gradual for all the Congo red feed concentrations used in this study. Therefore, 0.1(M) TX-100 may be considered as the optimum TX-100 dose for efficient CPE of Congo red up to 70 ppm. For dye concentration beyond 70 ppm, 0.2(M) TX-100 is the optimum dose.

8.4 Effects of Dye Concentration on Extraction

8.4.1 Chrysoidine

The effects of feed chrysoidine concentration on the dilute-phase chrysoidine concentrations have been shown in Fig. 8.7. Surfactants used are 0.1(M) of TX-100 and 0.075(M) of TX-114 for the feed chrysoidine concentrations of 12.5, 25, 50, 100, and 200 ppm. It has been observed from the figure that in both cases, the dilute-phase chrysoidine concentration increases sharply with the feed chrysoidine concentration. From Fig. 8.7, it may also be observed that for a particular feed dye concentration, dilute-phase dye concentration remains much higher when TX-100 is used. For example, for 100 ppm dye, about 92% of dye extraction is possible when 0.1(M) of TX-100 is used at 75 °C. For the same dye concentration, 0.075(M) of TX-114 at 40 °C increases the extraction up to about 97.5%.

As in Table 8.1, the fractional coacervate phase volume increases with feed dye concentration at constant temperature, surfactant concentration, and pH. It is well known that compounds, like urea, formamide, etc., increase CMC of the nonionic surfactants in the aqueous solution because of their disruption of the water structure (Rosen 2004). This may increase the hydration of the hydrophilic group of the

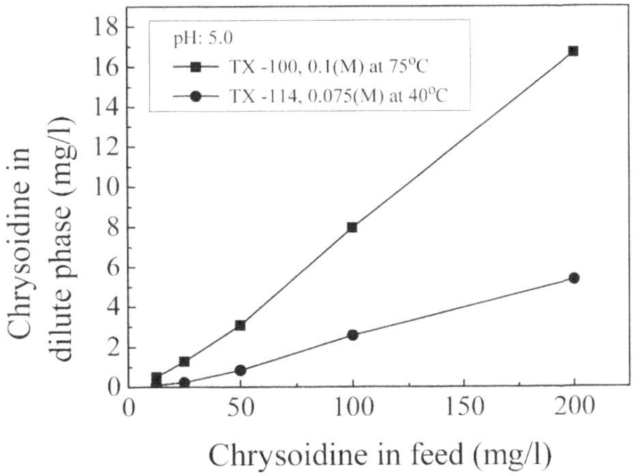

Fig. 8.7 Effect of feed dye concentration on the dilute-phase dye concentration (Reproduced from Purkait et al. (2006). With permission from Elsevier)

surfactant (Rosen 2004). In the present study, the chrysoidine dye has similar active functional group of urea. Therefore, it may be assumed that in the presence of dye (at constant surfactant concentration), CMC of the nonionic surfactant increases. This implies that the number concentration of the micelles decreases with dye concentration. Therefore, increasing feed dye concentration only results in an increase of unsolubilized dye, increasing its concentration in the dilute phase and hence reducing the extraction efficiency. The presence of dye increases the CMC resulting in lesser number of micelle and hence a lowering of the volume. But this decrease is offset by the hydrating effect of the dye resulting in the formation of larger molecules. The effective volume thus shows a small increase with the chrysoidine concentration.

8.4.2 Eosin and Congo Red

Variations of the extraction of TX-100 with feed eosin and TX-100 concentration are shown in Fig. 8.8 at 85 °C. It may be observed from the figure that extraction efficiency of the surfactant increases with initial surfactant concentration for all the dye concentrations. For example, at a feed surfactant concentration of 0.25(M), the surfactant is extracted in the coacervate phase in the range of 98.7–99.8% for all the eosin concentrations. Since, TX-100 concentration in the dilute phase remains almost constant (at lower surfactant concentration just above the CMC) at constant temperature and eosin concentration, the extraction efficiency of TX-100 increases with feed TX-100 concentration. Beyond a concentration of 0.2(M), the increase in extraction efficiency is gradual because of the fact that at higher surfactant

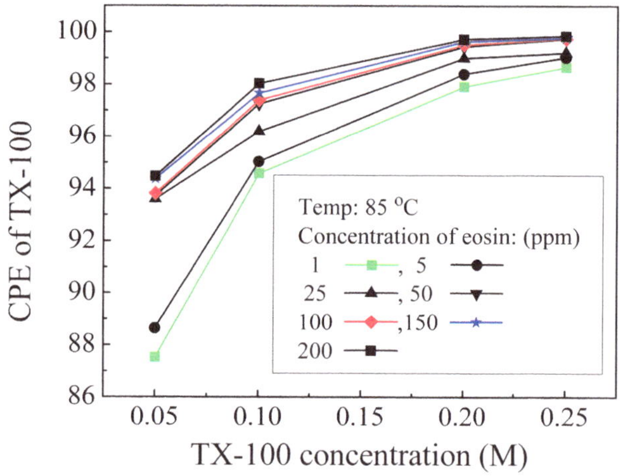

Fig. 8.8 Effect of concentration of TX-100 on the CPE of TX-100 at different feed eosin concentrations at 85 °C

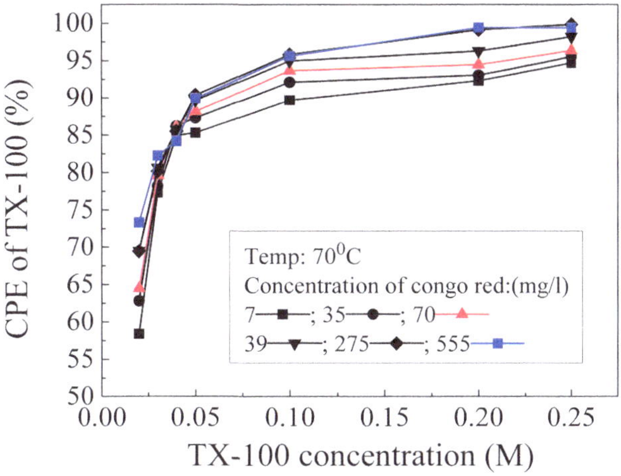

Fig. 8.9 Effect of concentration of TX-100 on the CPE of TX-100 at different feed Congo red concentrations at 70 °C

concentration (>0.2 M), the surfactant concentration in dilute phase increases with feed TX-100 concentration (>0.2 M). At constant surfactant concentration, TX-100 extraction increases with eosin concentration. As more eosin molecules are solubilized, intermicellar repulsion decreases, increasing the micellar size and hence the coacervate phase volume.

Same trend is observed for the CPE of Congo red and is shown in Fig. 8.9 at 70 °C. In this case, surfactant is extracted in the coacervate phase in the range of 94.7–99.41% for all Congo red concentrations at a feed surfactant concentration of 0.25(M).

8.5 Effects of Temperature on Extraction

8.5.1 Chrysoidine

The effects of temperature on the efficiency of chrysoidine extraction are shown in Fig. 8.10 for an initial chrysoidine concentration of 100 ppm at 0.03, 0.05, 0.10, 0.20, and 0.25(M) of TX-100. It is clear from the figure that the extraction of chrysoidine increases with temperature and TX-100 concentration. It may be observed that the extraction of chrysoidine (for 100 ppm of feed dye and 0.25 (M) of TX-100) increases from about 93 to 97.5%, when the temperature increases from 75 to 90 °C. Variation of chrysoidine extraction with temperature using TX-114 at different feed dye concentrations is shown in Fig. 8.11. Same trend of dye extraction with temperature has been observed when TX-114 is used instead of

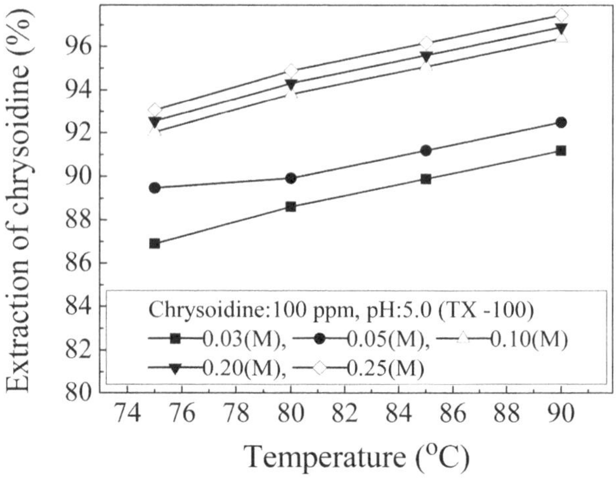

Fig. 8.10 Effect of temperature on the dye extraction at different TX-100 concentrations and at a dye concentration of 100 ppm (Reproduced from Purkait et al. (2006). With permission from Elsevier)

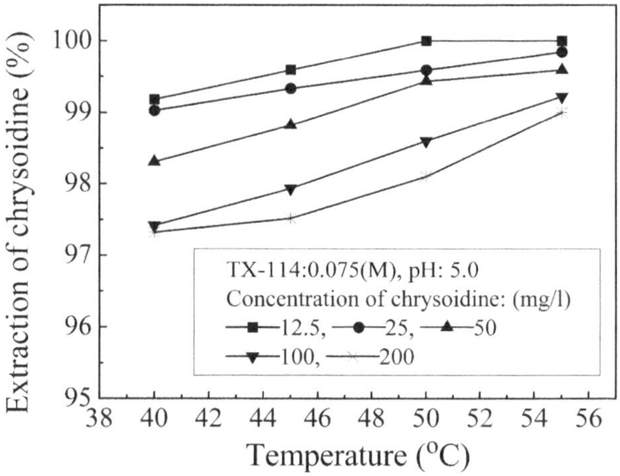

Fig. 8.11 Effect of temperature on the dye extraction at different dye concentrations and at a TX-114 concentration of 0.075(M) (Reproduced from Purkait et al. (2006). With permission from Elsevier)

TX-100. It may be noted that for TX-114, the extraction of dye is marginally improved with temperature. The extraction (for 100 ppm of feed dye and 0.25 (M) of TX-114) increases marginally from about 97.5 to only 98%, when the temperature increases from 40 to 55 °C. Therefore, the extraction should be carried out at an operating temperature of 40 °C for TX-114.

At higher temperature, CMC of nonionic surfactants decreases (Clint 1992). Moreover, nonionic surfactants appear relatively more hydrophobic at higher temperatures, due to an equilibrium shift favoring dehydration of the ether oxygens (Hiemenz and Rajagopalan 1997). This leads to an increase in the number concentration of micelles. Therefore, the solubilization capability of the micellar solution increases with temperature leading to an increase in the dye extraction. It is also evident from Table 8.1 that the volume of coacervate phase decreases with

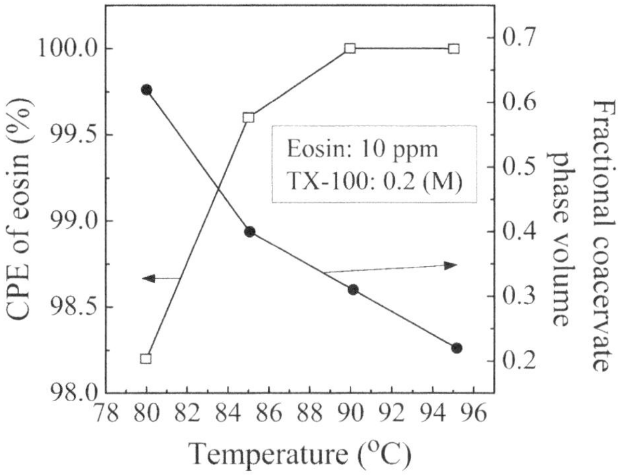

Fig. 8.12 Effect of temperature on the efficiency of CPE for 10 ppm of feed dye

temperature. For example, at 100 ppm of dye and 0.05(M) of TX-100, fractional volume of coacervate phase decreases from 0.10 to 0.07 when temperature is raised from 75 to 90 °C. At an elevated temperature, the interaction among the TX-100 micelles increases leading to dehydration from the external layers of micelles resulting in a decrease in volume of coacervate phase (Kimchuwanit et al. 2000).

8.5.2 Eosin and Congo Red

The effects of temperature on the efficiency of CPE are shown in Figs. 8.12 and 8.13 for initial eosin concentration of 10 and 150 ppm, respectively, using 0.2(M) of TX-100. It is clear from the figures that the extraction of eosin increases and the fractional coacervate phase volume decreases with temperature. From Fig. 8.12, it is observed that the extraction of dye increases from 98.2 to 100% and fractional coacervate phase volume decreases from 0.62 to 0.22, when the temperature increases from 80 to 95 °C. Eosin molecule is ionic, but due to the presence of three benzene rings, it shows an amphiphilic nature. The increase in extraction efficiency or solubilization in micelles with temperature is due to the fact that the polarity of eosin molecule decreases with temperature and shows preferentially hydrophobic nature at higher temperature. Therefore, eosin molecules get solubilized in the micelles. At higher temperature, this is augmented by the increase in aggregation number, leading to an increase in the extraction efficiency or solubilization. Similar trend is observed for higher feed concentration (150 ppm) of eosin as shown in Fig. 8.13.

Figure 8.14 shows the variations of extraction of Congo red with temperature for initial Congo red concentration of 139, 275, and 555 ppm using 0.04 and 0.05(M) of TX-100. This figure also shows that the extraction of Congo red increases with

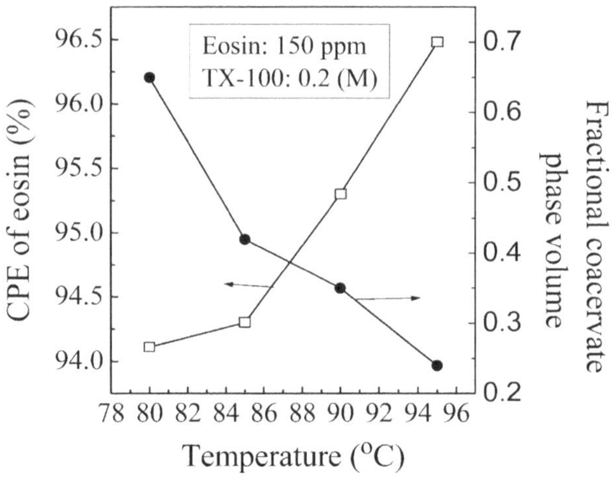

Fig. 8.13 Effect of temperature on the efficiency of CPE for 150 ppm of feed dye

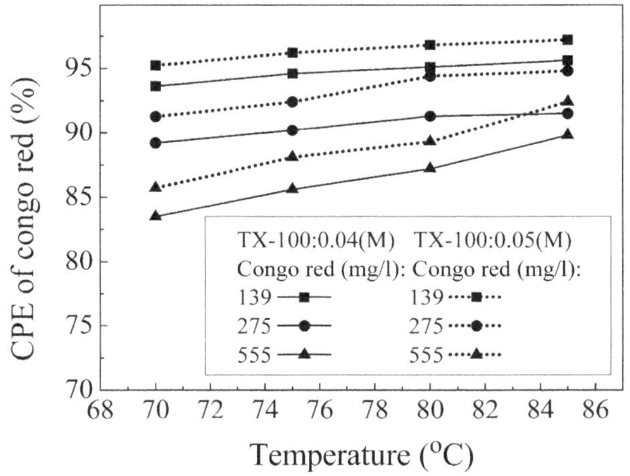

Fig. 8.14 Effect of temperature on the efficiency of CPE for 139, 275, and 555 ppm of CR at TX-100 concentrations of 0.04 and 0.05(M) (Reproduced from Purkait et al. (2004). With permission from Elsevier)

temperature and feed TX-100 concentration. It may be observed that the extraction of Congo red (for 139 ppm of feed dye and 0.04(M) of TX-100) increases from 93.62 to 95.6%, when the temperature increases from 70 to 85 °C. The reason is due to the interaction of the structural feature of the Congo red molecule as in the case of eosin molecule.

Figure 8.15 shows the variation of fractional coacervate phase volume with temperature. It is evident from Fig. 8.15 that the volume of coacervate phase decreases with temperature. For example, at 200 ppm of dye, fractional volume of coacervate phase decreases from 0.16 to 0.07 when the temperature is raised from 80 to 95 °C. At elevated temperature, the interaction among the TX-100 micelles increases leading to dehydration from the external layers of micelles resulting in a decrease in volume of coacervate phase. At a fixed temperature,

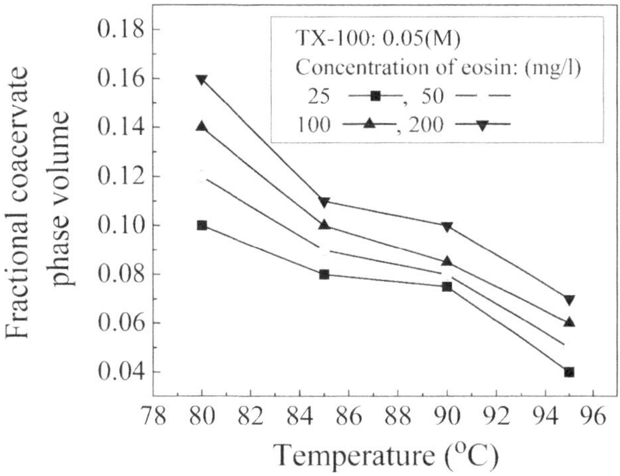

Fig. 8.15 Effect of temperature on the fractional coacervate phase volume at different feed eosin concentrations and at TX-100 concentration of 0.05(M)

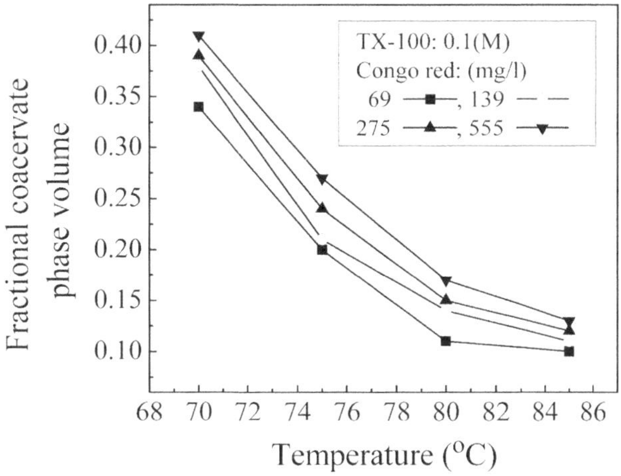

Fig. 8.16 Effect of temperature on the fractional coacervate phase volume at different feed CR concentrations and at TX-100 concentration of 0.1(M) (Reproduced from Purkait et al. (2004). With permission from Elsevier)

fractional volume of coacervate phase increases with dye concentration. As dye concentration increases, more dyes will be solubilized leading to an increase in coacervate phase volume. Similar trend is observed for CPE of Congo red and is shown in Fig. 8.16 for different Congo red concentrations using 0.1(M) of TX-100.

The variations of the extent of extraction of the surfactant with temperature are presented in Fig. 8.17 for different feed dye concentrations. It may be observed from the figure that the extraction of the surfactant increases initially with temperature and feed dye concentration. For example, at 139 ppm dye concentration, the efficiency of extraction of surfactant increases from 95.54 to 99.79% when temperature is raised from 70 to 75 °C at 0.25(M) of TX-100. On the other hand, at 70 °C, extraction efficiency increases with dye concentration as discussed earlier. Beyond 75 °C, the efficiency of TX-100 extraction is almost independent of

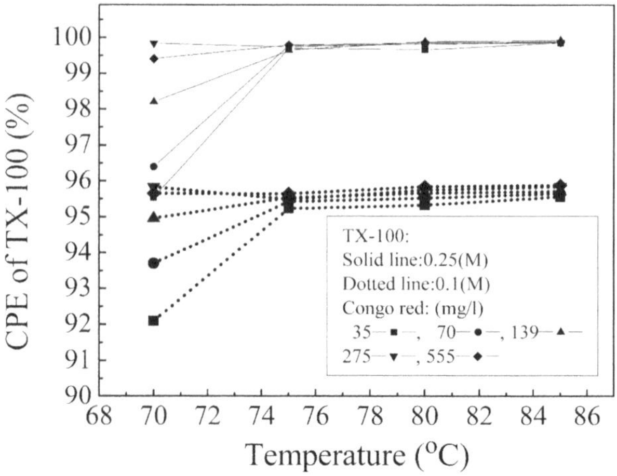

Fig. 8.17 Effect of temperature on CPE of TX-100 at different feed dye and TX-100 concentrations (Reproduced from Purkait et al. (2004). With permission from Elsevier)

temperature and dye concentration. The increasing extraction efficiency at higher temperature is due to the more micellar attraction. The same trend has been observed for TX-100 extraction with temperature and dye concentration at 0.1 (M) of surfactant. It may be observed from figure that the extraction efficiency is more for higher feed TX-100 concentration. Since all the feed TX-100 concentrations are much above the CMC and dilute phase remains slightly above the CMC, the ratio of the surfactant concentration of dilute phase to that of the feed decreases with increase in feed TX-100 concentration, and hence extraction efficiency decreases with feed surfactant concentration.

8.6 Effects of pH on Extraction

The effects of the pH of the solution on the extent of chrysoidine extraction are shown in Fig. 8.18 for 100 ppm of feed chrysoidine using 0.1(M) of TX-100 and 0.075(M) of TX-114 at 75 °C and 40 °C, respectively. Extraction of chrysoidine is less in acidic pH and increases with pH. The lower extraction at acidic pH may be due to the increasing ionic character of oxy group of nonionic surfactant, which increases the CMC (Clint 1992). This leads to a decrease in the micellar concentration and the aggregation number resulting in less solubilization of the dye. On the other hand, at basic pH, CMC is lowered due to increasing hydrophobicity of oxy groups that increase the size of the micelles as well as the aggregation number (Rosen 2004). Therefore, dye solubilization is more at basic pH values leading to an increase in the dye extraction and lower fractional coacervate phase volume (Table 8.1).

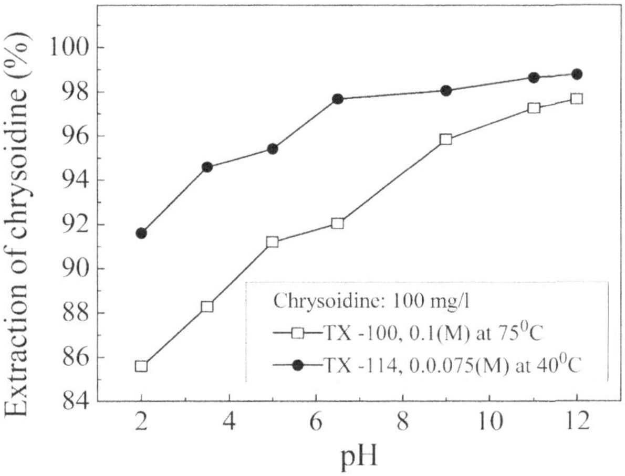

Fig. 8.18 Effect of pH on extraction of dye using TX-100 and TX-114 (Reproduced from Purkait et al. (2006). With permission from Elsevier)

8.7 Effects of Salt Concentration on Extraction

8.7.1 Chrysoidine

Figure 8.19 shows the variation of extraction efficiency with salt (NaCl and $CaCl_2$) concentration. The surfactants used are TX-114 at 40 °C and TX-100 at 75 °C. It may be observed from the figure that the extraction of chrysoidine increases from about 93 to 97% when concentration of $CaCl_2$ increases from 0.05(M) to 0.5(M) at a fixed initial dye concentration (100 ppm in this case) and TX-100 concentration (0.1 M). Beyond 0.3(M), the increase in efficiency becomes gradual. Same trend of chrysoidine extraction has been observed with NaCl. The trend with TX-114 is similar. The CMC of the nonionic surfactants decreases in the presence of electrolytes which are known to be capable of "salting out," e.g., NaCl, KCl, $CaCl_2$, etc. (Clint 1992). Therefore, the number of concentration and aggregation number of the micelles increase with the addition of NaCl and $CaCl_2$. This enhances the amount of solubilized dye. Beyond a salt concentration of 0.3(M), the increase in extraction efficiency is marginal. This may be due to the marginal change of aggregation number beyond this condition. From Fig. 8.19, it may be noted that at the same level of salt concentration, the dye extraction efficiency is more for $CaCl_2$ compared to NaCl. This is due to the increased micellar aggregation because of the enhanced salting out effect of $CaCl_2$ compared to NaCl in presence of divalent calcium salt.

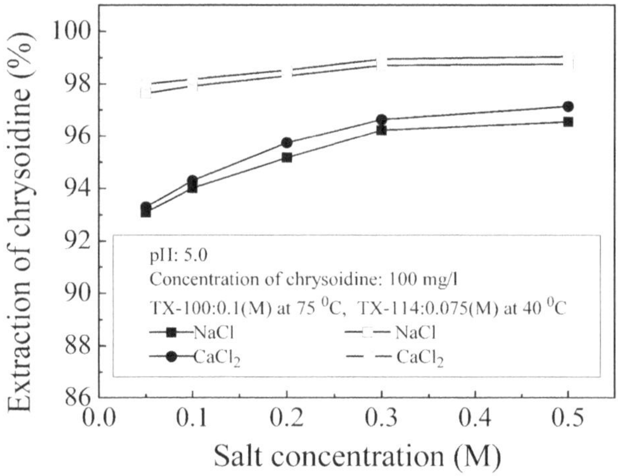

Fig. 8.19 Effect of NaCl and $CaCl_2$ concentration on extraction of chrysoidine (Reproduced from Purkait et al. (2006). With permission from Elsevier)

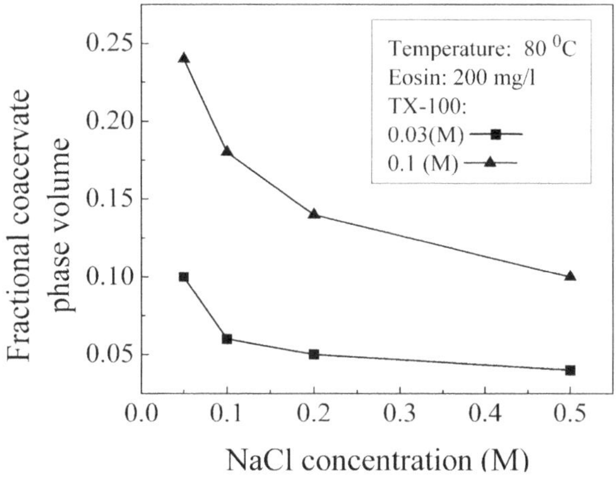

Fig. 8.20 Effect of NaCl concentration on fractional coacervate phase volume

8.7.2 Eosin

Variation of fractional coacervate phase volume with NaCl concentration is shown in Fig. 8.20. It is clear from the figure that the fractional coacervate phase volume decreases sharply (from 0.24 to 0.14 and 0.1 to 0.05 for 0.1(M) and 0.03(M) of feed TX-100, respectively) with increasing NaCl concentration up to 0.2(M). Beyond that, decrease in fractional coacervate phase volume is marginal. It has also been found that for constant eosin and salt concentration, fractional coacervate phase volume increases with TX-100 concentration. Decrease in fractional coacervate phase volume with increase in NaCl concentration is due to the decrease in intermicellar repulsion and increase in aggregation number. With the increase in NaCl concentration, intermicellar repulsion decreases and reaches minimum at 0.2(M). Beyond 0.2(M),

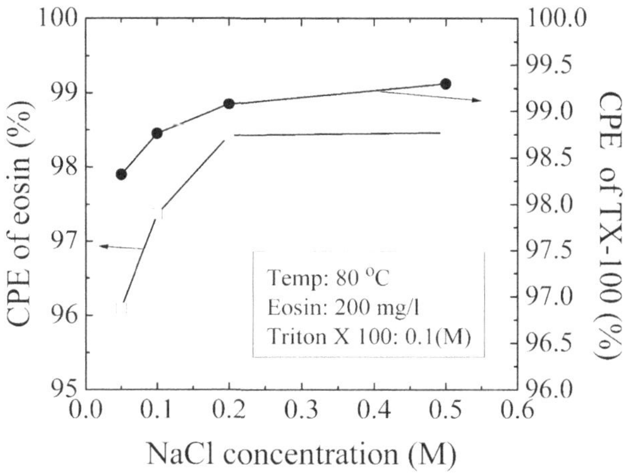

Fig. 8.21 Effect of NaCl concentration on the CPE for 200 ppm of feed eosin

the repulsion is marginal, and hence decrease in fractional coacervate phase volume is also very less.

Figure 8.21 shows the variation of extraction efficiency with NaCl concentration at 80 °C. It may be observed from the figure that the extraction of both eosin and TX-100 increases with NaCl concentration at a fixed initial eosin (200 ppm in this case) and TX-100 concentration (0.1 M). It is also found that the extraction of both eosin and TX-100 increases sharply (from 96.1 to 98.42% for dye and 98.32 to 99.08% for TX-100) up to 0.2(M) of NaCl, but beyond that, the increase in efficiency becomes gradual. The increase in extraction of dye with salt concentration has already been discussed earlier.

8.7.3 *Congo Red*

Figures 8.22 and 8.23 show the variation of extraction efficiency with $CaCl_2$ concentration at TX-100 concentration of 0.03(M) and 85 °C at initial Congo red concentration of 400 and 600 ppm, respectively. It may be observed from both the figures that the extraction efficiency of both Congo red and TX-100 increases with $CaCl_2$ concentration. From Fig. 8.22, it is found that the extraction of both dye and TX-100 increases sharply (from 89.45 to 93.99% for dye and from 80.14 to 85.78% for TX-100) up to 0.2(M) of $CaCl_2$, but beyond that, the increase in efficiency becomes gradual. Similar trend is observed in Fig. 8.23 for higher dye concentration (600 ppm).

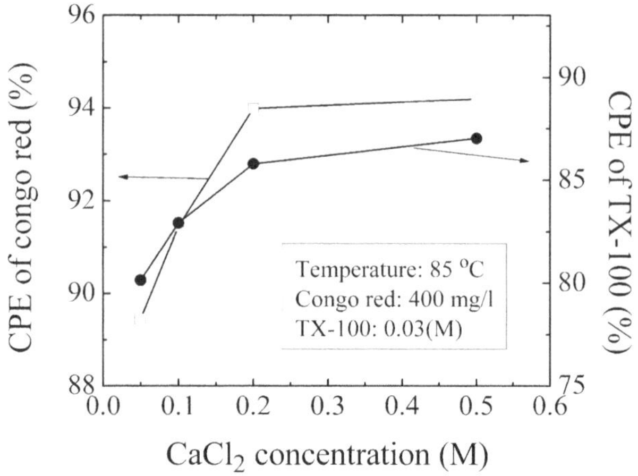

Fig. 8.22 Effect of $CaCl_2$ concentration on the efficiency of CPE for 400 ppm of feed CR at 0.03 (M) of TX-100 (Reproduced from Purkait et al. (2004). With permission from Elsevier)

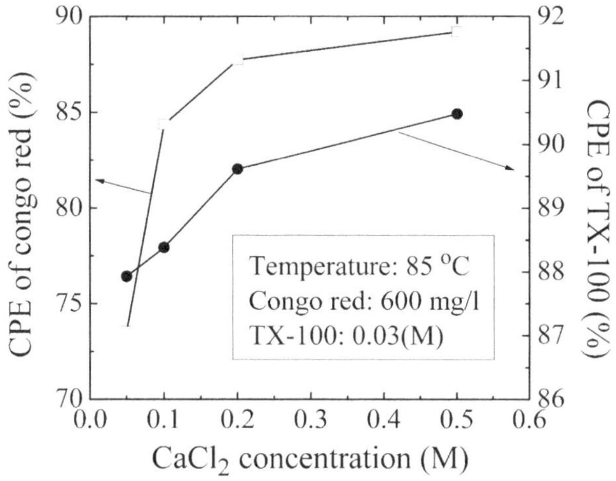

Fig. 8.23 Effect of $CaCl_2$ concentration on the efficiency of CPE for 600 ppm of feed CR at 0.03 (M) of TX-100 (Reproduced from Purkait et al. (2004). With permission from Elsevier)

8.8 Determination of Design Parameters for Cloud Point Extraction of Congo Red and Eosin Dyes Using TX-100

CPE has been successfully used for removal of Congo red and eosin dye using TX-100 as the nonionic surfactant (Namasivayam and Kavitha 2002; Purkait et al. 2004, 2005, 2006). In the present study, various design parameters of a CPE process have been estimated by developing correlations for dye solubilization and fractional coacervate phase volume with the operating conditions, namely, temperature, feed surfactant, and dye concentration. A method is presented to calculate the feed surfactant concentration required for the removal of dyes up to a level of 1.0 mg/L. The developed correlations may be useful to design a cloud point extractor of a desired efficiency.

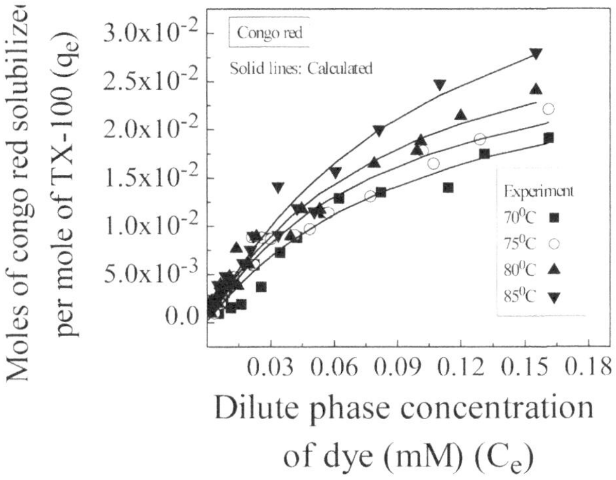

Fig. 8.24 Solubilization isotherm for Congo red at different temperatures using TX-100 (Reproduced from Purkait et al. (2006). With permission from Elsevier)

8.8.1 Solubilization Isotherm

In order to determine dye solubilization capacity of TX-100 at different temperatures, the experimental data are used to calculate the solubilization isotherms. These isotherm data are the basic requirements for the design of CPE system.

Figures 8.24 and 8.25 show the isotherms at different temperatures, for Congo red-TX-100 and eosin-TX-100 system, respectively. The Langmuir-type adsorption isotherm is successfully used in describing many adsorption processes. The same model has been used to explain the solubilization of two different dyes in TX-100. Equation (8.1) gives the expression of the well-known Langmuir model.

$$q_e = \frac{mnC_e}{1 + nC_e} \tag{8.1}$$

where q_e is the moles of dye solubilized per mole of surfactant. C_e is the dilute-phase equilibrium concentration of the dye. The constants m and n are the Langmuir constants signifying the solubilization capacity and energy of solubilization, respectively (Namasivayam and Kavitha 2002). Values of m and n for each operating temperatures are evaluated by regression analysis using the experimental data. The variations of m and n with temperature are fitted to a quadratic model as shown in Figs. 8.26 and 8.27. The expressions of m and n as function of temperature are given below:

For Congo red-TX-100 system

$$m = 0.6398 - 1.682 \times 10^{-2}T + 1.16 \times 10^{-4}T^2 \quad \left(r^2 = 0.996\right) \tag{8.2}$$

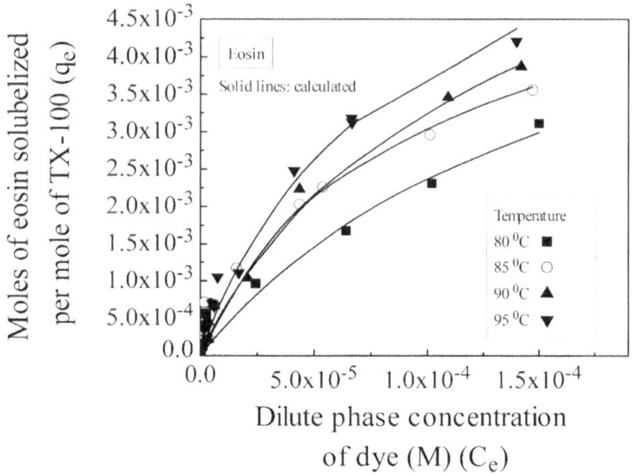

Fig. 8.25 Solubilization isotherm for eosin at different temperatures using TX-100 (Reproduced from Purkait et al. (2006). With permission from Elsevier)

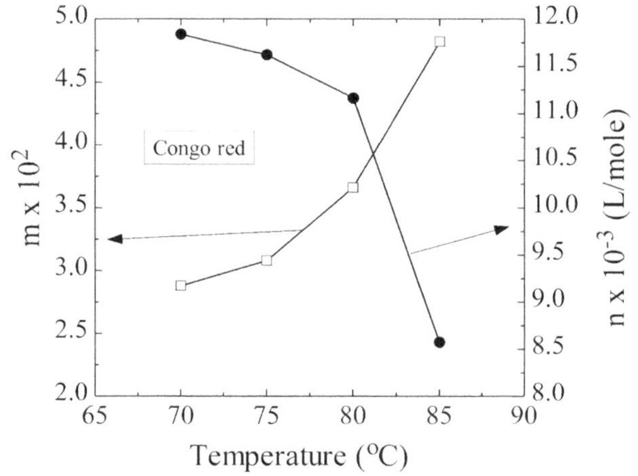

Fig. 8.26 Variation of the values of *m* and *n* with temperature for solubilization of Congo red in TX-100

$$n = -263438.5 + 7172.2T - 46.730.98T^2\left(r^2 = 0.990\right) \tag{8.3}$$

For eosin-TX-100 system

$$m = 2.376 \times 10^{-2} - 3.93 \times 10^{-4}T + 2.1 \times 10^{-6}T^2\left(r^2 = 0.992\right) \tag{8.4}$$

$$n = -3.774 \times 10^5 + 8384.7T - 44.69T^2\left(r^2 = 0.994\right) \tag{8.5}$$

where T is the temperature in $^\circ$C.

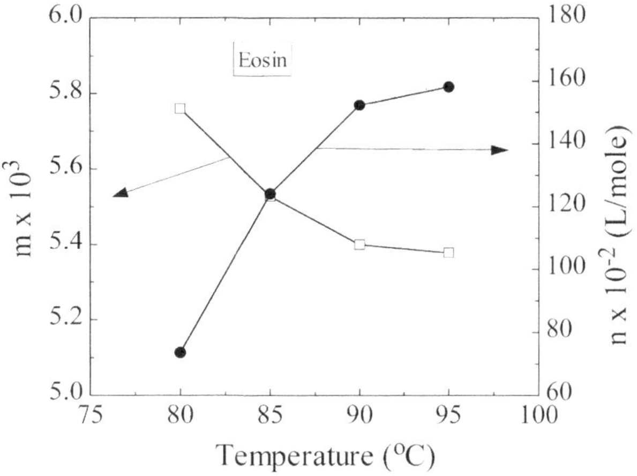

Fig. 8.27 Variation of the values of m and n with temperature for solubilization of eosin in TX-100

Presence of salts reduces the critical micellar concentration of the surfactant (Clint 1992). Thus, the solubilization capacity of the micelles increases as the number of micelles increases on addition of salts. Hence, the performance of CPE increases with salt concentration. This effect is more for a divalent salt (e.g., $CaCl_2$) compared to a monovalent one (e.g., NaCl). In the present study, divalent $CaCl_2$ is used as an external performance-enhancing agent for CPE of Congo red, and monovalent NaCl is used for CPE of eosin. To incorporate the concentration of salts in the isotherm, Eq. (8.1) is modified and expressed in Eqs. (8.6) and (8.7) for constant concentration of dye and surfactant.Isotherm for the solubilization of 600 mg/L of Congo red in 0.03(M) of TX-100 micelles at 70 °C in presence of $CaCl_2$ is as follows:

$$q_{es} = \frac{mnC_e}{1 + nC_e}\left[1 + 6.96\left(1 - \exp^{-10C_{\mathrm{salt}}}\right)\right] \tag{8.6}$$

The isotherm for the solubilization of 200 mg/L of eosin in 0.1(M) of TX-100 micelles at 85 °C in presence of NaCl is as follows:

$$q_{es} = \frac{mnC_e}{1 + nC_e}\left[1 + 163.97\left(1 - \exp^{-9.52C_{\mathrm{salt}}}\right)\right] \tag{8.7}$$

Using Eqs. (8.6) and (8.7), the solubilization isotherms at various salt concentrations are calculated and presented in Figs. 8.28 and 8.29 for Congo red using $CaCl_2$ at 70 °C and for eosin using NaCl at 85 °C, respectively. It is observed from both the figures that the extent of dye solubilization is significantly higher in the presence of salt.

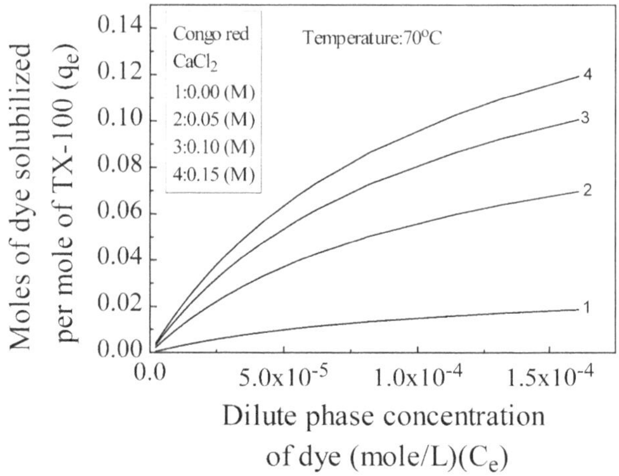

Fig. 8.28 Isotherm for the solubilization of 600 mg/L of Congo red in 0.03(M) of TX-100 micelles at 70 °C in presence of $CaCl_2$

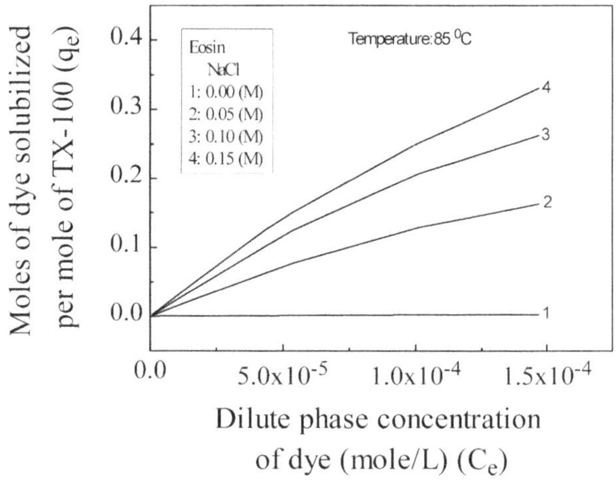

Fig. 8.29 Isotherm for the solubilization of 200 mg/L of eosin in 0.1(M) of TX-100 micelles at 85 °C in presence of NaCl

8.8.2 Variation of Fractional Coacervate Phase Volume

In order to calculate the performance of a CPE process, the variation of the fractional coacervate phase volume with concentration of the feed surfactant and the operating temperature needs to be studied. In this regard, the following correlation for the fractional coacervate phase volume with the feed surfactant concentration is proposed:

$$F_c = aC_s^b \tag{8.8}$$

where F_c is the fractional coacervate volume and C_s is the molar concentration of the feed surfactant solution. As mentioned in the experimental section, a wide range

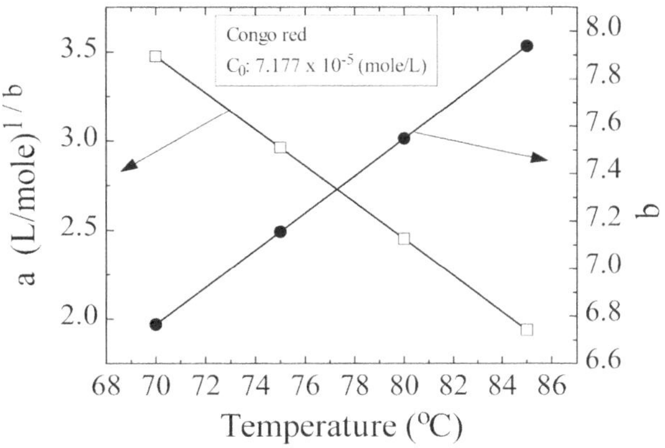

Fig. 8.30 Variation of the values of *a* and *b* with temperature for solubilization of Congo red in TX-100

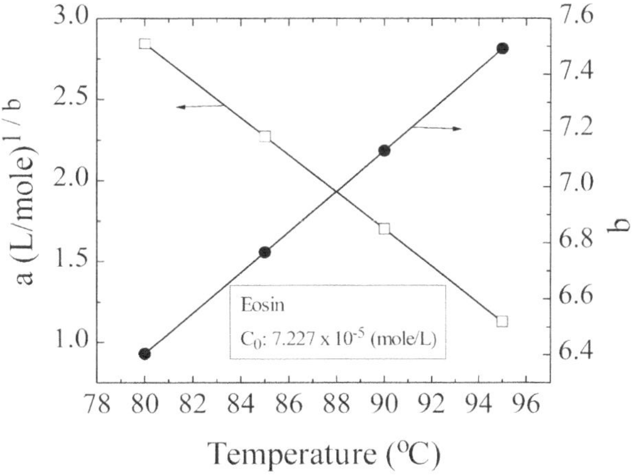

Fig. 8.31 Variation of the values of *a* and *b* with temperature for solubilization of eosin in TX-100

of feed surfactant concentration and dye concentration is used in the experiments at various operating temperatures. F_c is thereby correlated with C_s using Eq. (8.8). For fixed feed dye concentration, variations of the parameters a and b with temperature are shown in Figs. 8.30 and 8.31 for the two dyes. It is quite clear from the figures that a and b vary linearly with temperature and can be expressed as follows:

$$a = P + QT \tag{8.9}$$

$$b = R + ST \tag{8.10}$$

The parameters P, Q, R, and S are evaluated for various feed dye concentrations. The value of Q varies from -0.06 to -0.08 for Congo red and from -0.06 to -0.18 for eosin. Therefore, an average value of Q is considered for further calculations for both the dyes. They are -0.07 for Congo red and -0.12 for eosin. The variation of S for both the dyes remains within 0.07 to 0.08. For both the dyes, the average value

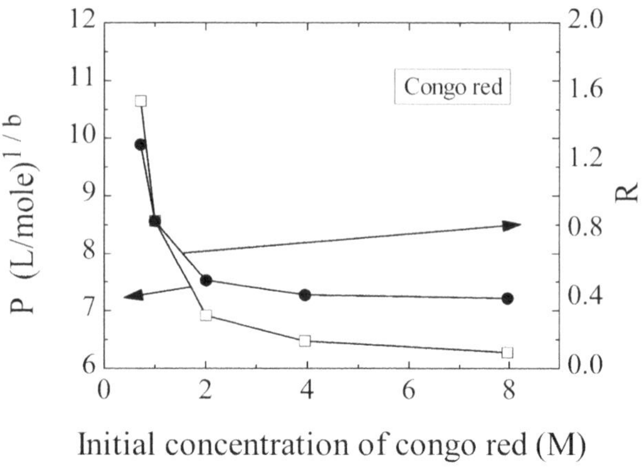

Fig. 8.32 Variation of the values of *P* and *R* with temperature for solubilization of Congo red in TX-100

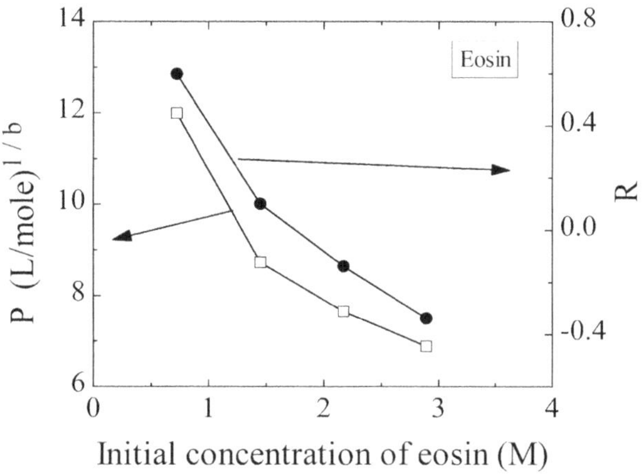

Fig. 8.33 Variation of the values of *P* and *R* with temperature for solubilization of eosin in TX-100

of *S* considered is 0.075. Typical variations of the parameters *P* and *R* with feed dye concentration for the two dyes are shown in Figs. 8.32 and 8.33. Variations of *P* and *R* for the two dyes are expressed by the following correlation with feed dye concentration:

For Congo red

$$P = 6.422 - 225.05C_0 - 2.185 \times 10^{-8}\frac{1}{C_0^2}\left(r^2 = 0.980\right) \tag{8.11}$$

$$R = 0.3917 + 7.015C_0 + 4.651 \times 10^{-9}\frac{1}{C_0^2}\left(r^2 = 0.996\right) \tag{8.12}$$

For eosin

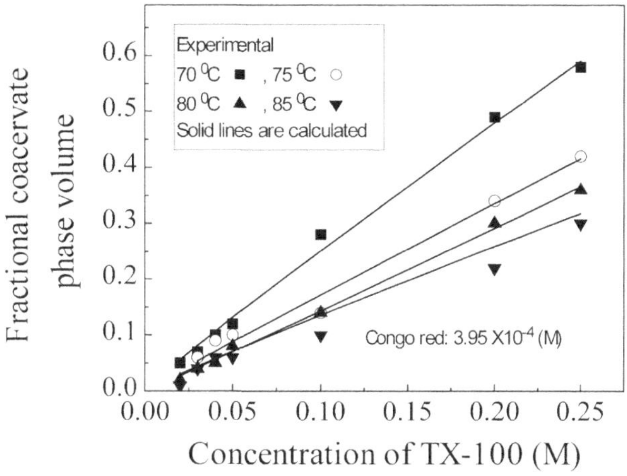

Fig. 8.34 Variation of fractional coacervate phase volume with initial TX-100 concentration at different temperatures for Congo red (Reproduced from Purkait et al. (2006). With permission from Elsevier)

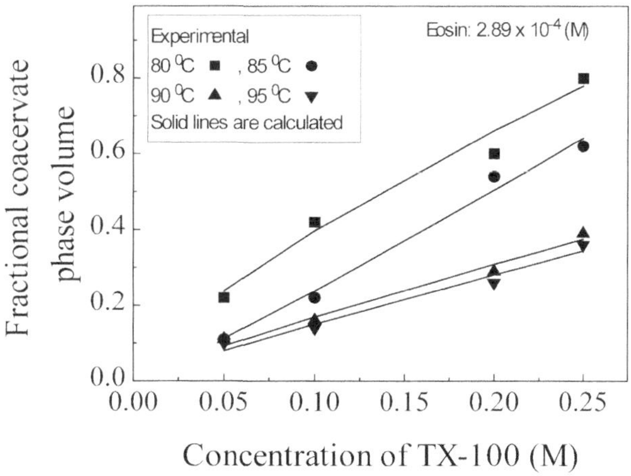

Fig. 8.35 Variation of fractional coacervate phase volume with initial TX-100 concentration at different temperatures for eosin (Reproduced from Purkait et al. (2006). With permission from Elsevier)

$$P = 9.02 - 8149.2C_0 + 1.862 \times 10^{-8} \frac{1}{C_0^2} \left(r^2 = 0.970\right) \tag{8.13}$$

$$R = 0.355 - 2487.6C_0 + 2.216 \times 10^{-9} \frac{1}{C_0^2} \left(r^2 = 0.976\right) \tag{8.14}$$

Variations of fractional coacervate phase volume with the feed surfactant concentration for two typical operating conditions for the two dyes are presented in Figs. 8.34 and 8.35. Using Eqs. 8.8, 8.9, 8.10, 8.11, 8.12, 8.13, and 8.14, the calculated trend of F_c is also presented in the same figures.

8.8.3 Determination of Surfactant Requirement for the Removal of Dye to a Desired Level Without Using Salts

Using the developed correlations (Eqs. 8.1, 8.2, 8.3, 8.4, 8.5, 8.8, 8.9, 8.10, 8.11, 8.12, 8.13, and 8.14), a calculation procedure is outlined to determine the amount of surfactant required for the removal of dye up to a desired level. The solubilization isotherm is defined as,

$$q_e = \frac{\text{Moles of dye solubilized}}{\text{Moles of TX}-100\ \text{used}} = \frac{A}{X} \tag{8.15}$$

Moles of dye solubilized can be obtained from mass balance,

$$A = V_0 C_0 - V_d C_e \tag{8.16}$$

where V_0 and V_d are the volume of the feed solution and that of the dilute phase after CPE. C_0 and C_e are the molar dye concentration in the feed and that remaining in the dilute phase after CPE. Hence, C_e is the desired concentration level of the dye for which the CPE is performed. In terms of fractional coacervate phase volume (F_c), Eq. (8.16) can be written as,

$$A = V_0[C_0 - C_e(1 - F_c)] \tag{8.17}$$

Using Eqs. (8.8), (8.15), and (8.17), the moles of surfactant required can be expressed as,

$$X = \frac{V_0}{q_e}\left[C_0 - C_e\left(1 - aC_s^b\right)\right] \tag{8.18}$$

If C_s is the concentration of surfactant initially in the feed, X in Eq. (8.18) can be expressed as,

$$X = C_s V_o \tag{8.19}$$

Equating Eqs. (8.18) and (8.19), the following governing equation of C_s is obtained:

$$C_s = \frac{1}{q_e}\left[C_0 - C_e\left(1 - aC_s^b\right)\right] \tag{8.20}$$

Expressing in terms of C_e from Eq. (8.1), the expression of C_s is obtained as,

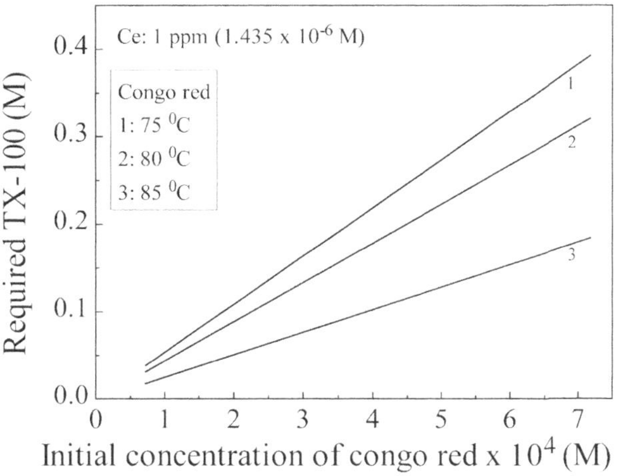

Fig. 8.36 Variation of the requirement of TX-100 concentration for different initial Congo red concentrations at three different temperatures to bring down its dilute-phase concentration to 1 ppm (1.435×10^{-3} mM)

$$C_s = \frac{(1 + nC_e)\left[C_0 - C_e\left(1 - aC_s^b\right)\right]}{mnC_e} \tag{8.21}$$

With the knowledge of feed dye concentration (C_0), the desired level of dye concentration in the dilute phase (C_e), isotherm constants m and n, and design parameters a and b, Eq. (8.21) can be solved by trial and error to obtain C_s. Solving Eq. (8.21) using some typical temperature conditions and fixing the desired concentration level of dye in the dilute phase as 1.0 mg/L, the surfactant concentrations required for various feed dye concentrations are calculated and plotted in Figs. 8.36 and 8.37 for Congo red and eosin dyes, respectively. It may be observed from the figures that the required surfactant concentration increases with feed dye concentration and is less at higher temperatures. Higher operating temperature requires higher energy input to the system. Therefore, there exists a trade-off between the feed surfactant dose and the operating temperature with respect to the feed dye concentration to effect a desired level of dye removal.

8.8.4 Surfactant Recovery by Solvent Extraction (SE)

To meet the environmental standards and the economy of the CPE process, it is necessary to recover the surfactant from both the coacervate and aqueous phases. Unlike ionic surfactant, precipitation method is not applicable for nonionic surfactant. For volatile solute, it is easy to recover surfactant from the coacervate phase by vacuum, steam, or gas stripping (Purkait et al. 2004). But problem arises for nonvolatile solute like dye, which is used in the present case. Although regeneration of surfactant from coacervate phase is not studied in this work, efficacy of SE is explored here to recover surfactant from the dilute phase.

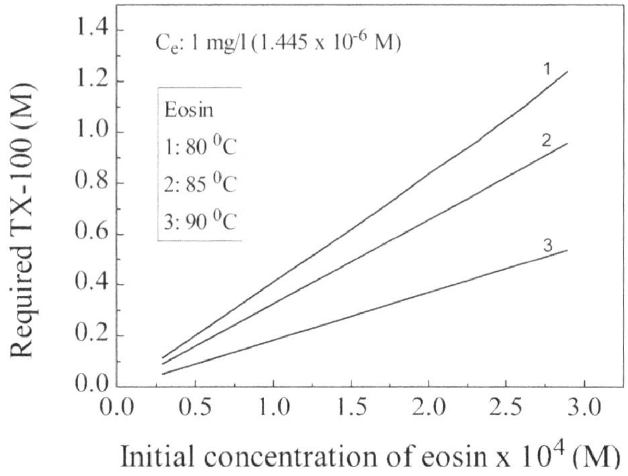

Fig. 8.37 Variation of the requirement of TX-100 concentration for different initial eosin concentrations at three different temperatures to bring down its dilute-phase concentration to 1 ppm (1.445×10^{-3} mM)

The experimental investigation on separation of TX-100 at 30 °C from aqueous solution is performed using heptane as extracting medium. All SE experiments are conducted in a cylindrical vessel. The composition of each set of experiment is prepared by adding different volumes (2, 5, 10, 15, 20, and 25 ml) of heptane to a fixed volume (30 ml) of TX-100 solution of 243, 479, 2079, and 5157 ppm. Dye concentration is zero, since the concentration of dye in the dilute phase is almost zero for a wide range of feed dye-TX-100 system. The mixture of TX-100 solution and heptane is vigorously shaken for 20 min using a stirrer (Type-RQ-123, Remi Motors Ltd., India) at 450 rpm to achieve equilibrium. After that, the mixture is transferred into a separating funnel and kept for 3 h for complete separation of two phases. The volume and concentration of aqueous and nonaqueous phase are measured. The experiments are conducted in batch mode.

The results are presented in Fig. 8.38. It is clear from the figure that for a fixed TX-100 concentration (e.g., 243 ppm), the surfactant recovery increases sharply from 42 to 88%, when the ratio of heptane to dilute phase increases from 0.067 to 0.5. With further increase of the above ratio up to 0.833, the surfactant recovery reaches up to 88%. On the other hand, at the same ratio, extraction efficiency decreases with increase in feed TX-100 concentration. The increase in recovery with the solvent to dilute-phase volume ratio is due to more distribution of TX-100 molecule in the heptane phase. The concentration distribution of the surfactant in both the solvent and dilute phases is constant at a fixed temperature (room temperature of 32 ± 2 °C). Therefore, fixing the ratio of solvent to dilute phase, increase in feed surfactant concentration results in lower surfactant recovery.

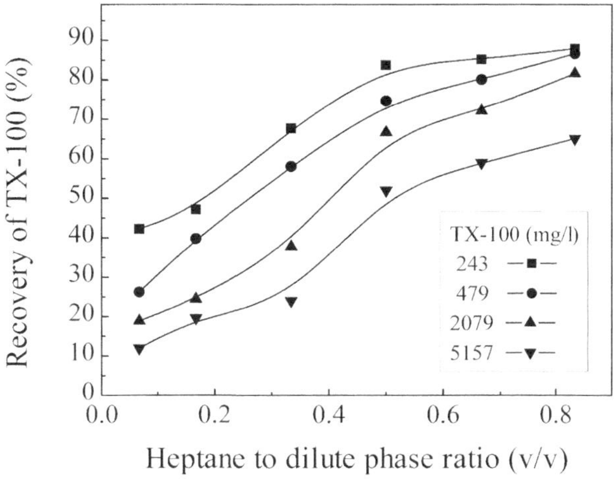

Fig. 8.38 Effect of heptane to aqueous phase volumetric ratio on the extraction of TX-100 at different feed TX-100 concentrations (Reproduced from Purkait et al. (2005). With permission from Elsevier)

References

Casero I, Sicilia D, Rubio S, Perez-Bendito D (1999) An acid induced phase cloud point separation approach using anionic surfactants for extraction and preconcentration of organic compounds. Anal Chem 71:4519–4526

Clint JH (1992) Surfactant aggregation. Blackie, Glasgow, p 154

Fernandez AE, Sosa-Ferrera Z, Santana-Rod- riguez JJ (1998) Determination of polychlorinated biphenyls by liquid chromatography following cloud-point extraction. Anal Chim Acta 358:145–155

Ferrer R, Beltran JL, Guiteras J (1996) Use of cloud point extraction methodology for the determination of PAHs priority pollutants in water samples by high performance liquid chromatography with fluorescence detection and wavelength programming. Anal Chim Acta 330:199–206

Garcia AL, Gonzalez EB, Alonso JG, Sanz-Medel A (1992) Potential of micelle-mediate procedures in the sample preparation steps for the determination of polynuclear aromatic hydrocarbons in waters. Anal Chim Acta 264:241–248

Heegaard NH, Jakobsen DR, Klattschou D (1997) Purification of wegener's granulomatosis autoantigen, proleinase 3, from neutrophils by triton X-114 extraction of azurophilic granules. Anal Bio- Chem 253:259–262

Hiemenz PC, Rajagopalan RH (1997) Principles of colloid and surface chemistry. Marcel Dekker, New York, p 377

Inoue T, Ohmura H, Murata D (2003) Cloud point temperature of polyoxyethylene-type nonionic surfactants and their mixtures. J Colloid Interf Sci 258:374–382

Kimchuwanit W, Osuwan S, Scamehorn JF, Harwell JH, Haller KJ (2000) Use of a micellar-rich Coacervate phase to extract Trichloroethylene from water. Sep Sci Technol 35:1991–2002

Lindman B, Wennerstrom H (1991) Nonionic micelles grow with increasing temperature. J Phys Chem 95:6053–6054

Liu CL, Nikas YJ, Blankschtein D (1996) Novel bioseparations using two-phase aqueous micellar systems. Biotechnol Bioeng 52:185–192

Namasivayam C, Kavitha D (2002) Removal of Congo Red from water by adsorption onto activated carbon prepared from coir pith and agricultural solid waste. Dyes Pigments 54:47–58

Pinto CG, Perez-Pavon JL, Moreno-Cordero B (1994) Cloud point preconcentration and high performance liquid chromatography determination of polycyclic aromatic hydrocarbons with fluorescence detection. Anal Chem 66:874–881

Purkait MK, Vijay SS, DasGupta S, De S (2004) Separation of congo red by surfactant mediated cloud point extraction. Dyes Pigments 63:151–159
Purkait MK, Banerjee S, Mewara S, DasGupta S, De S (2005) Cloud point extraction of toxic eosin dye using Triton X-100 as nonionic surfactant. Water Res 39:3885–3890
Purkait MK, Banerjee S, DasGupta S, De S (2006) Performance of TX-100 and TX-114 for the separation of chrysoidine dye using cloud point extraction. J Hazard Mater 137:827–835
Qiao L, Easteal AJ (1998) The interaction between triton X series surfactants and poly (ethylene glycol) in aqueous solutions. Colloid Polym Sci 276:313–320
Rosen MJ (2004) Surfactants and interfacial phenomena. Wiley, New York
Saitoh T, Hinze WL (1995) Use of surfactant mediated phase separation (cloud point extraction) with affinity ligands for the extraction of hydrophilic proteins. Talanta 42:119–127
Silva MAM, Frescura VLA, Aguilera FJN, Curtius AJ (1998) Determination of Ag and Au in geological samples by flame atomic absorption spectrometry after cloud point extraction. J Anal At Spectrom 13:1369–1373
Sirimanne SR, Patterson DG, Ma L, Justice JB (1998) Application of cloud point extraction reverse phase high performance liquid chromatography: a preliminary study of the extraction and quantification of vitamins A and E in human serum and whole blood. J Chromatogr B 716:129–137
Toerne K, Jackson R, Wandruska R (2003) POE chain length selectivity in the clouding of a triton surfactant. J Colloid Interf Sci 257:412–414
Wang Z, Zhao F, Li D (2003) Determination of solubilization of phenol at coacervate phase of cloud point extraction. Colloids Surf A 216:207–214
Watanabe H, Tanaka H (1978) A nonionic surfactant as new solvent for liquid-liquid extraction of zinc(II) with 1-(2-pyridylazo)-2-napthol. Talanta 25:585–589

Chapter 9
Electrocoagulation

Abstract The science of electrocoagulation is similar to an electrolysis. Here the anode continuously disintegrates due to the electrical current flowing though it produces cations which attract the pollutants in the wastewater. The cations released in water are produced in situ and acts as the electrocoagulant in this case. The dyes attached to the electrocoagulant either precipitates or flocculates, thereby separated. This is perhaps one of the advanced technologies where the system is very robust, efficient, and easily controllable requiring minimal maintenance. The chapter presents the details of the electrochemical science behind the electrocoagulation technique and further explains the engineering considerations in dealing with practical implementation.

Keywords Electrocoagulation • Electrochemical reaction • Electrolysis • Sacrificial anode • Flocculation

Electrochemistry is bridge between the two branches of science: chemistry and electricity. In the quest for more cleaner and green technologies for water purification, electrochemical means of water treatment stands a fair chance. It has a distinct advantage of the absence of the requirement of any chemical reagent for the process. The flexibility in the choice of the operating conditions (potential) and the electrode material makes the technology selective and robust. Primarily, electrochemistry is related to the interfacial charge transfer across an electrically conducting material (electrode) and an ionic conductor (electrolytes-liquid, melts, solid, etc.). Degradation of wastes using electrochemical means has several paybacks in terms of the operational simplicity, efficiency, and cost. For example, the process can be terminated by switching off the electrical circuit; operating at room temperature and normal environmental conditions, this reduces the chance of volatilization; there is no discharge of by-product streams. Separation of the undesired solutes (charged) is dependent on the choice of the electrode and operating conditions. However, separation (or removal) can also be facilitated using an intermediate membrane system to improve the selectivity (e.g., in proton exchange fuel cells). This has the benefit of wide applicability of several solutes (or mixtures) in water treatment. A schematic of the various physical mechanisms during electrocoagulation is illustrated in Fig. 9.1.

S. Mondal et al., *Advances in Dye Removal Technologies*, Green Chemistry and Sustainable Technology, DOI 10.1007/978-981-10-6293-3_9

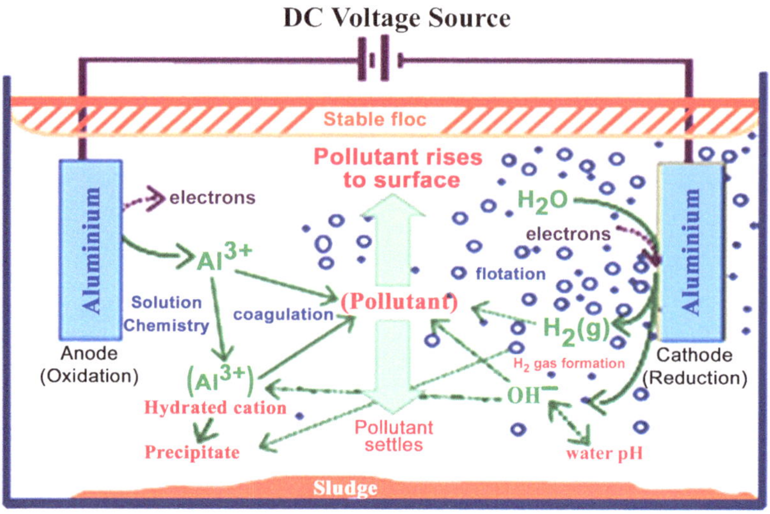

Fig. 9.1 Interactions within the electrocoagulation process

The phenomenon of electrocoagulation (EC) involves a continuously dissolving (sacrificial) anode due to the passage of electricity, releasing cations into the wastewater. Since the contaminant particulates are charged (by virtue of its surface potential), the cations are attracted toward it. This results in increasing agglomeration making the suspension inherently unstable. Similarly, the vapors generated by this hydrolysis forms miniature bubbles, which helps in floating the coagulated agglomerates. The principle of EC is based on three leading parameters – the ionic charge, particulate size, and droplet (or vapor) spatial density. The generation of the charged agglomeration involves three stages: (1) coagulant formation by electrolytic disintegration of the anode, (2) destabilization of the suspension, and (3) aggregation to form the floating material (flocs). The EC process is successful in removal of heavy metals, organic dyes, fine suspended particles, and oil and grease (or other heavier hydrocarbons) from diverse industrial effluents. Typically, the hydroxides and the oxyhydroxides make the available surface area necessary for the interaction or binding of the contaminant. The complete mechanism of the entire EC process can be succinctly described as:

(a) In the event of the electrical interaction of the cations (produced from the sacrificial anode) with the charge particulate contaminant, the diffused double layer is compressed.
(b) The presence of the counterions (due to the dissolution of the anode) neutralizes the excess species present in the solution. This decreases the interparticle electrostatic interactions significantly causing dominance of the van der Waals attraction, leading to a stable, electrically neutral coagulation.

(c) The gaseous bubbles (due to the hydrolysis, by-product of the process) entrap the remaining colloidal particles and produce floatation aiding in simple gravity separation.

The reactions taking place around the metal electrode (sacrificial anode) made of iron or aluminum are:

$$M \rightarrow M^{n+} + ne^{-} \tag{9.1}$$

$$2H_2O + 2e^{-} \rightarrow 2OH^{-} + H_2 \tag{9.2}$$

(d) The cations (M^{n+}) which can be either Fe/Al produced from the sacrificial anode subsequently hydrolyze to form the metal hydroxides (or oxyhydroxides) acting as the coagulating agents. Coagulation happens due to the interaction (which leads to combination) with the negatively charged solutes migrating toward the anode due to electrophoresis. This is sharp contrast to the conventional coagulation process, where the wastewater stream is treated by chemical precipitation added externally, whereas in the EC, the coagulant is produced in situ.
(e) Parallel to the above steps, water is hydrolyzed as a side reaction, which generates a pool of oxygen bubbles at the anode. This aids in flocculating the coagulated pollutants to the surface due to the buoyancy.

Besides, the following chemical reactions also occur in the electrochemical cell: (1) electro-levitation of the coagulated particulates by the hydrogen and oxygen bubbles near the electrodes, (2) cathodic reduction of the pollutants, and (3) cation reduction at the cathode. In short, it is clear that electrocoagulation has the capability to remove a large range of pollutants under a variety of conditions ranging from: suspended solids (Matteson et al. 1995) to heavy metals (Al-Shannag et al. 2015).

The electrode assembly is generally connected to a DC source, in an EC experiment. The quantity of metal dissolved from the sacrificial anode is proportional to the electrical charge (electricity) transferred through the electrolytic solution. Based on the Faraday's laws, the amount of electrode material dissolved (w, g/cm^2) is related to the mass of the electrode (M, g/cm^2) and the current density (i, A/cm^2):

$$w = \frac{i \times t \times M}{z \times \mathrm{F}} \tag{9.3}$$

where z is the oxidation number, t is the time (s), and F is the Faraday's constant (96,500 °C/mol). The above simple relationship does not take into account of the geometry of the electrode as well as environmental factors in electrolytic cell – temperature, pH, etc. Additionally, there is an uncertainty with the accuracy of estimating the cell potential. The measured potential is the applied overpotential (η_{AP}) and is combined of three different potentials:

$$\eta_{AP} = \eta_k + \eta_{Mt} + \eta_{IR} \quad (9.4)$$

where η_{IR} is the overpotential caused by solution or internal resistance drop, η_κ is the kinetic overpotential (V), and η_{Mt} is the concentration overpotential. The internal resistance drop (η_{IR}) is proportional to the distance between the electrodes (d), current (I), and inverse to the cathode surface area (A) (Vik et al. 1984):

$$\eta_{IR} = \frac{Id}{A\kappa} \quad (9.5)$$

where the specific conductivity (κ) is the proportionality constant. Looking into Eq. (9.2), it can be inferred that the potential drop due to the internal resistance can be reduced by either decreasing the interelectrode separation gap (d) or enhancing the cross-sectional surface area of the electrodes. The concentration overpotential (η_{Mt}), often referred to as the diffusion overpotential, is due to the gradient in the electrolyte concentration from the bulk near the electrode surface by the electrode reaction. This occurs when the electrochemical reaction rate is high enough to reduce the concentration of electroactive species at the electrode surface below the bulk solution. The concentration overpotential is negligibly small when the reaction timescale (or kinetics) is comparable or larger than the mass transfer rate. Essentially, η_{Mt} can be reduced by enhancing the rate of ion mass transport rate. This can be achieved by (1) increasing the amount of the metal ion transported from the anode to the bulk by enhancing the turbulence of the solution and (2) transferring the electrolyte solution at a higher velocity from anode to the cathode, by some mechanical options. The kinetic overpotential (η_κ) is originated by the barrier due to the activation energy in the electron transfer reactions. Both the kinetics and concentration overpotential increase with the current intensity (I). However, the physical and chemical characteristics of the species in the aqueous solution affect these changes. The implications of the electric field gradient and pH on the heterogeneous reactions near the electrode surface and homogeneous bulk reactions need to be explored in detail for optimization of the EC performance. A literature survey suggests that this has not been done extensively to characterize or relate the electrode degradation.

Coagulation Mechanism

The frequently used element in the sacrificial anode of the electrocoagulation unit is aluminum. The aluminum cation facilitates for (1) precipitation, (2) interaction with the pollutant present in the coagulator, and (3) hydrolysis in forming an aluminum-hydro complex. Thus, understanding the speciation is important for design and operation of an electrocoagulation phenomenon. For example, the dissociation and solubilization of aluminum ions are described in the following reactions (Eqs. 9.9, 9.10, 9.11, and 9.12). The principles of solution thermodynamics are often helpful in determining the stability of the process.

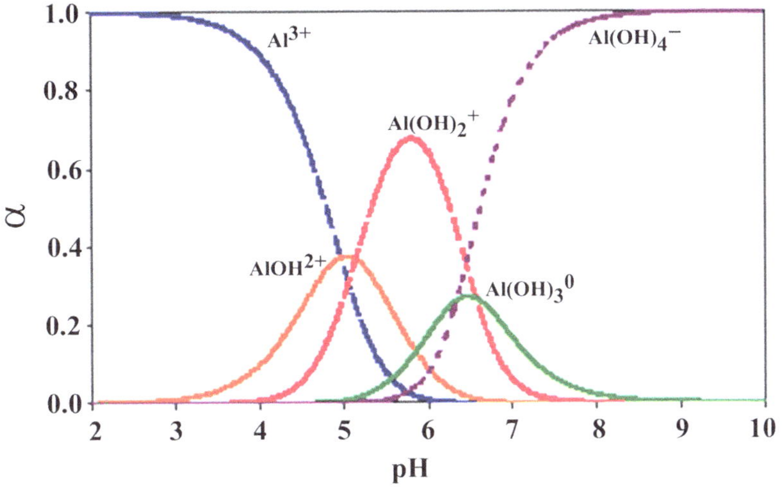

Fig. 9.2 Distribution diagram for Al-H_2O considering only mononuclear species

$$Al^{3+} + H_2O \rightarrow AlOH^{2+} + H^+ \quad (9.6)$$

$$AlOH^{2+} + H_2O \rightarrow Al(OH)_2{}^+ + H^+ \quad (9.7)$$

$$Al(OH)_2{}^+ + H_2O \rightarrow Al(OH)_3{}^0 + H^+ \quad (9.8)$$

$$Al(OH)_3{}^0 + H_2O \rightarrow Al(OH)_4{}^- + H^+ \quad (9.9)$$

Considering only mononuclear speciation, the total aluminum present in the solution (α) at a given pH can be calculated using the distribution diagram for aluminum-water system as shown in Fig. 9.2.

The boundary of solubilization denotes the thermodynamic equilibrium that exists between the dominant aluminum species in solution at a given pH and solid aluminum hydroxide. The minimum solubility occurs at pH 6.3(0.03 mg-Al/L), with solubility increasing as the solution pH shifts away from the minimum solubility pH (i.e., becoming either more acidic or alkaline), as demonstrated in Fig. 9.3.

The hydronium ions produced at the cathode react with the active counterions generated at the cathode (cations) forming an in situ coagulant (charged metal hydroxide). The charged (ionic) coagulant reacts or binds to the pollutant producing large agglomerates which precipitate or may flocculate with the help of the hydrogen bubbles. The coagulant typically remains as a soluble hydrolyzed species in the solution. Besides, the precipitate or the flocculants may further aid in subsequent precursor to pollutant removal by entrainment or can react with pollutant particles binding to the precipitate/flocculants. The solubility of the agglomerate can provide more insight on this mechanism of the pollutant removal. It may be noted that the solubility diagram is considered on the basis of mononuclear aluminum species

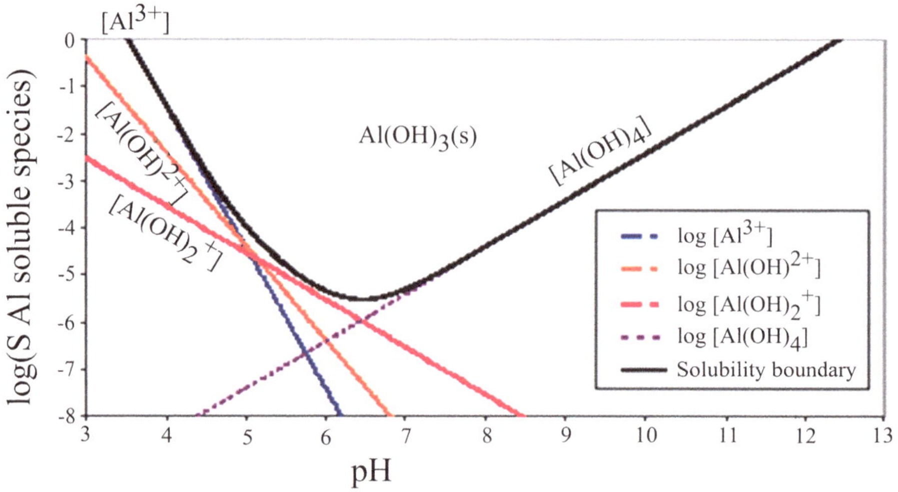

Fig. 9.3 The solubility diagram for aluminum hydroxide, $Al(OH)_3(s)$

which in reality is only a part of the scheme. With the elapse of time (aging), the species concentration in the solution increases triggering formation of polynuclear cation complexes which adds to the precipitation. For example, the reaction in Eq. (9.10) illustrates one of the many possibilities of such pathways and requires further understanding on the dissociation-association kinetics:

$$Al^{3+} \rightarrow Al(OH)_n^{(3-n)} \rightarrow Al_2(OH)_2^{4+} \rightarrow Al_3(OH)_4^{5+} \rightarrow Al_{13}\text{Complex} \rightarrow Al(OH)_3 \quad (9.10)$$

Coagulation and binding can occur in many different pathways and is very unique to the type of pollutant or cations present in the solution. The particle interaction can be via physisorption, chemisorption, and electrostatic interaction or may be coagulation sweeping. It is difficult to generalize the specific path taken by the coagulant for a specific pollutant, as it is sensitive to the concentration, pH, temperature, or presence of other pollutants. In most cases, there is a range of different pollutant species which makes them enormously complex. Modeling the solution thermodynamics can provide a possible explanation or insights into the probable coagulation pathways.

Eh-pH Diagrams

The information on the relationship between the state variables in the electrochemical and speciation process can be inferred from the Nernst equation. In practice, the Eh-pH plots the electric potential to pH, which identifies different region of the stability of the species in the solution environment (Pourbaix 1974). The particulars of both the electrochemical and chemical reactions are included, illustrating the intersection of the coagulation and electrochemical regions. Looking into Fig. 9.4 for the aluminum-water system, the areas passivation, corrosion, and immunity can

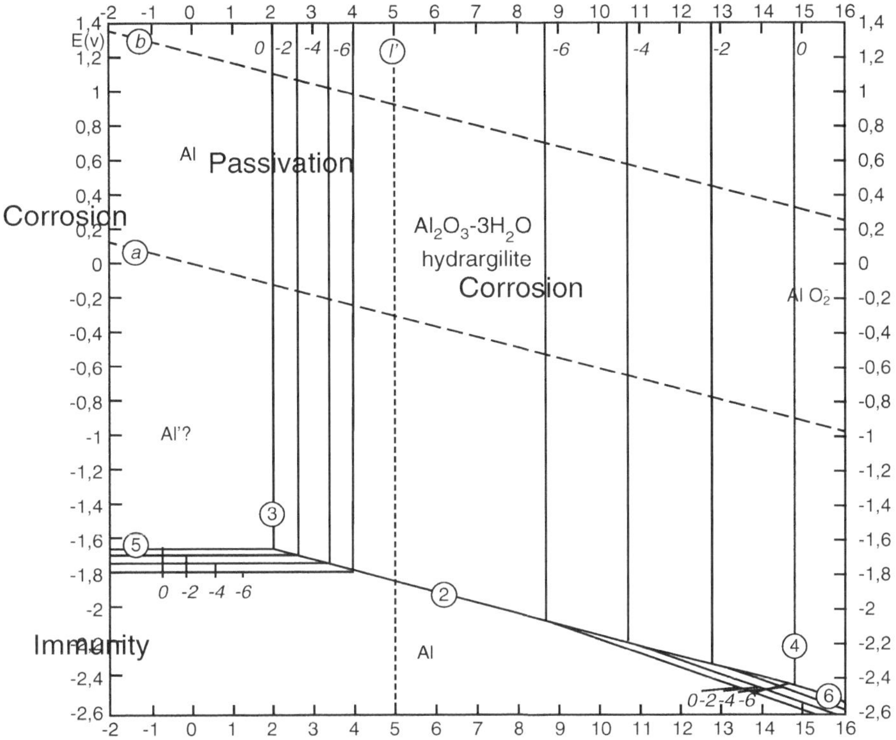

Fig. 9.4 Potential pH diagram for the system aluminum-water at 25 °C (Pourbaix 1974)

be observed. Passivation refers to a state of the surface being resistant to external degradation, self-happens by the formation of the oxide layer. The favorable conditions for corrosion, combination of the solution pH and potential under which aluminum degradation takes place, are obvious from such a plot. However, the utility of Eh-pH diagrams is restricted by the availability of the relevant thermodynamic data and its incapacity to consider for the kinetics, which are not to be considered separately.

It may be emphasized here that passivation moderates the performance of electrocoagulation by preventing dissolution of metal from the electrode and necessary electro transfers due to the oxide layer. Thus, on one hand, the passivation prevents excessive electrode degradation, while it also reduces the efficacy of the process. With continuous usage of the electrode, the passivation layer seriously limits the effectiveness of the electrocoagulation and thus obviates the need for mechanical cleaning. The existence of the chloride ion in the solution can help in decreasing passivation, thus enhancing the pollutant removal (Donini et al. 1994).

Floatation

One of the expected by-products of any electrolytic process is the generation of the gases (mostly hydrogen and oxygen dealing with aqueous solutions). In the present process of electrocoagulation, these gases help in the flotation of the agglomerates and also facilitate solution mixing. The method of production of the gaseous bubble and the variability in its size is the key difference with any the standard flotation technique used in separation technique. The rater and size of the bubble produced is related to the electrochemistry of the process. In the given situation, the smaller the bubble is, the more number of bubbles are generated, thereby raising the chances of contacting and removing the coagulant loaded pollutant aggregates. Apart from this, the likelihood of the contact of a bubble with a particle depends strongly on the residence time and bubble path line. Generally, the bubble density is influenced by the current intensity, while the path line is affected by the cell geometry. However, the parameters affecting the size of the bubble are not fully understood.

As described before the electrolytic bubbles flocculate the aggregates due to buoyancy increasing the mixing and promote local turbulence in the solution, although too much of shear in the solution can jeopardize the growth or development of the agglomerate. Thus, an optimum bubble density (which determines the current intensity indirectly) producing a moderate bubble density, causing a mild agitation, is ideal for the application. Electrochemistry, coagulation, and flotation or precipitation are the three fundamental phenomena and each of them is a well-researched topic in itself.

Electrochemistry

In principle any electrocoagulation reactors are electrochemical cells. The metal ions from the electrode dissociate into the polluted stream forming the coagulant on site. The potential difference in the system drives the electrocoagulation reaction. The operational potential required for the process can be obtained from the electrochemical half-cell reactions for a specific pH. Apart from aluminum, stainless steel and platinum are reported to have been good electrode materials. The electrochemical reactions are only dependent on the type of the anodic material, which we have considered as an example in the following dissociation reactions:

$$Al^{3+} + 3e^- \leftrightarrow Al \qquad E_A^0 = -1.66V \tag{9.11}$$

At the anode, there is also a possibility of evolution of oxygen (Eq.9.12):

$$4OH^- \rightarrow O_2 + 2H_2O + 4e^- \qquad E_A^0 = -0.4V \tag{9.12}$$

Similarly at the cathode, the generation of hydrogen is dependent on the pH. At alkaline or at natural pH, hydrogen is produced as described in Eq. (9.13):

$$2H_2O + 2e^- \rightarrow 2OH^- + H_2 \qquad E_C^0 = -0.83V \tag{9.13}$$

While under acidic conditions, Eq. (9.14) best describes the hydrogen evolution at the cathode:

$$2H^+ + 2e^- \rightarrow H_2 \qquad E_C^0 = 0V \tag{9.14}$$

The mass of electrode decomposition is related to the current intensity (I) by the Faraday's Law within the reactor and hence in this case provides a theoretical estimate of the amount of aluminum going into solution (Eq. 9.3):

$$Al + 3H_2O \rightarrow Al(OH)_3 + 1.5H_2 \tag{9.16}$$

$$Al + 4H_2O + e^- \rightarrow Al(OH)_4^- + 2H_2 \tag{9.17}$$

The half-cell reactions for the aluminum electrode are stated in Eqs. (9.16 and 9.17), aluminum anodes and inert cathodes. The equilibrium potential in any half-cell reaction is calculated by the Nernst equation (Eq. 9.18):

$$E = -\frac{\Delta G^0}{nF} - \frac{RT}{nF}\sum v_i \ln c \tag{9.18}$$

The electrochemical potential on the whole is estimated as the addition of the cathodic (E_c), anodic (E_a), solution (E_{solution}), and the loss (E_{loss}) potentials as shown in Eq. (9.19). The solution potential is dependent on the distance between the electrodes (b), current density (i_c), and the electrical conductivity of the solution (s), described in Eq. 9.20. The potential energy loss to overcome the resistance due to the passivation layer is included in E_{loss}. The last two terms suggest that in addition to the cathodic and anodic potential requirement, the reactor potential is influenced by the electrode configuration (geometry), solution character, and operating conditions.

$$E_{\text{Cell}} = E_c - E_a - E_{\text{solution}} - E_{\text{loss}} \tag{9.19}$$

$$E_{\text{solution}} = \frac{bi_c}{s} \tag{9.20}$$

Electrochemical Kinetics

The electrochemical reactions near the electrodes generally take place in the interfacial boundary layer with the electrode and the solution. For the illustration purpose, let us consider a generalized electron transfer scheme, O + ne ↔ R, which is detailed in Fig. 9.5. Due to the diffusion of the oxidant ions (O_{bulk}) in the interfacial region (from the solution to the surface, O_{surf}), a concentration gradient develops which eventually leads to the potential gradient between the electrode and the solution. The heterogeneous reaction at the electrode surface forms a reaction intermediate (O′), which reduces at the surface to form R'_{ads}. Likewise, a similar counter phenomenon occurs for the reductant. The overall electrochemical kinetics is limited wither by the diffusion timescale or the electron charge transfer timescale.

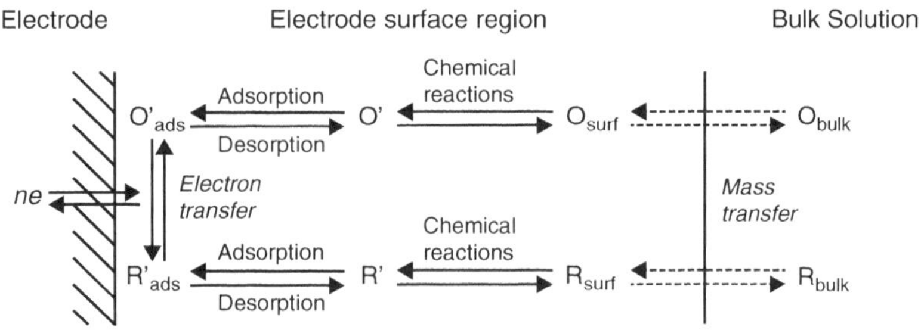

Fig. 9.5 Pathway of general electrode reaction

9.1 Design of Electrocoagulation Unit

In designing an efficient electrochemical cell, the following may be taken into consideration:

1. Minimizing the potential drop due to the internal resistance between the electrodes.
2. Resistance to mass transfer in the interelectrode region to be as low as possible.
3. Reduction in the buildup of the oxygen and hydrogen gases at the electrode surfaces.

The potential drop due to IR is related to the (a) separation distance between the electrodes, (b) electrical conductivity of the solution, and (c) electrode shape or structure.

The evolution of the gaseous bubbles at the electrodes acts as a partially insulating layer which increases internal resistance of the cell. So, bubble accumulation has to be minimized to improve the overall energy (electrical current) efficiency. On the other hand, the released bubbles aid in flotation of the electrocoagulated particles, helpful in preferential separation of the flocculent. So, a balance between these two counteracting phenomena can help in deciding an optimum level of bubble density. The mass transport rate in the electrochemical cell can be enhanced by promoting the local turbulence, e.g., increasing the flow rate. This facilitates in reducing the passivation later as well as sweeping the bubbles formed near the electrode.

Physical Design Issues

The electrocoagulation process is typically combined with many other unit operation process including dissolved air flotation (DAF), sand filtration, microfiltration, and electro-flotation. As understood, the efficacy of the pre- and post-water treatment technique impacts the performance of the electrocoagulation reactor significantly. The design of the electrochemical reactor depends on the following physical factors:

- Operational capacity of the treatment process
- Continuous or batch operation

- Reactor geometry
- Operating current density

Geometry
The geometrical configuration of the reactor affects the physical parameters such as the floc formation, bubble path line, fluid flow regime, flotation effectiveness, and mixing/settling characteristics. Traditionally, the most common approach is to use electrodes plates (iron or aluminum) and continuous operation. Water is continuously dosed with dissolved metal ions as it is introduced into the electrocoagulation cell. Often, a downstream unit is essential to separate pollutant and water.

Scale-Up Issues
Identifying the important scale-up parameters between the laboratory and full-scale unit is a challenging nontrivial engineering task. In the electrocoagulator, the ratio of the surface area to volume (S/V) is one of the most significant parameters to consider in scaling up. The current density, rate of cation dosing, and dynamics of bubble transport are influenced by the electrode area. In 1998, Mameriet et al. reported that the optimal current intensity is decreased with S/V. The values of the S/V ratio for other instances are reported in Table 9.1 and are not widely available in literature reports, justifying the need for exploration of the scaling aspect of the electrocoagulation reactors based on electrode area.

Current Density
The rate of electrochemical metal dosing to the water and the bubble density generation is impacted by the current density. Typically the current density ranges from 10 to 150 A/m^2 depending on the situation. Small current densities are appropriate for electrocoagulators which are integrated with conventional sand filters, while relatively large current densities produce desirable results for separation processes involving flotation cells or large settling tanks which can handle large amount of bubbles.

Electrode Material
It is pertinent by now that the electrode material impacts significantly the performance and operation of the electrocoagulation reactor. There are several reports on the choice of anode material with the performance of the reactor. As mentioned before, the most common choices are iron plates or aluminum (Vik et al. 1984). In another study by Do and Chen (1994), the performance of the aluminum and iron was compared for its efficiency in treatment of dye solutions. They have found that

Table 9.1 Comparison of surface to volume ratio

Reference (Author)	S/V (m^2/m^3)
Osipenko and Pogorelyi (1977)	18.8
Novikova et al. (1982)	42.5
Amosov et al. (1976)	30.8

the operating conditions for similar range of optimal performance vary with the choice of the anode material and are determined by:

- Pollutant type and concentration
- Rate of mechanical agitation

Passivation
Electrode passivation is an inevitable consequence of electrocoagulation, and one of the major operational issues, which affects the durability and smooth running of the process.

The passivation of electrodes is concern for the longevity of the process. Various methods of preventing and/or controlling electrode passivation are:

- Hydromechanical or mechanical cleaning of the electrodes
- Alternating polarity of the electrode application of chemical inhibiting agents

In most of the cases, the reliable and efficient method of electrode maintenance was to mechanically clean the electrodes regularly which for a continuous large-scale system is a nontrivial issue.

Solution pH
Typically the solution pH decides the state and speciation of the cations and also affects the product solubility. Thus, solution pH influences the overall effectiveness and efficiency of the process. The pH of the solution can easily be altered, which makes it necessary to maintain its level during the course of the process, e.g., using an in situ neutral buffer. Generally, an optimal pH exists for a given pollutant, ranging from 6.5 to 7.5.

9.2 Removal of Dyes Using Electrocoagulation

Decolorization of crystal violet dye (CI 42555) in a batch electrocoagulation unit with aluminum electrodes is reported in this section. Important parameters (like effect of current density, interelectrode distance, solution conductivity, type and quantity of salts, initial pH, and initial dye concentration) that affect on the extent of removal of crystal violet were studied. There were several experiments conducted to find the optimum conditions for the removal of dyes. The operating conditions at which the experiments were conducted are presented in Table 9.2.

Table 9.2 Operating conditions for the EC of crystal violet dye

Parameters	Values
pH	4.5, 5.5, 6.5, 7.5, 8.5, and 10
Current density (A/m^2)	3.625, 6.125, 8.625, and 11.125
Crystal violet concentration (mg/L)	50, 60, 80,100, and120
Dose of electrolyte ($\times$ 10^{-3}, kg/L)	2, 4, 6, 8, and 10
Interelectrode distance ($\times$ 10^{-2}, m)	0.5, 1,1.5, and 2

9.2.1 *Effect of Current Density*

Normally, in electrocoagulation, duration of electrolysis, and current density (A/m^2) are important operational parameters (Pouet and Grasmick 1995) which define the overall energy consumption dependent on the process throughput. However, in contrast to this, some researchers have reported that the current density does not influence treatment efficiency significantly (Chen et al. 2000), which makes it unclear as to the role of the current density on the removal efficiency. Nevertheless, it is an important parameter to consider in upscaling. Moreover, the choice of the electrode material also affects the cell voltage, since the oxidation potential is different on the type of material and hence the separation efficiency.

In our work, aluminum was selected as the sacrificial electrode material because of its low cost, ready availability, and non-harmfulness, and it requires comparatively less oxidation potential. Figure 9.6 shows the aqueous phase crystal violet concentration as a function of time for four different current densities. It may be seen from the figure that a sharp decrease in dye concentration just at the beginning of the process for all current intensity occurs and gradually decreases thereafter. It is also realized that the dye removal efficiency is not sensitive to the current density, but the process rate (and hence the throughput) is directly proportional to the current density.

In any electrocoagulation process, as the isoelectric point is reached, the cations account for the electrical charge neutralization of the pollutant species. Subsequently, the sorption mechanism occurs leading to the formation of agglomerates (in situ electrocoagulant). With the elapse of time, further addition of cation results in precipitation of the amorphous metal hydroxide, which facilitates pollutant binding due to sweep coagulation. In the final moment, coagulated aggregate flocculates with the bubbles to the surface or settles down. As observed from the figure, there is a sharp reduction in the concentration due to the nascent electrode surface at the start which slowly reduces (with decelerating rate) further with generation of metal hydroxide as an electrocoagulant for the pollutant particles.

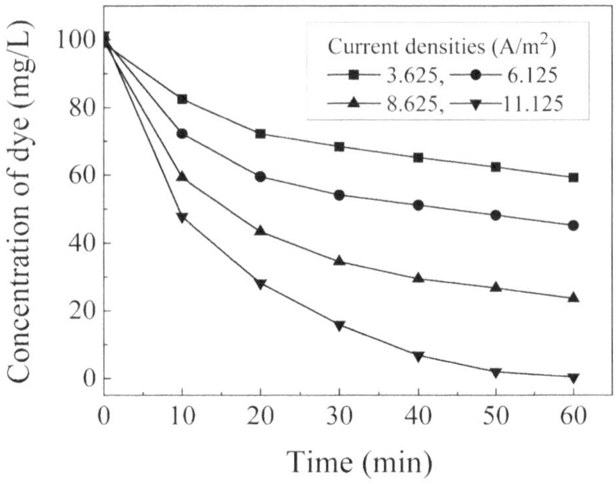

Fig. 9.6 Variation of concentration of crystal violet dye with time at different current densities. Interelectrode distance, 0.005 m; initial dye concentration, 100 mg/L; pH, 8.5; conductivity, 1.613 S/m

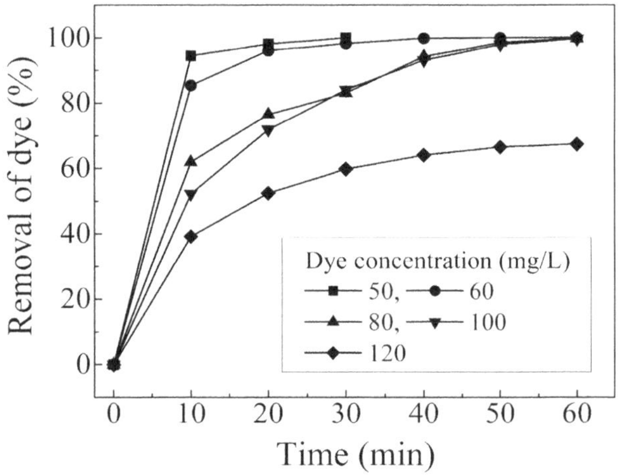

Fig. 9.7 Variation of extent of dye removal with time for different initial dye concentrations. Interelectrode distance, 0.005 m; current density, 11.125 A/m^2; pH, 8.5; conductivity, 1.613 S/m

9.2.2 Effect of Initial Dye Concentration

To estimate the separation efficiency with varying dye concentrations, the dye solutions with different initial concentrations of 50, 60, 80, 100, and 120 mg/L were treated by electrocoagulation using a current density of 11.125 A/m^2. The removal efficiencies of the dye are shown in Fig. 9.7. It is evident from the figure that the fraction of dye removed increases with time. At the end of an hour, complete dye removal is observed for concentrations up to 100 mg/L. The removal is drastically reduced when initial concentration is increased to 120 mg/L, because of the formation of insufficient number of aluminum hydroxide complexes generated by the electrode to coagulate the dye molecules. Therefore, it is quite clear that

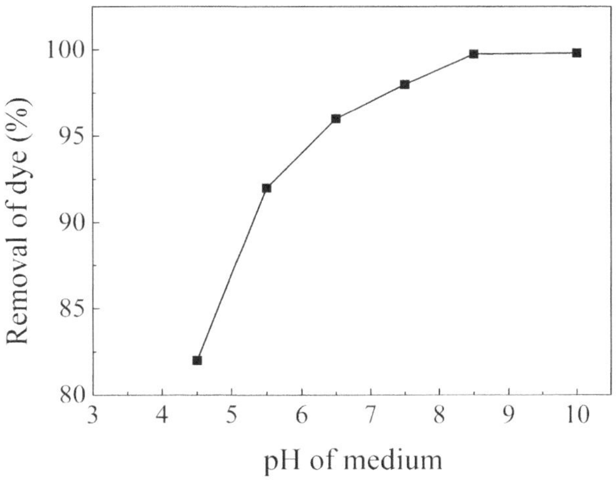

Fig. 9.8 Effects of solution pH on the dye removal. Interelectrode distance, 0.005 m; current density, 11.125 A/m^2; conductivity, 1.613 S/m; initial dye concentration, 100 mg/L; time, 60 min

under the present experimental conditions, the lower is the dye concentration the better is the removal efficiency.

9.2.3 *Effect of Initial pH*

The process of electrocoagulation is sensitive to the solution pH. To understand its effect, the dye-containing solutions are balanced for the specific pH either by addition of sodium hydroxide (to make it alkaline) or sulfuric acid (to make it acidic). It can be observed from Fig. 9.8 that the efficiency of dye removal is affected significantly by the initial solution pH, which decreases on reducing the pH (acidic). It is apparent from the figure that for a dye (crystal violet) concentration of 100 mg/L, complete removal (100%) is achieved at a pH 8.5. Therefore, the optimum pH for separation of crystal violet using electrocoagulation is 8.5. The role of pH on the speciation (hence the process mechanism) can be explained as: at low pH 2–3, $Al^{3+}$and $Al(OH)^2+$ are dominant, and various polymeric species such as $Al_{13}O_4(OH)_{24}{}^{7+}$are formed and precipitate as $Al(OH)_3(s)$ for pH in the range of 4 and 9. At low pH, the hydroxide ions generated at the cathode are neutralized by the hydronium ions and reduce the formation of the aluminum hydroxide complexes, which are responsible for removal of dye molecules. Contrastingly, at high pH, this process is just facilitated as number of OH^- ions in the solution are more.

9.2.4 *Effect of Interelectrode Distance*

The electrode assembly configuration is important to obtain the required surface area of the electrode and the interelectrode distance. The internal resistance

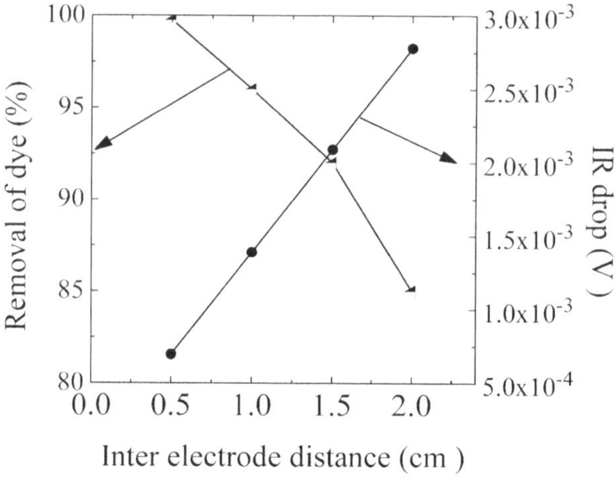

Fig. 9.9 Effects of interelectrode distance on the dye removal. Current density, 11.125 A/m^2; conductivity, 1.613 S/m; initial dye concentration, 100 mg/L; time, 60 min; pH, 8.5

increases with the interelectrode separation gap, leading to enhanced cell voltage, and adversely affects the removal efficiency. In Fig. 9.9, the effect of interelectrode distance on the removal efficiency and the consequent IR drop is illustrated. As expected, the removal process is enhanced by reducing the interelectrode gap. It can be inferred from Eq. (9.2) that for given surface area of the anode and the conductivity of the solution, the potential drop is proportional to the interelectrode distance. There exists an optimum electrode separation distance considering the energy consumption, removal efficiency with respect to the practicalities of fabrication. The optimum interelectrode distance is found to be 0.5 mm to attain complete removal of the dye with the concentration level up to 100 mg/L.

9.2.5 *Effect of Conductivity*

Typically, the ionic strength of a textile effluent is high due to the post-processing steps after dyeing leading to the contamination of salt and high conductivity of the effluent. For a fixed cell voltage, the current density increases with the ionic strength. Similarly, for a fixed current density, the cell voltage decreases with increasing effluent electrical conductivity. Thus, this obviates the need for investigation of the effect of effluent conductivity on the performance of the electrocoagulation process. The variation in the solution conductivity depends on the electrolyte concentration and type. In Table 9.3, the dependencies of conductivity, removal efficiency, and the applied voltage using the same concentration (8×10^{-3} kg/L) but different types of electrolyte (NaCl, $BaCl_2$, KCl, and KI) are illustrated in Table 9.3. It is observed from Table 9.3 that sodium chloride (NaCl) exhibits high conductivity and highest removal capacity (99.75%) with small

Table 9.3 Effect of different types of electrolyte on dye removal

Electrolyte	Conductivity ($\times 10^{-1}$, S/m)	Applied voltage (V)	Dye removal (%)
NaCl	16.13	6.75	99.75
$BaCl_2$	9.67	10.75	75
KCl	18.75	15.75	94
KI	9.18	26.75	98

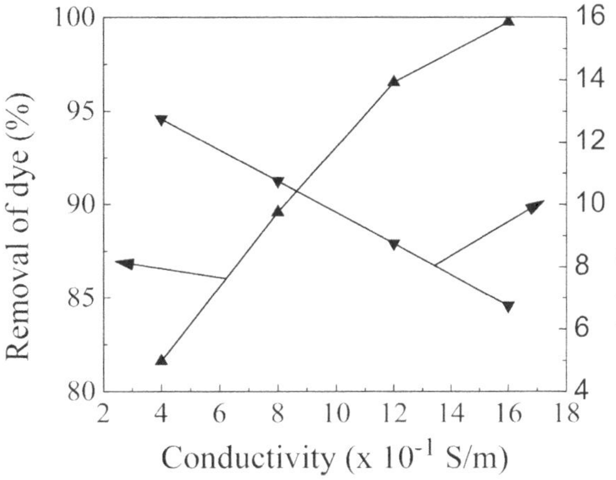

Fig. 9.10 Effects of solution conductivity over dye removal and cell voltage. Current density, 11.125 A/m^2; conductivity, 1.613 S/m; initial dye concentration, 100 mg/L; time, 60 min; pH, 8.5; interelectrode distance, 0.005 m

applied voltage in contrast to other electrolytes. As such, NaCl seems to be a superior electrolyte and viable in large-scale systems.

Figure 9.10 describes the quantitative dependence of the dye removal with conductivity of the solution. It is observed that the dye removal efficiency rises with the solution conductivity but decreasing cell voltage. For a dye concentration of 100 mg/L, the optimum conductivity is found to be around 16.1 S/m. Consequently, to achieve a certain current density, the voltage required is reduced, thus decreasing the electrical energy consumption. Moreover, with NaCl, the presence of chlorides in the solution accounts for release of nascent Cl_2 and OCl^-, which are strong oxidants. This in turn helps in oxidation of the dye molecules present in the effluent solution. So, the added NaCl not only increases the conductivity but also contributes as strong oxidizing agents.

9.2.6 Energy Consumption

With increasing focus on low-energy processes for environmental issues, current efficiency and electricity utilization and key economical parameters. This is usually determined for a simple equation (Vorobiev et al. 2003):

$$E = V \times I \times t_{EC} \tag{9.24}$$

where t_{EC} is the electrocoagulation time (in s), I is the current (in A), V is the cell voltage (in volts), and E is the electrical energy (in kWh) per unit kilogram of dye removed. As shown in Fig. 9.14 the color removal efficiency increases with the current density, so does the energy consumption (from Eq. 9.24). So, there exists an optimum current density with respect to the overall energy consumption and the removal efficiency.

The current efficiency (φ) of EC process can be calculated using Eq. (9.25),

$$\varphi = \frac{\Delta M_{\text{Exp}}}{\Delta M_{\text{Theo}}} \times 100 \tag{9.25}$$

This is based on the relative mass loss of the sacrificial aluminum electrodes (M_{exp}) during the electrocoagulation process and the theoretical amount based on the same operating and environmental condition (M_{theo}) from the Faraday's law, given by

$$\Delta M_{\text{theo}} = \frac{MIt_{EC}}{nF} \tag{9.26}$$

where n the number of electron moles, t_{EC} is the time of the electrocoagulation process, F is the Faraday constant, and M is the molecular weight of the aluminum (g/mol). The specific electrical energy consumption (S_{eec}) is related to the mass of aluminum electrodes lost (consumed) during the electrocoagulation process (in terms of kWh/kg Al) (Mollah et al. 2004; Pons et al. 2005) expressed as

$$S_{\text{eec}} = \frac{n \times F \times U}{3.6 \times 10^3 \times M \times \varphi} \tag{9.27}$$

It is observed in Fig. 9.11 that the specific electrical energy consumption and the dye removal efficiency enhance with the density of current. As current density increases, it facilitates formation of aluminum hydroxide complexes, increasing the removal efficiency. Consequently, the rate of dissolution of the aluminum electrode is more, thereby increasing the specific electrical energy.

9.2.7 Characterization of Treated Dye Solution and By-Products Obtained from EC Bath

The absorption spectrum shown in Fig. 9.12 is for the treated dye solution at every 10-min intervals. At the end of an hour, almost complete removal is achieved, without any change in the λ_{max} corresponding to 584 nm. Figures 9.13 and 9.14 show the SEM image and EDAX spectra of solid precipitate (by-product of EC), respectively. The SEM image describes the morphology of the precipitate of

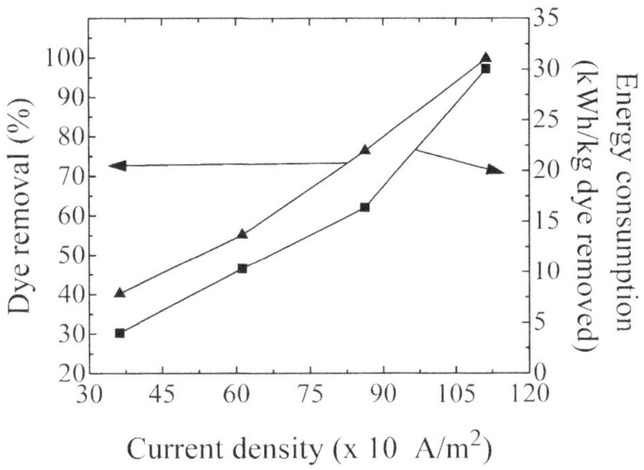

Fig. 9.11 Effect of current density on percentage removal of dye and energy consumption. Initial dye concentration, 100 mg/L; conductivity, 16.13×10^{-1} S/m; pH, 8.5; interelectrode distance, 0.5×10^{-2} m

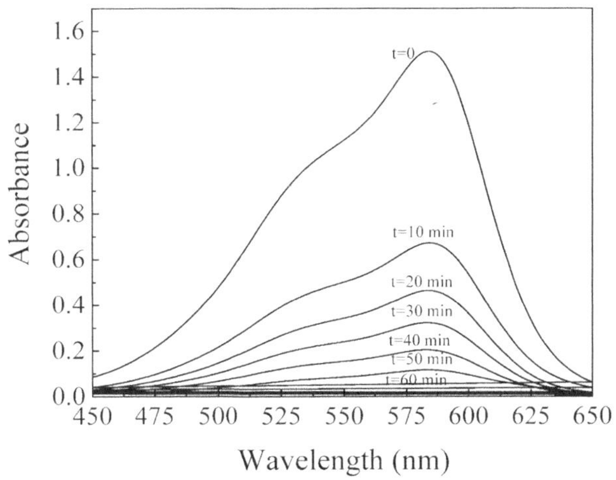

Fig. 9.12 Absorption spectra of EC-treated samples with time (taken after every 10 min)

various aluminum hydroxides with the modified dye molecule as confirmed by the elemental analysis using EDAX. The presence of peaks and trenches in the EDAX depicts that the by-product of electrocoagulation consists of oxygen (O) and carbon from the remains of the destroyed dye molecule and aluminum (Al) from the electrode. While sodium (Na) and chlorine (Cl) are from the added sodium chloride salt and.

9.2.8 Operation Cost

The operational cost in an electrocoagulation process is due to the electrical energy demand, like any other electrical process. In addition to this, the operational cost also includes material costs for the electrode, sludge dewatering, as well as running

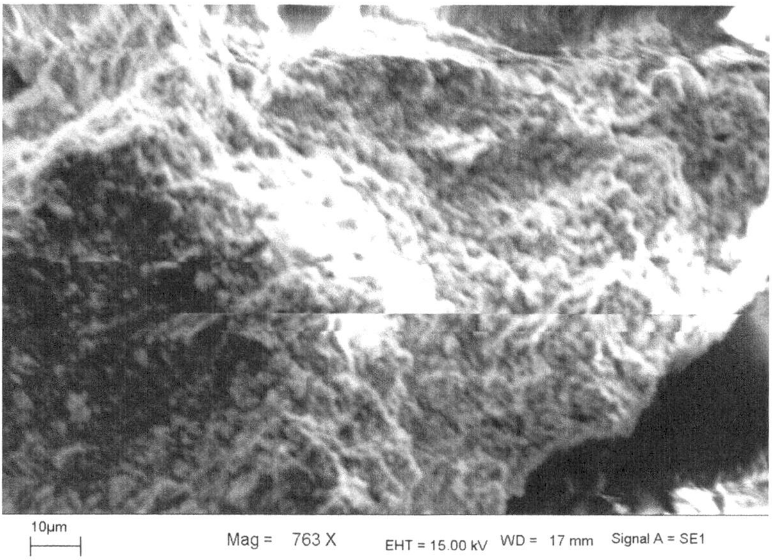

Fig. 9.13 SEM image of by-products obtained from EC bath

Fig. 9.14 Elemental analysis of the by-products obtained from EC bath

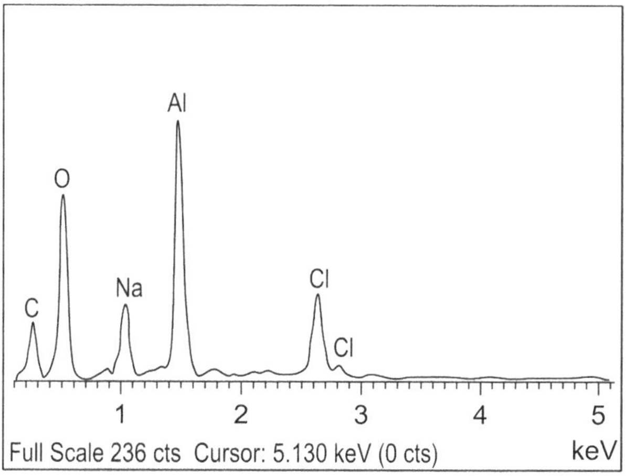

maintenance and labor (Donini et al. 1994). In the following preliminary analysis, energy and the cost associated with the electrode material are taken into consideration as major limiting costs in the calculation of the operation cost (USD/m^3 of solution):

$$\text{Operating cost} = a\,C_{\text{energy}} + b\,C_{\text{electrode}} \tag{9.28}$$

where $C_{electrode}$ (kg Al/m^3 of dye solution) and C_{energy} (kWh/m^3 of dye solution) are consumption quantities for the dye removal, which are obtained experimentally. Here the parameters, *a* and *b*, are representative of the Indian market in June 2017: *a* electrical energy price 0.1 USD/kWh and *b* is the electrode material price 1.3 USD/kg. The cost due to the electrical energy (KWh/m^3 dye solution) is calculated as

$$C_{\text{energy}} = \frac{U \times I \times t_{\text{EC}}}{v} \tag{9.29}$$

where U is the cell voltage (V), I is current (A), t_{EC} is the time of electrolysis (s), and v is the volume (m^3) of dye solution. Cost for electrode (Kg Al/m^3 dye solution) is calculated by the Faraday's Law:

$$C_{\text{electrode}} = \frac{I \times t_{\text{EC}} \times M_{\text{w}}}{z \times \text{F} \times v} \tag{9.30}$$

where z is the number of electron transferred ($z = 3$), M_W is molecular mass of aluminum (26.98 g/mol), and F is Faraday's constant (Kobya et al. 2006). The cost due to electrical energy consumption as well as electrode assembly is calculated for different initial dye concentration (up to 100 mg/L) and shown in Fig. 9.15 for 100% dye removal at optimum operating conditions. It is apparent from the illustration that both the cost (electrode material and electrical energy) increases with the dye concentration. Similar trend is also observed for the operating cost as well.

9.3 Benefits and Drawbacks of Electrocoagulation

Electrocoagulation is one of the potential alternative options for an eco-friendly and economical colored wastewater treatment. The principle of hydroxide adsorption on mineral surface "in situ" is at least two orders of magnitude higher than over pre-precipitated hydroxides obtained from chemical coagulants. The start-up and the operational costs are relatively low, more because it does not involve any moving parts. The energy requirement is comparatively small than any pumping costs involved in hydrodynamic process and can even run on the renewable sources (solar, wind, etc.). As there is no addition of any chemical in the process, it does not produce any secondary pollution or require any secondary stage of removing the added chemicals, producing minimum sludge. The efficiency of removal of small-sized particles is quite high relative to the chemical and biological options as the electrical field controls the electrophoretic velocity of the micro-pollutants and also facilitates its coagulation. Besides, the generation of the gas bubbles aids in the flocculation of the coagulated pollutants. Apart from the advantages, there are quite a few disadvantages with the electrocoagulation process. The sacrificial anode

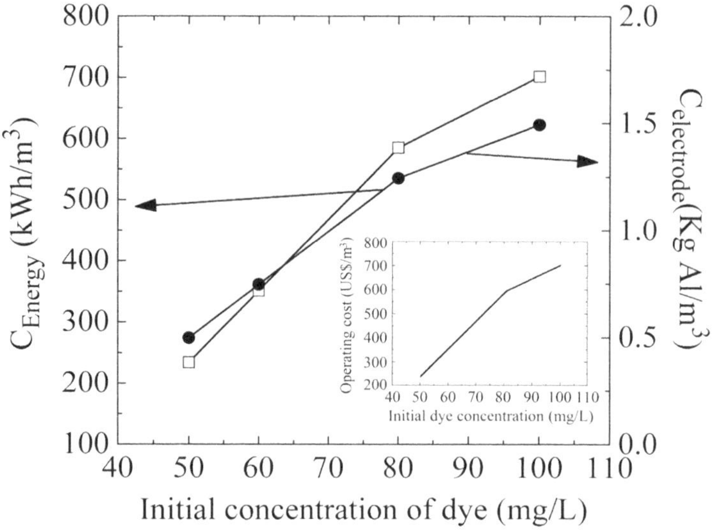

Fig. 9.15 Cost for the treatment of dye solution containing different concentration of crystal violet. Current density, 11.125 A/m^2; conductivity, 1.613 S/m; pH, 8.5; interelectrode distance, 0.005 m

requires to be replaced periodically, thus needs continuous monitoring and maintenance. In dealing with organic compounds, the by-products formed can contain toxic chlorinated organics, if chloride is present in the solution. Finally, the overall cost can be high in regions where cost of electrical energy is high.

Advantages of EC

To summarize, the following list of advantages are not exhaustive but certainly some of the favorable considerations in choosing it as a water treatment option.

1. Electrocoagulation needs simple electrical equipments and does not require skilled operators for maintenance.
2. The wastewater treated removes color and odor, lowers TDS levels, and balances the salinity. Since the TDS level is low, this facilitates the option for its recycled usage.
3. The floc produced by electrocoagulation is larger, stable, and acid resistant and contains less water compared to chemical flocculation, leading to faster separation in the downstream process.
4. The sludge produced in the process settles easily and simple to de-water as it is mostly metal hydroxides. In any case, it is low sludge-producing process.
5. It does not involve any use of chemicals (as the electrocoagulant is produced in situ), so the chance of secondary pollution is negligible. Also there is no issue with neutralizing excess chemicals for safe disposal.

6. The process is almost independent of the particle size and is particularly effective for micro-pollutants, because of the electric field which sets it in a faster motion, accelerating coagulation.
7. Since the overall electrical energy required is low, this is a viable alternative where conventional electrical supply is not available, and energy from other renewable sources can be utilized.

Disadvantages of EC

The factors which determine the overall economics of the process are (in descending order) labor, electrode material, electrical energy, capital investment, and disposal of solid residuals. Generally, the electrical energy requirement is dependent on the process capacity, conductivity of the wastewater, and configuration (as well as the electrode material used) of the reactor. Salt can be added to the solution as required to boost the effect of the current density. Some of the limitations of the process can be outlined here as:

1. The sacrificial anode needs to be replaced regularly and needs continuous monitoring of its activity.
2. Loss of efficiency due to the inevitable formation of the impermeable oxide layer on the cathode.
3. Depending on the regional electricity price, it can be expensive.
4. Not a major disadvantage but generally high electrical conductivity of the wastewater is necessary for treatment, which can be obtained by adding salt.

References

Al-Shannag M, Al-Qodah Z, Bani-Melhem K, RasoolQtaishat M, Alkasrawi M (2015) Heavy metal ions removal from metal plating wastewater using electrocoagulation: kinetic study and process performance. Chem Eng J 260:749–756

Amosov VV, Zilberman AG, Kucheryavykh EI, Sorkin EI, Tsarik LY, Eppel SA, Timoshek VE, Titov IP (1976) Experience in local treatment of wastewaters from petrochemical production. Chem Technol Fuels Oils 12:850–852

Chen X, Chen G, Po LY (2000) Separation of pollutants from restaurant wastewater by electrocoagulation. Sep Purif Technol 19:65–76

Do JS, Chen ML (1994) Decolourization of dye-containing solutions by electrocoagulation. J Appl Electrochem 24:785–790

Donini JC, Kan J, Szynkarczuk J, Hassan TA, Kar KL (1994) Operating cost of electrocoagulation. Can J Chem Eng 72:1007–1012

Kobya M, Demirbas E, Can OT, Bayramoglu M (2006) Treatment of levafix orange textile dye solution by electrocoagulation. J Hazard Mater 132:183–186

Mameri N, Yeddou AR, Lounici H, Belhocine D, Grib H, Bariou B (1998) Defluoridation of septentrional Sahara water of North Africa by electro-coagulation process using bipolar aluminum electrode. Water Res 32:604–612

Matteson MJ, Dobson RL, Glenn RW, Kukunoor NS, Waits WH, Clayfield EJ (1995) Electrocoagulation and separation of aqueous suspensions of ultrafine particles. Colloids Surf A Physicochem Eng Asp 104:101–109

Mollah MYA, Morkovsky P, Gomes JAG, Kesmez M, Parga J, Cocke DL (2004) Fundamentals, present and future perspectives of electrocoagulation. J Hazard Mater B 114:199–210
Novikova SP, Shkorbatova TL, Sokol EY (1982) Purification of effluents from the production of synthetic detergents by electrocoagulation. Soviet J Water Chem Technol 4:353–357
Osipenko VD, Pogorelyi PI (1977) Electrocoagulation neutralization of chromium containing effluent. Metallurgist 21:628–630
Pons MN, Alinsafi A, Khemis M, Leclerc JP, Yaacoubi A, Benhammou A, Nejmeddine A (2005) Electro-coagulation of reactive textile dyes and textile wastewater. Chem Eng Process 44:461–470
Pouet MF, Grasmick A (1995) Urban wastewater treatment by electrocoagulation and flotation. Water Sci Technol 31:275–283
Pourbaix M (1974) Atlas of electrochemical equilibria in aqueous solutions. National Association of Corrosion Engineers Science, Houston
Vik EA, Carlson DA, Eikum AS, Gjessing T (1984) Electrocoagulation of potable water. Water Res 18:1355–1360
Vorobiev E, Larue O, Vu C, Durand B (2003) Electrocoagulation and coagulation by iron of latex particles in aqueous suspensions. Sep Purif Technol 31:177–192

Chapter 10
Emulsion Liquid Membrane

Abstract Emulsion liquid membranes are widely used for recovery of metal ions and organics due to the fast extraction and are a single-stage operation of stripping-extraction. Essentially emulsion liquid membranes are double emulsions which are made stable with surfactants. Separation of dyes using such an emulsion technique is quite novel and new, being researched only for the last one or two decades. The chapter introduces the details of the technique and the efficacy of the method in removing dye-contaminated wastewater.

Keywords Liquid emulsion • Surfactant • Dye • Droplet • Alkali

Emulsion liquid membranes are demonstrated to have significant potential for treatment of various industrial wastes since their invention (Li 1968, 1971). Many studies have been carried out using emulsion liquid membrane (ELM) for recovery of metal ions (Uddin and Kathiresan 2000; Kargari et al. 2006), phenol (Ng et al. 2010), organic acids (Itoh et al. 1990; Hone and Yang 1994), cephalexin from dilute solution (Sahoo and Dutta 1998), bioactive materials (Thien and Hatton 1988), and aniline (Datta et al. 2003). A schematic showing a typical liquid emulsion membrane droplet is shown in Fig. 10.1. Recovery of textile dye from an aqueous solution has been studied by supported liquid membrane using vegetable oil as liquid membrane (Muthuraman and Palanivelu 2006). Liquid emulsion membranes are essentially double emulsion, i.e., water/oil/water (w/o/w) system or oil/water/oil (o/w/o) system. The main advantages of liquid surfactant membranes are (1) fast extraction rates due to availability of high specific surface area, (2) extraction and stripping in one stage only so that the product can be separated and concentrated simultaneously, and (3) possibility of extraction from very dilute solutions.

Stability of w/o/w emulsions is generally understood as the resistance of the individual globules against coalescence (Hou and Papadopoulos 1996). The breakdown of w/o/w-type dispersions is described through several possible mechanisms (Florence and Whitehill 1981) that include (1) coalescence of the internal aqueous droplets into larger internal droplets, (2) coalescence of the oil droplets suspended in the aqueous phase, (3) the expulsion of the internal droplets following rupture of the thin oil films during the interaction of the internal and external aqueous phases (Li et al. 1988; Sanyal et al. 1998), and (4) swelling or contraction due to water permeation through the oil membrane by diffusion (Xuan-cai and Fu-quan 1991;

S. Mondal et al., *Advances in Dye Removal Technologies*, Green Chemistry and Sustainable Technology, DOI 10.1007/978-981-10-6293-3_10

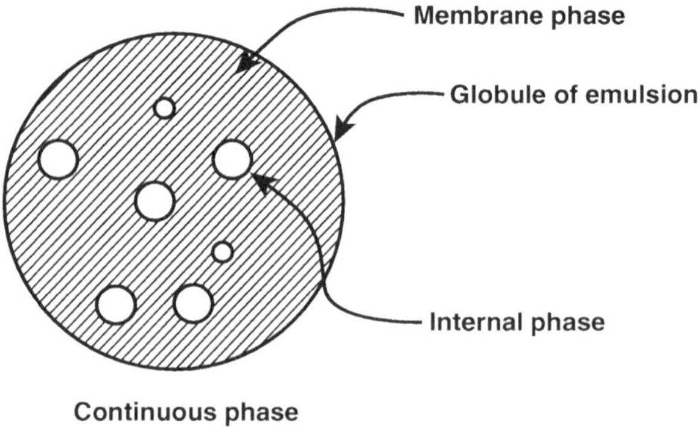

Fig. 10.1 Liquid emulsion membrane droplet

Zihao et al. 1996; Wan and Zhang 2002). In emulsion-type liquid membrane process, surfactant plays a very important role. It influences the emulsion stability and the transport rate of the solute. With the increase of surfactant concentration, emulsion stability improves; however, the extraction rate decreases due to presence of more surfactant molecules at the reaction site, aqueous-organic interface. This problem may be resolved by use of a new type of surfactant, known as bifunctional surfactant, which acts as an emulsifier and an extractant as well (Jee et al. 1985; Wodzki et al. 1990; Uddin and Kathiresan 2000).

Methylene blue (MB) and crystal violet (CV) are removed from aqueous solution using ELM technique. Removal of both single component and their binary mixture is investigated. The effects of concentration of surfactant span 80, NaOH, stirring speed, and feed dye concentration are presented herewith.

10.1 Emulsion Preparation

20 ml n-heptane was added to span 80 (concentration varying from 1 to 8 (wt/wt) %) in a beaker. NaOH concentration was varied from 0.01(M) to 0.5 (M) and added drop by drop, while the system was stirred at 2500 rpm. The final volume was 80 ml. The system was stirred for about 20 min resulting in stable emulsion. 200 ml feed containing dye was taken in a beaker, and the solution was stirred using a magnetic stirrer in the range of 300–600 rpm. 40 ml (1/5th of feed) of stable liquid emulsion was added to the feed. The emulsion slowly gets dispersed. Samples were drawn at regular intervals up to 60 min and were collected in clean and dry test tubes. Each sample was subjected to gravity settling for 25 min. The emulsion being lighter was collected at the top, and clear solution was present at the bottom in each test tube.

10.2 Effect of Surfactant Concentration

Surfactant concentration is an important factor as it directly affects the stability, swelling, and breakup of ELM. Figure 10.1 presents the variation of percentage extraction of CV for various span 80 concentrations. It is observed from this figure that the percentage extraction of CV increases up to 5% of span 80 concentration and decreases thereafter. At lower surfactant concentration (less than 5%), emulsions break easily leading to poor extraction. At higher surfactant concentration (beyond 5%), although the membrane stability increases, mass transfer resistance also increases due to presence of more surfactant at aqueous-organic phase interface, resulting in less transfer of dye molecules to internal phase reducing dye extraction. It is observed from Fig. 10.2 that dye extraction is maximum (~95%) at span 80 concentration of 5%. It may be observed that the dye extraction is slightly reduced after 35 min of separation. This may be due to diffusion of sodium salt of CV from internal phase. However, this effect is marginal.

Effect of surfactant concentration on the extraction of MB is presented in Fig. 10.3. In case of MB, almost 100% extraction is achieved at the surfactant concentration of 5%. Below 5%, the extraction is reduced as discussed earlier. It may be observed from this figure that extraction of MB reaches a constant value after 45 min, due to reduction in driving force (i.e., the concentration gradient of MB between external and internal phase). Comparing this figure with Fig. 10.2, two trends are apparent. First, extraction profile of CV is faster than MB. Second, unlike CV, MB extraction does not fall below the maximum level at any of the surfactant concentration. It is observed that, in MB, the reacting chloride ion is surrounded by three benzene rings providing steric hindrance, and the huge electron cloud associated with these rings makes the reaction difficult. On the other hand, although CV molecule is bigger in size, less hindered chloride ion reacts easily. Therefore, reaction of CV in internal phase favors faster transport of CV. Thus, the extraction

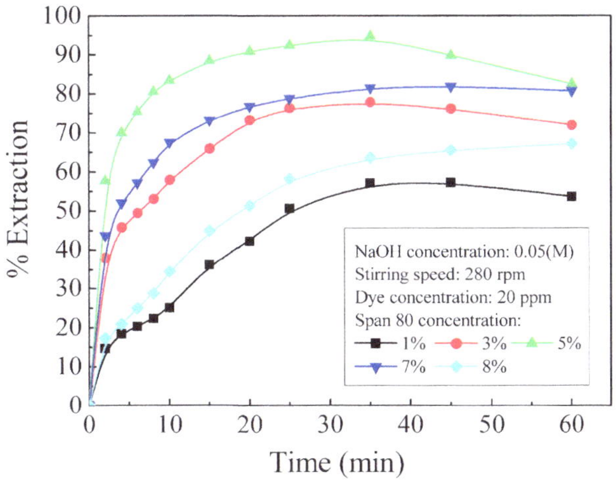

Fig. 10.2 Variation of extraction of CV with time for different span 80 concentrations (Reproduced from Das et al. (2008). With permission from Elsevier)

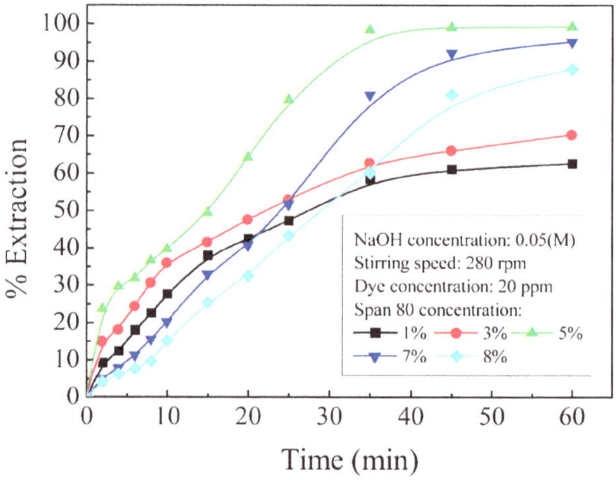

Fig. 10.3 Variation of % extraction of MB with time for different span 80 concentrations (Reproduced from Das et al. (2008). With permission from Elsevier)

profile of CV stabilizes at an earlier time of operation. The amount of NaOH used is approximately equal to that required according to the stoichiometric reaction with the respective dyes. As the reaction with CV is very fast, NaOH present in the inner phase gets depleted at an earlier time as compared the case with MB. Therefore, the chance of leakage of MB into the external phase is small. Ho and Li presented a concept where the solute can transfer from the internal phase of an ELM system to the external phase by two mechanisms: diffusive transport and breakage (Ho and Li 1984). But, solute can transfer from the external phase to the internal phase only by diffusive transport. Our experiments indicate that for CV, the breakage and subsequent leakage of CV into the external phase, over time, manifests itself as a drop in percentage extraction, whereas this phenomenon is absent for experiments with MB for the time of operation used herein.

Extraction profile of the dye mixture is presented in Fig. 10.4 for various concentrations of surfactant. The qualitative observations are similar to those of one-component system as discussed earlier. Two distinctly new observations are evident from this figure. Firstly, the extraction values of both components are slightly less than the single component system. For example, at 5% surfactant concentration, percentage extraction of CV and MB in mixture is 90% and 97%, respectively, compared to 95 and 99% in their single component system. This is due to competitive transport and reaction of two dyes. Secondly, percentage extraction of both CV and MB is quicker compared to one component system. For example, maximum extraction of CV and MB takes place at 35 min and 45 min, respectively, for single dye system, whereas maximum extraction of these dyes takes place at 5 and 35 min, respectively, in case of mixture. As discussed previously, the reaction of CV is preferred over MB due to absence of pi clouds, leading to development of extraction profiles faster in CV compared to MB. In the mixture, unlike CV, MB extraction does not fall below the maximum level for reasons already discussed.

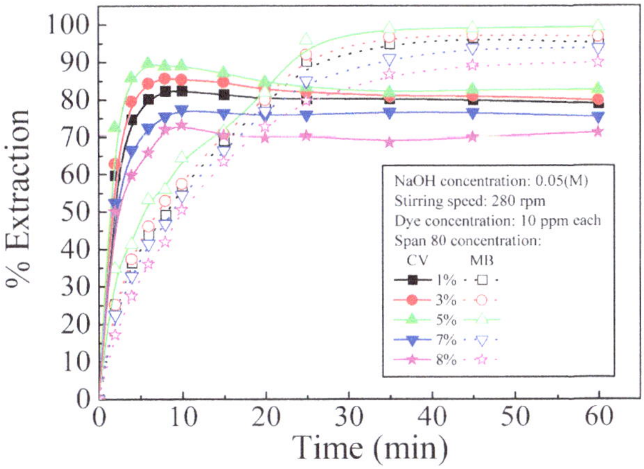

Fig. 10.4 Effect of span 80 concentration in binary mixture (Reproduced from Das et al. (2008). With permission from Elsevier)

10.3 Effect of NaOH Concentration

Effect of NaOH concentration in the internal phase on the extraction of CV is shown in Fig. 10.5. It is observed that maximum extraction (95%) occurs at NaOH concentration of 0.05 (M). At higher concentration (beyond 0.05 M), extraction of dye decreases. At lower concentration (below 0.05 M) of NaOH, almost all reactant (NaOH) is consumed in the internal phase resulting in lower extraction efficiency. At concentration above 0.05 M, excess NaOH causes swelling of emulsion leading to destabilization of liquid membrane system, causing reduction in percentage extraction.

Effect of NaOH concentration on the extraction of MB is presented in Fig. 10.6. The observations are similar to those in case of CV (Fig. 10.5). In this case also, the optimum value of NaOH concentration is found to be 0.05 M. Effect of NaOH concentration on the extraction of CV-MB mixture is shown in Fig. 10.7. The optimum NaOH concentration is found to be 0.05 M in this case as well. Compared with one component system (refer to Figs. 10.5 and 10.6), it is observed that percentage extraction of each dye is slightly less than the single component system as discussed earlier. It is interesting to note that time required for stabilization of percentage extraction profile is much less in case of dye mixture compared to single component system. For example, 35 min and 45 min are required for stabilization for CV and MB, respectively, in single dye system, whereas only 5 and 35 min are required for these dyes in their mixture. This is due to preferential reaction of CV with NaOH in the internal phase as discussed earlier.

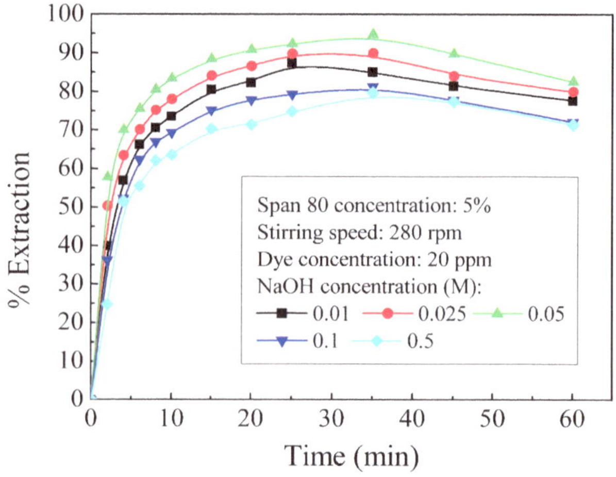

Fig. 10.5 Variation of extraction of CV with time for different NaOH concentrations (Reproduced from Das et al. (2008). With permission from Elsevier)

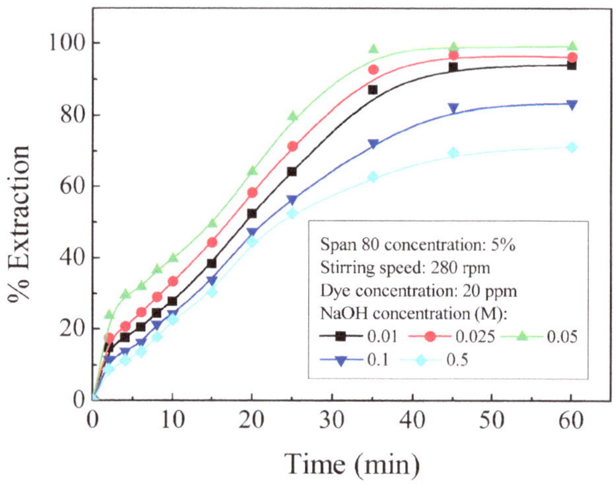

Fig. 10.6 Variation of % extraction of MB with time for different NaOH concentrations (Reproduced from Das et al. (2008). With permission from Elsevier)

10.4 Effect of Stirring Speed

Stirring speed during extraction is an important factor. Effects of stirring speed on the extraction of CV are shown in Fig. 10.8. It is observed that at earlier period of operation, extraction is more at higher stirring speed. This trend is observed during initial 15 min. At higher stirring speed, smaller-sized emulsion droplets are formed leading to more surface area for mass transfer. But at the same time, higher stirring speed adversely affects the stability of emulsion globules leading to breakage. Therefore, percentage extraction decreases in the long run. It is observed from Fig. 10.8 that maximum extraction (95%) occurs at 35 min and with a stirring speed of 280 rpm. Beyond 15 min, percentage extraction decreases with rpm. Similar observations are made in case of extraction of MB from Fig. 10.9. In this case,

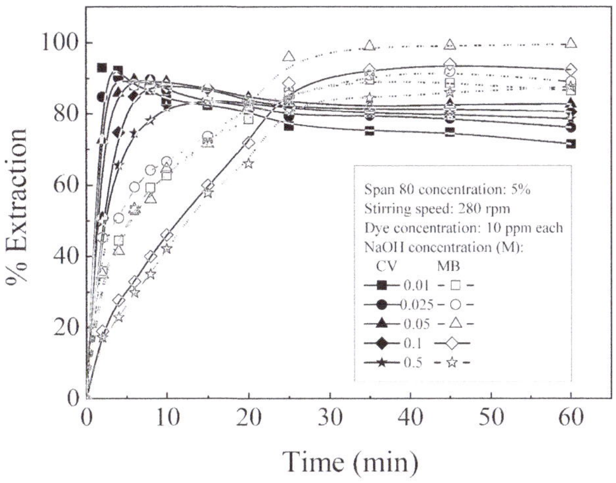

Fig. 10.7 Effect of NaOH concentration in binary mixture (Reproduced from Das et al. (2008). With permission from Elsevier)

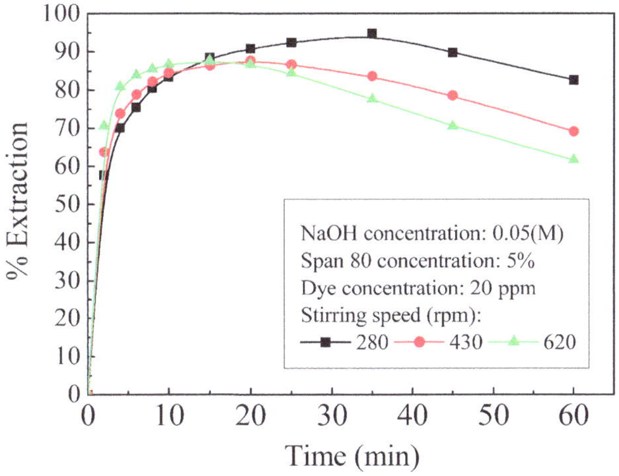

Fig. 10.8 Variation of extraction of CV with time for different stirring speeds (Reproduced from Das et al. (2008). With permission from Elsevier)

percentage extraction increases with rpm up to about 35 min. Beyond that, breakage of emulsion occurs, and extraction of MB decreases with rpm. Maximum extraction observed is about 99% at 280 rpm at the end of operation. Effects of rpm on extraction of dye mixture are presented in Fig. 10.10. The trends are qualitatively similar to the single component system (Fig. 10.8 and 10.9). In the case of mixture, the maximum extraction decreases compared to single component system due to competitive transport of dyes. Maximum extraction for CV is found to be 90% and that for MB is about 97%.

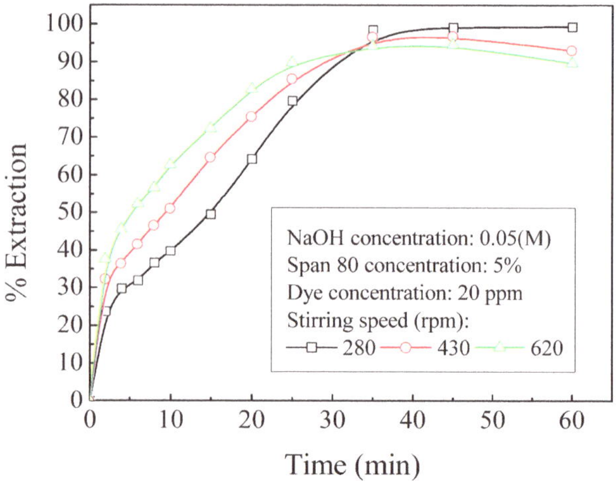

Fig. 10.9 Variation of extraction of MB with time for different stirring speeds (Reproduced from Das et al. (2008). With permission from Elsevier)

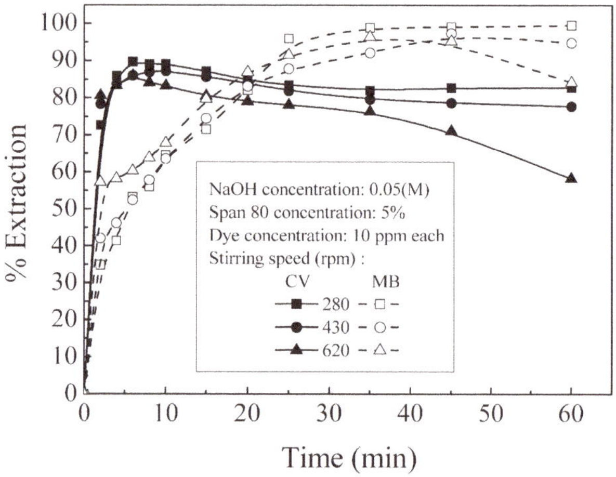

Fig. 10.10 Effect of stirring speed volume in binary mixture (Reproduced from Das et al. (2008). With permission from Elsevier)

10.5 Effect of Feed Concentration

Effects of feed concentrations of dye on the liquid membrane system are shown in Fig. 10.11, in case of CV. It is observed from this figure that dye concentration in the external phase decreases sharply up to 15 min and then remains almost constant up to 35 min and increases slightly thereafter. This effect is prominent at higher dye concentration. As dye concentration in external phase increases, more dye is extracted in the internal phase due to enhanced driving force and after 35 min leakage of dye to external phase occurs due to breakage of internal phase. Effects of feed concentration on ELM system in case of MB are shown in Fig. 10.12. In case of MB, external phase dye concentration decreases up to 40 min and remains almost constant thereafter. As discussed earlier, there is no leakage of MB, and hence, dye

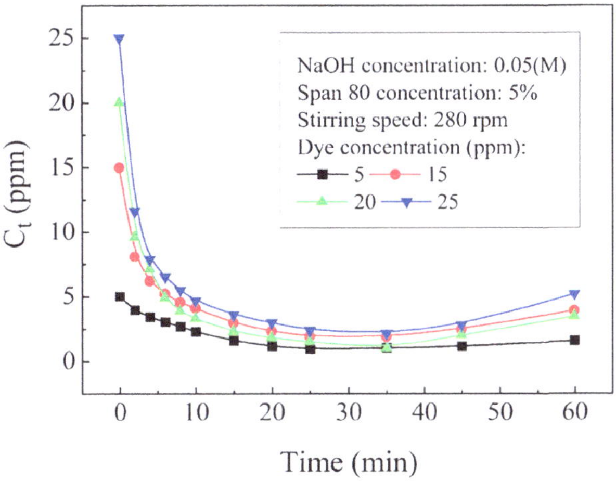

Fig. 10.11 Variation of CV concentration with time for different feed CV concentrations (Reproduced from Das et al. (2008). With permission from Elsevier)

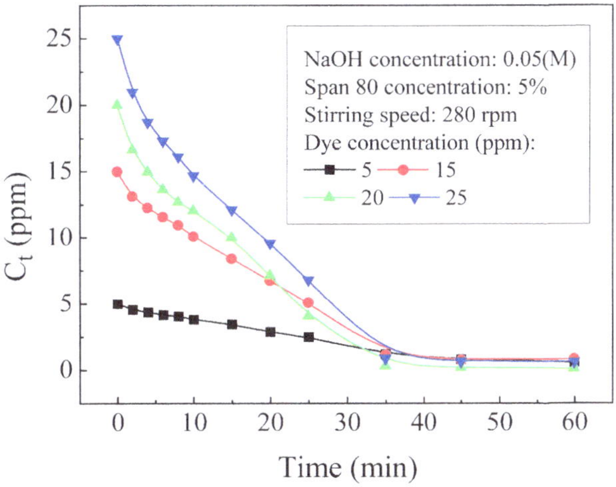

Fig. 10.12 Variation of MB concentration with time for different feed MB concentrations (Reproduced from Das et al. (2008). With permission from Elsevier)

concentration in external phase remains constant almost at zero level indicating almost 99% extraction. Effects of feed concentration in dye mixture are presented in Fig. 10.13. The trends obtained are similar to single component system. Compared to single component system, the dye concentration in the external phase at the end of operation is slightly higher (indicating lower percentage extraction) due to competitive transport of dyes as discussed earlier.

A study of recovery of two dyes, namely, CV and MB, separately and simultaneously by emulsion liquid membrane is undertaken. The effects of concentration of span 80, concentration of NaOH, stirring speed, and composition of feed solution are studied both for single and binary system. The optimum span 80 and NaOH concentrations are found to be 5% and 0.05(M), respectively, for a feed concentration of 20 ppm. The optimum stirring speed for extraction is 280 rpm. Maximum

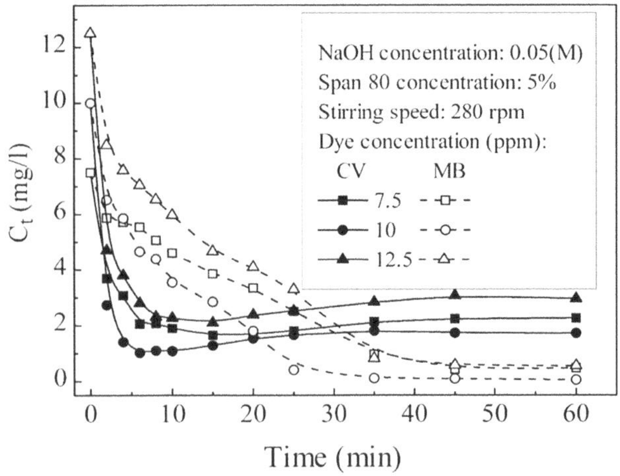

Fig. 10.13 Effect of dye concentration in binary mixture (Reproduced from Das et al. (2008). With permission from Elsevier)

extraction of MB is found to be 99%, and that for CV is about 95% in single component system. In binary mixture, these values are 97% and 90%, respectively. Maximum extraction of CV and MB takes place at 35 min and 45 min, respectively, for single dye system, whereas maximum extraction of these dyes takes place at 5 and 35 min, respectively, in the mixture. Unlike CV, MB extraction does not fall below the maximum level for any of the parameters at its optimum conditions. This technology seems to have a promising future in the treatment of aqueous effluents containing dyes in moderate concentrations, allowing their recovery and reuse.

References

Das C, Rungta M, Arya G, DasGupta S, De S (2008) Removal of dyes and their mixtures from aqueous solution using liquid emulsion membrane. J Hazard Mater 159:365–371

Datta S, Bhattacharya PK, Verma N (2003) Removal of aniline from aqueous solution in a mixed flow reactor using emulsion liquid membrane. J Membr Sci 226:185–201

Florence AT, Whitehill D (1981) Some features of breakdown in water-in-oil-in-water multiple emulsions. J Colloid Interface Sci 79:243–256

Ho WS, Li NN (1984) Modeling of liquid membrane extraction processes. In: Bautista RG (ed) Hydrometallurgical process fundamentals. Plenum Press, New York, pp 555–597

Hone SA, Yang JW (1994) Process development of amino acid concentration by a liquid emulsion membrane technique. J Membr Sci 86:181–192

Hou W, Papadopoulos KD (1996) Stability of water-in-oil-in-water type globules. Chem Eng Sci 51:5043–5051

Itoh H, Thien MP, Hatton TA, Wang DIC (1990) Water transport mechanism in liquid emulsion membrane process for the separation of amino acids. Biotechnol Bioeng 35:853–860

Jee CF, Long TB, Xia XM, Jin QQ, Yang ZL (1985) A study on a two-component liquid membrane system. J Membr Sci 23:137–154

Kargari A, Kaghazchi T, Mardangahi B, Soleimani M (2006) Experimental and modeling of selective separation of gold (III) ions from aqueous solutions by emulsion liquid membrane system. J Membr Sci 279:389–393

Li NN (1968) Separating hydrocarbons with liquid membranes. US patent 3,410,794
Li NN (1971) Permeation through liquid surfactant membranes. AICHE J 17:459–463
Li NN, Borwankar RP, Chan CC, Wassan DT, Kurzeja RM, Gu ZM (1988) Analysis of the effect of internal phase leakage on liquid membrane separations. AICHE J 34:753–762
Muthuraman G, Palanivelu K (2006) Transport of textile dye in vegetable oils based supported liquid membrane. Dyes Pigments 70:99–104
Ng YS, Jayakumar NS, Hashim MA (2010) Performance evaluation of organic emulsion liquid membrane on phenol removal. J Hazard Mater 184:255–260
Sahoo GC, Dutta NN (1998) Studies on emulsion liquid membrane extraction of cephalexin. J Membr Sci 145:15–26
Sanyal SK, Bandyopadhyaya R, Bhowal A, Datta S (1998) A new model of batch-extraction in emulsion liquid membrane: simulation of globule-globule interaction and leakage. Chem Eng Sci 53:2799–2807
Thien MP, Hatton TA (1988) Liquid emulsion membranes and their applications in bio-chemical processing. Sep Sci Technol 23:819–853
Uddin MS, Kathiresan M (2000) Extraction of metal ions by emulsion liquid membrane using bi-functional surfactant: equilibrium and kinetic studies. Sep Purif Technol 19:3–9
Wan Y, Zhang X (2002) Swelling determination of W/O/W emulsion liquid membranes. J Membr Sci 196:185–201
Wodzki R, Wyszynska R, Narebska A (1990) Two-component emulsion liquid membranes with macromolecular carriers of divalent ions. Sep Sci Technol 25:1175–1187
Xuan-cai D, Fu-quan X (1991) Study of the swelling phenomena of liquid surfactant membranes. J Membr Sci 59:183–188
Zihao W, Yuanli J, Jufu F (1996) The entrainment swelling of emulsion during lactic acid extraction by LSMs. J Membr Sci 109:25–34

Printed by Printforce, the Netherlands